D0548016

# A Beautiful Question

ALSO BY FRANK WILCZEK

*The Lightness of Being*
*Fantastic Realities*
*Longing for the Harmonies*

*This work was prepared especially for* A Beautiful Question *by He Shuifa, a modern master of traditional Chinese art and calligraphy. He is renowned for the vigor and subtlety of his brushwork and for the spiritual depth of his depictions of flowers, birds, and nature. A simple translation of the inscription is this: "Taiji double fish is the essence of Chinese culture. This image was painted by He Shuifa on a lake in early winter." The playful "double fish" aspect of Taiji comes to life in He Shuifa's image. The yin and yang resemble two carp playing together, and there are hints of their eyes and fins. In Henan, on the Yellow River, there is a waterfall called Dragon's Gate. Yulong carp attempt to jump the cataract, although it is very difficult for them. Those that succeed transform into lucky dragons. With a sense of humor, we may associate this event with the transformation of virtual into real particles, an essential quantum process that is now thought to underlie the origin of structure in the Universe (see plates XX and AAA). Alternatively we may identify ourselves with the carp, and their strivings with our quest for understanding.*

FRANK WILCZEK

# A Beautiful Question

## *Finding Nature's Deep Design*

ALLEN LANE
*an imprint of*
PENGUIN BOOKS

ALLEN LANE

UK | USA | Canada | Ireland | Australia
India | New Zealand | South Africa

Allen Lane is part of the Penguin Random House group of companies whose addresses can be found at global.penguinrandomhouse.com.

First published in the United States of America by Penguin Press, an imprint of Penguin Random House LLC 2015
First published in Great Britain by Allen Lane 2015
001



Illustration credits appear on page 411

Printed in Great Britain by Clays Ltd, St Ives plc

A CIP catalogue record for this book is available from the British Library

ISBN: 978-1-846-14701-2

www.greenpenguin.co.uk

Penguin Random House is committed to a sustainable future for our business, our readers and our planet. This book is made from Forest Stewardship Council® certified paper.

TO MY FAMILY AND FRIENDS:

BEAUTIFUL ANSWERS OF THE SECOND KIND

# CONTENTS

USER'S MANUAL *xi*

THE QUESTION *1*

PYTHAGORAS I: THOUGHT AND OBJECT *17*

PYTHAGORAS II: NUMBER AND HARMONY *27*

PLATO I: STRUCTURE FROM SYMMETRY—PLATONIC SOLIDS *37*

PLATO II: ESCAPING THE CAVE *55*

NEWTON I: METHOD AND MADNESS *77*

NEWTON II: COLOR *91*

NEWTON III: DYNAMIC BEAUTY *101*

MAXWELL I: GOD'S ESTHETICS *117*

MAXWELL II: THE DOORS OF PERCEPTION *141*

PRELUDE TO SYMMETRY *165*

QUANTUM BEAUTY I: MUSIC OF THE SPHERES *169*

SYMMETRY I: EINSTEIN'S TWO-STEP *199*

QUANTUM BEAUTY II: EXUBERANCE *209*

SYMMETRY II: LOCAL COLOR *221*

QUANTUM BEAUTY III: BEAUTY AT THE CORE OF NATURE *225*

SYMMETRY III: EMMY NOETHER—TIME, ENERGY, AND SANITY *279*

QUANTUM BEAUTY IV: IN BEAUTY WE TRUST *295*

A BEAUTIFUL ANSWER? *321*

*Acknowledgments 329*

*Timelines 331*

*Terms of Art 339*

*Notes 397*

*Recommended Reading 407*

*Illustration Credits 411*

*Index 413*

# USER'S MANUAL

- The "Timelines" are mainly focused on events mentioned or alluded to in the book. They do what timelines do. They are not intended to be complete histories of anything, and they aren't.
- The "Terms of Art" section contains explanatory definitions and discussions of key terms and concepts that occur in the main text. As you can infer from its length, it is rather more than a standard glossary. It contains alternative perspectives on many ideas in the text, and develops a few in new directions.
- The "Notes" section contains material that might, in an academic setting, have gone into footnotes. It both qualifies the text and provides some more technical references on particular points. You'll also find a pair of poems in there.
- The brief "Recommended Reading" section is not a routine list of popularizations, nor of textbooks, but a carefully considered set of recommendations for further exploration in the spirit of the text, emphasizing primary sources.

I hope you've already enjoyed the cover art and the frontispiece, which set the tone for our meditation beautifully.

There's also a "User's Manual"—but you knew that.

# THE QUESTION

This book is a long meditation on a single question:

Does the world embody beautiful ideas?

Our Question may seem like a strange thing to ask. Ideas are one thing, physical bodies are quite another. What does it mean to "embody" an "idea"?

Embodying ideas is what artists do. Starting from visionary conceptions, artists produce physical objects (or quasi-physical products, like musical scores that unfold into sound). Our Beautiful Question, then, is close to this one:

Is the world a work of art?

Posed this way, our Question leads us to others. If it makes sense to consider the world as a work of art, is it a successful work of art? Is the physical world, considered as a work of art, beautiful? For knowledge of the physical world we call on the work of scientists, but to do justice to our questions we must also bring in the insights and contributions of sympathetic artists.

## SPIRITUAL COSMOLOGY

Our Question is a most natural one, in the context of spiritual cosmology. If an energetic and powerful Creator made the world, it could be that what moved Him—or Her, or Them, or It—to create was precisely an impulse to make something beautiful. Natural though it may be, this is assuredly *not* an orthodox idea, according to most religious traditions. Many motivations have been ascribed to the Creator, but artistic ambition is rarely prominent among them.

In Abrahamic religions, conventional doctrine holds that the Creator set out to embody some combination of goodness and righteousness, and to create a monument to His glory. Animistic and polytheistic religions have envisaged beings and gods who create and govern different parts of the world with many kinds of motives, running the gamut from benevolence to lust to carefree exuberance.

On a higher theological plane, the Creator's motivations are sometimes said to be so awesome that finite human intellects can't hope to comprehend them. Instead we are given partial revelations, which are to be believed, not analyzed. Or, alternatively, God is Love. None of those contradictory orthodoxies offers compelling reasons to expect that the world embodies beautiful ideas; nor do they suggest that we should strive to find such ideas. Beauty can form part of their cosmic story, but it is generally regarded as a side issue, not the heart of the matter.

Yet many creative spirits have found inspiration in the idea that the Creator might be, among other things, an artist whose esthetic motivations we can appreciate and share—or even, in daring speculation, that the Creator is *primarily* a creative artist. Such spirits have engaged our Question, in varied and evolving forms, across many centuries. Thus inspired, they have produced deep philosophy, great science, compelling literature, and striking imagery. Some have produced works that combine several, or all, of those features. These works are a vein of gold running back through our civilization.

Galileo Galilei made the beauty of the physical world central to his own deep faith, and recommended it to all:

> The greatness and the glory of God shine forth marvelously in all His works, and is to be read above all in the open book of the heavens.

. . . as did Johannes Kepler, Isaac Newton, and James Clerk Maxwell. For all these searchers, finding beauty embodied in the physical world, reflecting God's glory, was the goal of their search. It inspired their work, and sanctified their curiosity. And with their discoveries, their faith was rewarded.

While our Question finds support in spiritual cosmology, it can also stand on its own. And though its positive answer may inspire a spiritual interpretation, it does not require one.

We will return to these thoughts toward the end of our meditation, by which point we will be much better prepared to appraise them. Between now and then, the world can speak for itself.

## HEROIC VENTURES

Just as art has a history, with developing standards, so does the concept of the world as a work of art. In art history, we are accustomed to the idea that old styles are not simply obsolete, but can continue to be enjoyed on their own terms, and also offer important context for later developments. Though that idea is much less familiar in science, and in science it is subject to important limitations, the historical approach to our Question offers many advantages. It allows us—indeed, forces us—to proceed from simpler to more complex ideas. At the same time, by exploring how great thinkers struggled and often went astray, we gain perspective on the initial strangeness of ideas that have become, through familiarity, too "obvious" and comfortable. Last but by no means least, we humans are especially adapted to think in story and narrative, to associate ideas with names and faces, and to find tales of conflicts and their resolution compelling, even when they are conflicts of ideas, and no blood gets spilled. (Actually, a little does . . .)

For these reasons we will sing, to begin, songs of heroes: Pythagoras, Plato, Filippo Brunelleschi, Newton, Maxwell. (Later a major heroine, Emmy Noether, will enter too.) Real people went by those names—very interesting ones! But for us they are not merely people, but also legends and symbols. I've portrayed them, as I think of them, in that style, emphasizing clarity and simplicity over scholarly nuance. Here biography is a means, not an end. Each hero advances our meditation several steps:

- *Pythagoras* discovered, in his famous theorem about right-angled triangles, a most fundamental relationship between numbers, on the one hand, and sizes and shapes, on the other. Because Number is the purest product of Mind, while Size is a primary characteristic of Matter, that discovery revealed a hidden unity between Mind and Matter.

  Pythagoras also discovered, in the laws of stringed instruments, simple and surprising relationships between numbers and musical harmony. That discovery completes a trinity, Mind-Matter-Beauty, with Number as the linking thread. Heady stuff! It led Pythagoras to surmise that All Things Are Number. With these discoveries and speculations, our Question comes to life.
- *Plato* thought big. He proposed a geometric theory of atoms and the Universe, based on five symmetrical shapes, which we now call the Platonic solids. In this audacious model of physical reality, Plato valued beauty over accuracy. The details of his theory are hopelessly wrong. Yet it provided such a dazzling vision of what a positive answer to our Question might look like that it inspired Euclid, Kepler, and many others to brilliant work centuries later. Indeed, our modern, astoundingly successful theories of elementary particles, codified in our Core Theory (see page 8), are rooted

in heightened ideas of symmetry that would surely make Plato smile. And when trying to guess what will come next, I often follow Plato's strategy, proposing objects of mathematical beauty as models for Nature.

Plato was also a great literary artist. His metaphor of the Cave captures important emotional and philosophical aspects of our relationship, as human inquirers, with reality. At its core is the belief that everyday life offers us a mere shadow of reality, but that through adventures of mind, and sensory expansion, we can get to its essence—and that the essence is clearer and more beautiful than its shadow. He imagined a mediating *demiurge,* which can be translated as *Artisan,* who rendered the realm of perfect, eternal Ideas into its imperfect copy, the world we experience. Here the concept of the world as a work of art is explicit.

- *Brunelleschi* brought new ideas to geometry from the needs of art and engineering. His *projective geometry,* in dealing with the actual appearance of things, brought in ideas—relativity, invariance, symmetry—not only beautiful in themselves, but pregnant with potential.
- *Newton* brought the mathematical understanding of Nature to entirely new levels of ambition and precision.

A common theme pervades Newton's titanic work on light, the mathematics of calculus, motion, and mechanics. It is the method he called Analysis and Synthesis. The method of Analysis and Synthesis suggests a two-stage strategy to achieve understanding. In the analysis stage, we consider the smallest parts of what we are studying their "atoms," using the word figuratively. In a successful analysis, we identify small parts that have simple properties that we can summarize in precise laws. For example:

- In the study of light, the atoms are beams of pure spectral colors.

- In the study of calculus, the atoms are infinitesimals and their ratios.
- In the study of motion, the atoms are velocity and acceleration.
- In the study of mechanics, the atoms are forces.

(We'll discuss these in more depth later.) In the synthesis stage we build up, by logical and mathematical reasoning, from the behavior of individual atoms to the description of systems that contain many atoms.

When thus stated broadly, Analysis and Synthesis may not seem terribly impressive. It is, after all, closely related to common rules of thumb, e.g., "to solve a complex problem, divide and conquer"—hardly an electrifying revelation. But Newton demanded precision and completeness of understanding, saying,

> 'Tis much better to do a little with certainty & leave the rest for others that come after than to explain all things by conjecture without making sure of any thing.

And in these impressive examples, he achieved his ambitions. Newton showed, convincingly, that Nature herself proceeds by Analysis and Synthesis. There really is simplicity in the "atoms," and Nature really does operate by letting them do their thing.

Newton also, in his work on motion and mechanics, enriched our concept of what physical laws are. His laws of motion and of gravity are *dynamical* laws. In other words, they are laws of change. Laws of this kind embody a different concept of beauty than the static perfection beloved of Pythagoras and (especially) Plato.

Dynamical beauty transcends specific objects and phenomena, and invites us to imagine the expanse of possibilities. For example, the sizes and shapes of actual planetary orbits are not simple. They are neither the (compounded) circles of Aristotle, Ptolemy, and

Nicolaus Copernicus, nor even the more nearly accurate ellipses of Kepler, but rather curves that must be calculated numerically, as functions of time, evolving in complicated ways that depend on the positions and masses of the Sun and the other planets. There is great beauty and simplicity here, but it is only fully evident when we understand the deep design. The appearance of particular objects does not exhaust the beauty of the laws.

- *Maxwell* was the first truly modern physicist. His work on electromagnetism ushered in both a new concept of reality and a new method in physics. The new concept, which Maxwell developed from the intuitions of Michael Faraday, is that the primary ingredients of physical reality are not point-like *particles*, but rather space-filling *fields*. The new method is *inspired guesswork*. In 1864 Maxwell codified the known laws of electricity and magnetism into a system of equations, but discovered the resulting system was inconsistent. Like Plato, who shoehorned five perfect solids into four elements plus the Universe, Maxwell did not give up. He saw that by adding a new term he could both make the equations appear more symmetric and make them mathematically consistent. The resulting system, known as the Maxwell equations, not only unified electricity and magnetism, but derived light as a consequence, and survives to this day as the secure foundation of those subjects.

By what is the physicist's "inspired guesswork" inspired? Logical consistency is necessary, but hardly sufficient. Rather it was beauty and symmetry that guided Maxwell and his followers—that is, all modern physicists—closer to truth, as we shall see.

Maxwell also, in his work on color perception, discovered that Plato's metaphorical Cave reflects something quite real and specific: the paltriness of our sensory experience, relative to available reality. And his work, by clarifying the limits of perception, allows us to transcend those limits. For the ultimate sense-enhancing device is a searching mind.

## QUANTUM FULFILLMENT

The definitive answer "yes" to our Question came only in the twentieth century, with the development of quantum theory.

The quantum revolution gave this revelation: we've finally learned what Matter is. The necessary equations are part of the theoretical structure often called the Standard Model. That yawn-inducing name hardly does the achievement justice, and I'm going to continue my campaign, begun in *The Lightness of Being*, to replace it with something more appropriately awesome:

Standard Model → Core Theory

This change is more than justified, because

1. "Model" connotes a disposable makeshift, awaiting replacement by the "real thing." But the Core Theory is already an accurate representation of physical reality, which any future, hypothetical "real thing" must take into account.
2. "Standard" connotes "conventional," and hints at superior wisdom. But no such superior wisdom is available. In fact, I think—and mountains of evidence attest—that while the Core Theory will be supplemented, its core will persist.

The Core Theory embodies beautiful ideas. The equations for atoms and light are, almost literally, the same equations that govern musical instruments and sound. A handful of elegant designs support Nature's exuberant construction, from simple building blocks, of the material world.

Our Core Theories of the four forces of Nature—gravity, electromagnetism, and the strong and weak forces—embody, at their heart, a common principle: *local symmetry*. As you will read, this principle both fulfills and transcends the yearnings of Pythagoras and Plato for harmony and conceptual purity. As you will *see,* this principle both builds upon and transcends the artistic geometry of Brunelleschi and the brilliant insights of Newton and Maxwell into the nature of color.

The Core Theory completes, for practical purposes, the analysis of matter. Using it, we can *deduce* what sorts of atomic nuclei, atoms, molecules—and stars—exist. And we can reliably orchestrate the behavior of larger assemblies of these elements, to make transistors, lasers, or Large Hadron Colliders. The equations of the Core Theory have been tested with far greater accuracy, and under far more extreme conditions, than are required for applications in chemistry, biology, engineering, or astrophysics. While there certainly are many things we don't understand—I'll mention some important ones momentarily!—we do understand the Matter we're made from and that we encounter in normal life (even if we're chemists, engineers, or astrophysicists).

Despite its overwhelming virtues, the Core Theory is imperfect. Indeed, precisely because it is such a faithful description of reality, we must, in pursuit of our Question, hold it to the highest esthetic standards. So scrutinized, the Core Theory reveals flaws. Its equations are lopsided, and they contain several loosely connected pieces. Furthermore, the Core Theory does not account for so-called dark matter and dark energy. Although those tenuous forms of matter are negligible in our immediate neighborhood, they persist in the interstellar and intergalactic voids, and thereby come to dominate the overall mass of the Universe. For those and other reasons, we cannot remain satisfied.

Having tasted beauty at the heart of the world, we hunger for more. In this quest there is, I think, no more promising guide than beauty itself. I shall show you some hints that suggest concrete possibilities for improving our description of Nature. As I aspire to inspired guesswork, beauty is my inspiration. Several times it's worked well for me, as you'll see.

## VARIETIES OF BEAUTY

Different artists have different styles. We don't expect to find Renoir's shimmering color in Rembrandt's mystic shadows, or the elegance of Raphael in either. Mozart's music comes from a different world entirely, the Beatles' from another, and Louis Armstrong's from yet another. Likewise, the beauty embodied in the physical world is a particular kind of beauty. Nature, as an artist, has a distinctive style.

To appreciate Nature's art, we must enter her style with sympathy. Galileo, ever eloquent, expressed it this way:

> Philosophy [Nature] is written in that great book which ever is before our eyes—I mean the universe—but we cannot understand it if we do not first learn the language and grasp the symbols in which it is written. The book is written in mathematical language, and the symbols are triangles, circles, and other geometrical figures, without whose help it is impossible to comprehend a single word of it; without which one wanders in vain through a dark labyrinth.

Today we've penetrated much further into the great book, and discovered that its later chapters use a more imaginative, less familiar language than the Euclidean geometry Galileo knew. To become a fluent speaker in it is the work of a lifetime (or at least of several years in graduate school). But just as a graduate degree in art history is not a prerequisite for engaging with the world's best art and finding that a deeply rewarding experience, so I hope, in this book, to help you engage with Nature's art, by making her style accessible. Your effort will be rewarded, for as Einstein might have said,

> Subtle is the Lord, but malicious She is not.

Two obsessions are the hallmarks of Nature's artistic style:

- Symmetry—a love of harmony, balance, and proportion
- Economy—satisfaction in producing an abundance of effects from very limited means

Watch for these themes as they recur, grow, and develop throughout our narrative and give it unity. Our appreciation of them has evolved from intuition and wishful thinking into precise, powerful, and fruitful methods.

Now, a disclaimer. Many varieties of beauty are underrepresented in Nature's style, as expressed in her fundamental operating system. Our delight in the human body and our interest in expressive portraits, our love of animals and of natural landscapes, and many other sources of artistic beauty are not brought into play. Science isn't everything, thank goodness.

## CONCEPTS AND REALITIES; MIND AND MATTER

Our Question can be read in two directions. Most obviously, it is a question about the world. That is the direction we've emphasized so far. But the other direction is likewise fascinating. When we find that *our* sense of beauty is realized in the physical world, we are discovering something about the world, but also something about ourselves.

Human appreciation of the fundamental laws of Nature is a recent development on evolutionary or even historical time scales. Moreover, those laws reveal themselves only after elaborate operations—looking through sophisticated microscopes and telescopes, tearing atoms and nuclei apart, and processing long chains of mathematical reasoning—that do not come naturally. Our sense of beauty is not in any very direct way adapted to Nature's fundamental workings. Yet just as surely, our sense of beauty is excited by what we find there.

What explains that miraculous harmony of Mind and Matter? Without an explanation of that miracle, our Question remains mysterious. It is an issue our meditation will touch upon repeatedly. For now, two brief anticipations:

1. We human beings are, above all, visual creatures. Our sense of vision, of course, and in a host of less obvious ways our deepest modes of thought, are conditioned by our interaction with light. Each of us, for example, is born to become an accomplished, if unconscious, practitioner of projective geometry. That ability is hardwired into our brain. It is what allows us to interpret the two-dimensional image that arrives on our retinas as representing a world of objects in three-dimensional space.

   Our brains contain specialized modules that allow us to construct, very quickly and without conscious effort, a dynamic worldview based on three-dimensional objects located in three-dimensional space. We do this beginning from two two-dimensional images on the retinas of our eyes (which, in turn, are the product of light rays emitted or reflected from the surfaces of external objects, which propagate to us in straight lines). To work back from the images we receive to the objects that cause them is a tricky problem in inverse projective geometry. In fact, as stated, it is an impossible problem, because there's not nearly enough information in the projections to do an unambiguous reconstruction. A basic problem is that even to get started we need to separate objects from their background (or foreground). We exploit all kinds of tricks based on typical properties of objects we encounter, such as their color or texture contrast and distinctive boundaries, to do that job. But even after that step is accomplished, we are left with a difficult geometrical problem, for which Nature has helpfully provided us, in our visual cortex, an excellent specialized processor.

Another important feature of vision is that light arrives to us from very far away, and gives us a window into astronomy. The regular apparent motion of stars and the slightly less regular apparent motion of planets gave early hints of a lawful Universe, and provided an early inspiration and testing ground for the mathematical description of Nature. Like a good textbook, it contains problems with varying degrees of difficulty.

In the more advanced, modern parts of physics we learn that light itself is a form of matter, and indeed that matter in general, when understood deeply, is remarkably light-like. So again, our interest in and experience with light, which is deeply rooted in our essential nature, proves fortunate.

Creatures that, like most mammals, perceive the world primarily through the sense of smell would have a much harder time getting to physics as we know it, even if they were highly intelligent in other ways. One can imagine dogs, say, evolving into extremely intelligent social creatures, developing language, and experiencing rich lives full of interest and joy, but devoid of the specific kinds of curiosity and outlook, based on visual experience, that lead to our kind of deep understanding of the physical world. Their world would be rich in reactions and decays—they'd have great chemistry sets, elaborate cuisines, aphrodisiacs, and, à la Proust, echoing memories. Projective geometry and astronomy, maybe not so much. We understand that smell is a chemical sense, and we are beginning to understand its foundation in molecular events. But the "inverse" problem of working from smell back to molecules and their laws, and eventually to physics as we know it, seems to me hopelessly difficult.

Birds, on the other hand, are visual creatures, like us. Beyond that, their way of life would give them an extra advantage over humans, in getting started on physics. For birds, with their freedom of flight, experience the essential symmetry of three-dimensional space in an intimate way that we do not. They also experience the basic regularities of motion, and especially the role of inertia, in their everyday lives, as they operate in a nearly frictionless environment. Birds are born, one might say, with intuitive knowledge of classical mechanics and Galilean relativity, as well as of geometry. If some species of bird evolved high abstract intelligence—that is, if they ceased being birdbrains—their physics would develop rapidly. Humans, on the other hand, have to unlearn the friction-laden Aristotelean mechanics they use in everyday life, in order to achieve deeper understanding. Historically that involved quite a struggle!

Dolphins, in their watery environment, and bats, with their echolocation, give us other interesting variations on these themes. But I will not develop those here.

A general philosophical point, which these considerations illustrate, is that the world does not provide its own unique interpretation. The world offers many possibilities for different sensory universes, which support very different interpretations of the world's significance. In this way our so-called Universe is already very much a multiverse.

2. Successful perception involves sophisticated inference, because the information we sample about the world is both very partial and very noisy. For all our innate powers, we must also learn how to see by interacting with the world, forming expectations, and comparing our predictions with reality. When we form expectations that turn out to be correct, we experience pleasure and satisfaction. Those reward mechanisms encourage successful learning. They also stimulate—indeed, at base they *are*—our sense of beauty.

Putting those observations together, we discover an explanation of why we find interesting phenomena (phenomena we can learn from!) in physics beautiful. An important consequence is that we especially value experience that is surprising, but not too surprising. Routine, superficial recognition will not challenge us, and may not be rewarded as active learning. On the other hand, patterns whose meaning we cannot make sense of at all will not offer rewarding experience either; they are noise.

And here we are lucky too, in that Nature employs, in her basic workings, symmetry and economy of means. For these principles, like our intuitive understanding of light, promote successful prediction and learning. From the appearance of part of a symmetric object we can predict (successfully!) the appearance of the rest; from the behavior of parts of natural objects we can predict (sometimes successfully!) the behavior of wholes. Symmetry and economy of means, therefore, are exactly the sorts of things we are apt to experience as beautiful.

## NEW IDEAS AND INTERPRETATIONS

Together with new appreciations of some very old and some less old ideas, you will find in this book several essentially new ones. Here I'd like to mention some of the most important.

My presentation of the Core Theory as geometry, and my speculations about the next steps beyond it, are adaptations of my technical work in fundamental physics. That work builds, of course, on the work of many others. My use of color fields as an example of extra dimensions, and my exploitation of the possibilities they open up for illustrating local symmetry, are (as far as I know) new.

My theory that promotion of learning underlies, and is the evolutionary cause of, our sense of beauty in important cases, and the application of that theory to musical harmony, which offers a rational explanation for

Pythagoras's discoveries in music, form a constellation of ideas I've entertained privately for a long time but present here for the first time publicly. Caveat emptor.

My discussion of the expansion of color perception draws on an ongoing program of practical research that I hope will lead to commercial products. Patents have been applied for.

I'd like to think that Niels Bohr would approve of my broad interpretation of complementarity, and might even acknowledge his paternity—but I'm not sure he would.

# PYTHAGORAS I: THOUGHT AND OBJECT

## THE SHADOW PYTHAGORAS

There was a person named Pythagoras who lived and died around 570–495 BCE, but very little is known about him. Or rather a lot is "known" about him, but most of it is surely wrong, because the documentary trail is littered with contradictions. It combines the sublime, the ridiculous, the unbelievable, and the just plain weird.

Pythagoras was said to be the son of Apollo, to have a golden thigh, and to glow. He may or may not have advocated vegetarianism. Among his most notorious sayings is an injunction not to eat beans, because "beans have a soul." Yet several early sources explicitly deny that Pythagoras said or believed anything of the sort. More reliably, Pythagoras believed in, and taught, the transmigration of souls. There are several stories—each, to be sure, dubious—that corroborate this. According to Aulus Gellius, Pythagoras remembered four of his own past lives, including one as a beautiful courtesan named Alco. Xenophanes recounts that Pythagoras, upon hearing the cries of a dog who was being beaten, rushed to halt the beating, claiming to recognize the voice of a departed friend. Pythagoras also, like Saint Francis centuries later, preached to animals.

The *Stanford Encyclopedia of Philosophy*—a free and extremely valuable online resource, by the way—sums it up as follows:

> The popular modern image of Pythagoras is that of a master mathematician and scientist. The early evidence shows, however, that, while Pythagoras was famous in his own day and even 150 years later in the time of Plato and Aristotle, it was not mathematics or science upon which his fame rested. Pythagoras was famous
>
> 1. As an expert on the fate of the soul after death, who thought that the soul was immortal and went through a series of reincarnations
> 2. As an expert on religious ritual
> 3. As a wonder-worker who had a thigh of gold and who could be two places at the same time
> 4. As the founder of a strict way of life that emphasized dietary restrictions, religious ritual and rigorous self discipline

A few things do seem clear. The historical Pythagoras was born on the Greek island of Samos, traveled widely, and became the inspiration for and founder of an unusual religious movement. His cult flourished briefly in Crotone, in southern Italy, and developed chapters in several other places before being everywhere suppressed. The Pythagoreans formed secretive societies, on which the initiates' lives centered. These communities, which included both men and women, promoted a kind of intellectual mysticism that seemed marvelous, yet strange and threatening, to most of their contemporaries. Their worldview centered on worshipful admiration of numbers and musical harmony, which they saw as reflecting the deep structure of reality. (As we'll see, they were on to something.)

## THE REAL PYTHAGORAS

Here again is the *Stanford Encyclopedia:*

The picture of Pythagoras that emerges from the evidence is thus not of a mathematician, who offered rigorous proofs, or of a scientist, who carried out experiments to discover the nature of the natural world, but rather of someone who sees special significance in and assigns special prominence to mathematical relationships that were in general circulation.

Bertrand Russell was pithier:

A combination of Einstein and Mary Baker Eddy.

To scholars of factual biography, it is a major problem that later followers of Pythagoras ascribed their own ideas and discoveries to Pythagoras himself. In that way they hoped both to give their ideas authority and, by enhancing Pythagoras's reputation, to promote their community—the community he founded. Thus magnificent discoveries in different fields of mathematics, physics, and music, as well as an inspiring mysticism, a seminal philosophy, and a pure morality were all portrayed as the legacy of a single godlike figure. That awesome figure is, for us, the *real* Pythagoras.

It is not altogether inappropriate to assign the (historical) shadow Pythagoras credit for the real Pythagoras, because the latter's great achievements in mathematics and science emerged from the way of life the former inspired, and the community he founded.

(Those so inclined might draw parallels to the differing careers in life, and afterward, of other major religious figures . . .)

Thanks to Raphael, we know what thc real Pythagoras looked like. In plate B* he is captured deep in concentration as he writes in a great book, surrounded by admirers.

## ALL THINGS ARE NUMBER

It is difficult to make out what Pythagoras is writing, but I like to pretend it is some version of his most fundamental credo:

All Things Are Number

It is also difficult to know, at this separation in time and space, exactly what Pythagoras meant by that. So we get to use our imagination.

## PYTHAGORAS'S THEOREM

For one thing, Pythagoras was mightily impressed by Pythagoras's theorem. So much so that when he discovered it, in a notable lapse from vegetarianism, he offered a hecatomb—the ritual sacrifice of one hundred oxen, followed by feasting—to the Muses, in thanks.

Why the fuss?

Pythagoras's theorem is a statement about right triangles; that is, triangles that contain a 90-degree angle, or, in other words, a square corner. The theorem tells you that if you erect squares on the different sides of such a triangle, then the sum of the areas of the two smaller squares adds up to the area of the largest square. A classic example is the 3-4-5 right triangle, shown in figure 1:

* Color plates, which are denoted by letters of the alphabet, appear in the inserts following pages 146 and 274. Black-and-white figures, which are numbered, appear throughout the main text.

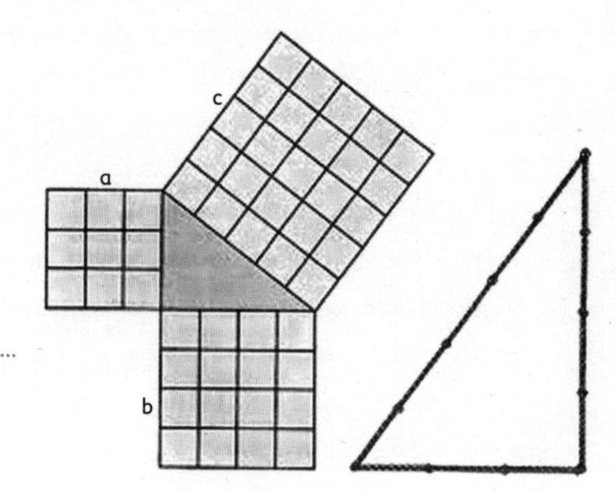


FIGURE 1. THE 3-4-5 RIGHT TRIANGLE, A SIMPLE CASE OF PYTHAGORAS'S THEOREM.

The areas of the two smaller squares are $3^2 = 9$ and $4^2 = 16$, as we can see, in the spirit of Pythagoras, by *counting* their subunits. The area of the largest square is $5^2 = 25$. And we verify $9 + 16 = 25$.

By now Pythagoras's theorem is familiar to most of us, if only as a dim memory from school geometry. But if you listen to its message afresh, with Pythagoras's ears, so to speak, you realize that it is saying something quite startling. It is telling you that the *geometry* of objects embodies hidden *numerical* relationships. It says, in other words, that Number describes, if not yet everything, at least something very important about physical reality, namely the sizes and shapes of the objects that inhabit it.

Later in this meditation we will be dealing with much more advanced and sophisticated concepts, and I'll have to resort to metaphors and analogies to convey their meaning. The special joy one finds in precise mathematical thinking, when sharply defined concepts fit together perfectly, is lost in translation. Here we have an opportunity to experience that special joy. Part of the magic of Pythagoras's theorem is that one can prove it with minimal preparation. The best proofs are unforgettable, and their memory lasts a lifetime. They've inspired Aldous Huxley and Albert Einstein—not to mention Pythagoras!—and I hope they'll inspire you.

## *Guido's Proof*

"So simple!"

That is what Guido, the young hero of Aldous Huxley's short story "Young Archimedes," says, as he describes his demonstration of Pythagoras's theorem. Guido's proof is based on the shapes displayed in plate C.

## *Guido's Plaything*

Let's spell out what was obvious to Guido at a glance.

Each of the two large tiled squares contains four colored triangles that are matched in the other large square. All the colored triangles are right triangles, and all are the same size. Let's say the length of the smallest side is $a$, the next smallest $b$, and the longest (the hypotenuse) $c$. Then it's easy to see that the sides of both large (total) squares have length $a + b$, and in particular that those two squares have equal areas. So the non-triangular parts of the large squares must also have equal areas.

But what are those equal areas? In the first large square, on the left, we have a blue square with side $a$, and a red square with side $b$. They have areas $a^2$ and $b^2$, and their combined area is $a^2 + b^2$. In the second large square, on the right, we have a gray square with side $c$. Its area is $c^2$. Recalling the preceding paragraph, we conclude that

$$a^2 + b^2 = c^2$$

. . . which is Pythagoras's theorem!

## *Einstein's Proof(?)*

In Einstein's *Autobiographical Notes* he recalls,

> I remember that an uncle told me about the Pythagorean theorem before the holy geometry booklet had come into my hands. After much effort I succeeded in "proving" this theorem on the basis of the similarity of triangles; in doing so it seemed to me "evident" that the relations of the sides of the right-angled triangles would have to be determined by one of the acute angles.

There is not really enough detail in that account to reconstruct Einstein's demonstration with certainty, but here, in figure 2, is my best guess. That guess deserves to be right, because this is the simplest and most beautiful proof of Pythagoras's theorem. In particular, this proof makes it brilliantly clear why the *squares* of the lengths are what's involved in the theorem.

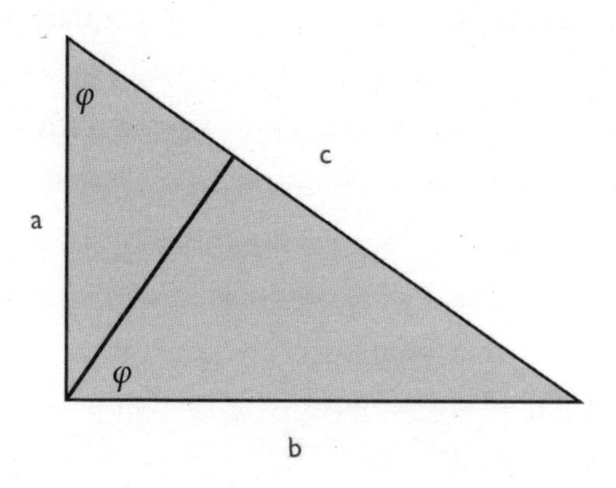


FIGURE 2. A PLAUSIBLE RECONSTRUCTION OF EINSTEIN'S PROOF, FROM *AUTOBIOGRAPHICAL NOTES.*

## *A Polished Jewel*

We start from the observation that right triangles that include a common angle $\varphi$ are all similar to one another, in the precise sense that you can get from any one to any other by an overall rescaling (magnification or shrinking). Also: if we rescale the length of the triangle by some factor, then we will rescale the area by the square of that factor.

Now consider the three right triangles that appear in figure 2: the total figure, and the two sub-triangles it contains. Each of them contains the angle $\varphi$, so they are similar. Their areas are therefore proportional to $a^2$, $b^2$, $c^2$, going from smallest to largest. But because the two sub-triangles add up to the total triangle, the corresponding areas must also add up, and therefore

$$a^2 + b^2 = c^2$$

. . . Pythagoras's theorem pops right out!

## *A Beautiful Irony*

It is a beautiful irony that Pythagoras's theorem can be used to undermine his doctrine All Things Are Number.

That scandalous result is the one discovery of the Pythagorean school that was not attributed to Pythagoras, but rather to his pupil Hippasus. Shortly after his discovery, Hippasus drowned at sea. Whether his death should be attributed to the wrath of the gods, or the wrath of the Pythagoreans, is a debated point.

Hippasus's reasoning is very clever, but not overly complicated. Let's waltz through it.

We consider isosceles right triangles with two equal sides—in other words, $a = b$. Pythagoras's theorem tells us that

$$2 \times a^2 = c^2$$

Now let's suppose that the lengths $a$ and $c$ are both whole numbers. If *all* things are numbers, they'd better be! But we'll find that it's impossible.

If both $a$ and $c$ are even numbers, we can consider a similar triangle of half the size. We can keep halving until we reach a triangle where at least one of $a$, $c$ is odd.

But whichever choice we make, we quickly derive a contradiction.

First let's suppose that $c$ is odd. Then so is $c^2$. But $2 \times a^2$ is obviously even because it contains a factor 2. So we can't have $2 \times a^2 = c^2$, as Pythagoras's theorem tells us. Contradiction!

Alternatively, suppose that $c$ is even, say $c = 2 \times p$. Then $c^2 = 4 \times p^2$. Then Pythagoras's theorem tells us, after we divide both sides by 2, that $a^2 = 2 \times p^2$. And so $a$ can't be odd, by the same reasoning as before. Contradiction!

So all things can't be whole numbers, after all. There cannot be an atom of length, such that all possible lengths are whole number multiples of that atom's length.

It doesn't seem to have occurred to the Pythagoreans that one might draw a different conclusion, saving All Things Are Number. After all, one *can* imagine a world where space is constructed from many identical atoms. Indeed, my friends Ed Fredkin and Stephen Wolfram advocate models of our world based on cellular automata, which have exactly this property. And your computer screen, based on atoms of light we call pixels, shows that such a world can look pretty realistic! Logically, the correct conclusion to draw is that in such a world, one cannot construct exact isosceles right triangles. Something has to go slightly wrong. The "right" angle might fail to be exactly 90 degrees, or the two shorter sides might fail to be perfectly equal or—as on the computer screen—the sides of your triangles might fail to be exactly straight.

This is not the option Greek mathematicians chose. Rather, they considered geometry in its more appealing continuous form, where we allow exact right angles and exact equality of sides to coexist. (This is also the choice that has proved most fruitful for physics, as we'll learn from Newton.) To do this, they had to prioritize geometry over arithmetic, because—as we've seen—the whole numbers are inadequate to describe even very simple geometric figures. Thus they abandoned the letter, though not the spirit, of All Things Are Number.

## THOUGHT AND OBJECT

For the true essence of Pythagoras's credo is not a literal assertion that the world must embody whole numbers, but the optimistic conviction that the world should embody *beautiful concepts*.

The lesson for which Hippasus paid with his life is that we must be willing to learn from Nature what those concepts are. In this enterprise, humility is mandatory. Geometry is not less beautiful than arithmetic. Indeed, it is more naturally suited to our highly visual brains, and most people prefer it. And geometry is no less conceptual, no less a pure world of Mind, than arithmetic. Much of ancient Greek mathematics, epitomized in Euclid's *Elements*, was devoted to showing precisely this: that geometry is a system of *logic*.

As we continue our meditation, we'll find that Nature is inventive in her language. She stretches our imagination with new kinds of numbers, new kinds of geometry—and even, in the quantum world, new kinds of logic.

# PYTHAGORAS II: NUMBER AND HARMONY

The essence of all stringed instruments, whether ancient lyre or modern guitar, cello, or piano, is the same: they produce sound from the motion of strings. The exact quality of sound, or timbre, depends on many complex factors, including the nature of the material that makes the string, the shapes of the surfaces—"sounding boards"—that vibrate in sympathy, and the way in which the string is plucked, bowed, or hammered. But in all instruments there is a principal tone, or pitch, that we recognize as the note being played. Pythagoras—the real one—discovered that the pitch obeys two remarkable rules. Those rules make direct connections among numbers, properties of the physical world, and our sense of harmony (which is one face of beauty).

The drawing that follows, not by Raphael, shows Pythagoras in action, performing experiments on harmony:

FIGURE 3. AN ETCHING FROM MEDIEVAL EUROPE DEPICTING PYTHAGORAS AT WORK ON MUSICAL HARMONY. WE CAN INFER FROM THE FIGURE THAT PYTHAGORAS LISTENED TO HOW THE SOUNDS PRODUCED BY HIS INSTRUMENT CHANGED AS HE VARIED TWO DIFFERENT THINGS. BY HOLDING A STRING DOWN FIRMLY AT DIFFERENT POINTS, HE COULD VARY THE EFFECTIVE LENGTH OF THE VIBRATING PART. AND BY CHANGING THE WEIGHT THAT STRETCHES A STRING, HE COULD VARY ITS TENSION.

## HARMONY, NUMBER, AND LENGTH: AN ASTONISHING CONNECTION

Pythagoras's first rule is a relationship between the length of the vibrating string and our perception of its tone. The rule says that two copies of the same type of string, both subject to the same tension, make tones that sound good together precisely when the lengths of the strings are in ratios of small whole numbers. Thus, for example, when the ratio of lengths is 1:2, the tones form an octave. When the ratio is 2:3, we hear the dominant fifth; when the ratio is 3:4, the major fourth. In musical notation (in the key of C) these correspond to playing two Cs, one above the other, together, a C-G, or C-F, respectively. People find those tone combinations appealing. They are the main building blocks of classical music, and of most folk, pop, and rock music.

In applying Pythagoras's rule, the length that we must consider is of course the effective length, that is, the length of the portion of the string that actually vibrates. By clamping down on the string, creating a dead zone, we can change the tone. Guitarists and cellists exploit that possibility when they "finger" with their left hands. As they do so they are, whether or not they know it, reincarnating Pythagoras. In the drawing, we see Pythagoras adjusting the effective length using a pointed clamp, which is a technique conducive to accurate measurement.

When tones sound good together, we say they are in harmony, or that they are concordant. What Pythagoras discovered, then, is that the perceived harmonies of tones reflect relationships in what might seem to be an entirely different world—the world of numbers.

## HARMONY, NUMBER, AND WEIGHT: AN ASTOUNDING CONNECTION

Pythagoras's second rule involves the tension of the string. The tension can be adjusted, in a controlled and readily measurable way, by burdening the string with different amounts of weight, as shown in figure 3. Here the result is even more remarkable. The tones are in harmony if the tensions are ratios of *squares* of small whole numbers. Higher tensions correspond to higher pitches. Thus a 1:4 ratio of tensions produces the octave, and so forth. When string musicians tune their instruments prior to a performance, stretching or relaxing the strings by winding their pegs, Pythagoras returns.

This second relationship is even more impressive than the first as evidence that Things are hidden Numbers. The relationship is better hidden because the numbers must be processed—squared, to be exact—before the relationship becomes evident. The shock of discovery is accordingly greater. Also, the relationship brings in weight. And weight, more unmistakably than length, links us to Things in the material world.

## DISCOVERY AND WORLDVIEW

Now we've discussed three major Pythagorean discoveries: the Pythagorean theorem on right triangles, and two rules of musical consonance. Together, they link shape, size, weight, and harmony, with the common thread being Number.

For the Pythagoreans, that trinity of discoveries was more than enough to anchor a mystic worldview. Vibration of strings is the source of musical sound. These vibrations are nothing but periodic motions; that is, motions which repeat themselves at regular intervals. We also see the Sun and planets move in periodic motions across the sky, and infer their periodic motion in space. So they too must emit sound. Their sounds form the Music of the Spheres, a music that fills the cosmos.

Pythagoras was fond of singing. He also claimed actually to hear the Music of the Spheres. Some modern scholars speculate that the historical Pythagoras suffered from tinnitus, or ringing in the ears. The real Pythagoras, of course, did not.

In any case, the larger point is that All Is Number, and Number supports Harmony. The Pythagoreans, drunk on mathematics, inhabited a harmony-filled world.

## THE FREQUENCY IS THE MESSAGE

Pythagoras's musical rules deserve, I think, to be considered the first quantitative laws of Nature ever discovered. (Astronomical regularities, beginning with the regular alternation of night and day, were of course noticed much earlier. Calendar-keeping and casting of horoscopes, using mathematics to predict or reconstruct the positions of the Sun, Moon, and planets, were significant technologies before Pythagoras was born. But empirical observations about specific objects are quite different from general laws of Nature.)

It is ironic, therefore, that we still don't fully understand why they are true. Today we have a much better understanding of the physical processes involved in the production, transmission, and reception of sound, but the connection between that knowledge and the perception of "notes that sound good together" has so far been elusive. I think there is a promising set of ideas about that. These ideas are close to the central concern of our meditation, because (if true) they elucidate an important origin of our sense of beauty.

Our account of the *why* of Pythagoras's rules has three parts. The first part starts with the vibrating string and proceeds to our eardrums. The second part starts with the eardrum and proceeds to primary nerve impulses. The third part starts with primary nerve pulses and proceeds to perceived harmony.

The vibration of a string goes through several transformations before arriving to our minds as a message. The vibration disturbs the surrounding air directly, simply by pushing it. The hum of an isolated string is quite weak, however. Practical musical instruments employ sounding boards, which respond to the string's vibration with stronger vibrations of their own. The motion of the sounding board pushes air around more robustly.

The disturbance of air near the string or sounding board then takes on a life of its own, becoming a propagating disturbance: a sound wave that spreads outward in all directions. Any sound wave is a recurring cycle of compression and decompression. The vibrating air in each region of space exerts pressure on neighboring regions and sets them into vibration. Eventually a portion of this sound wave, funneled by the complicated geometry of the ear, arrives at a membrane called the eardrum a few centimeters within. Our eardrums serve as inverse sounding boards, where now vibrations of air induce mechanical motion, instead of the opposite.

The eardrum vibrations set off more reactions, as we'll discuss momentarily. Before that, however, we should make a simple but fundamental observation. This long series of transformations can seem bewildering, and one may wonder how a meaningful signal, reflecting what that string was doing, can be extracted far down the line. The point is that throughout all these transformations there is a property that remains unchanged. The rate of the vibrations in time or, as we say, their *frequency*, whether they are vibrations in string, in sounding board, in air, or in eardrum—or in the ossicles, cochlear fluid, basilar membrane, and hair cells farther down the line—remains the same. For at each transformation, the pushes and pulls of one stage induce the compressions and decompressions of the next, one for one, and so the different kinds of disturbances are synchronized or, as we say, "in time." We can anticipate, therefore, and will find, that the useful things to monitor, if we want our perception to reflect a property of the initial vibration, is the frequency of vibrations it eventually sets up in our heads.

The first step toward understanding Pythagoras's rules, therefore, is to cast them in terms of frequency. Today we have reliable equations of mechanics that allow us to calculate how the frequency of vibration of a string changes as we vary its length or tension. Using those equations, we find that the frequency falls proportionally to the length, and rises proportionally to the square root of the tension. Therefore Pythagoras's rules, translated into frequency, both make the same simple statement. They both state that notes sound good together if their frequencies are in ratios of small whole numbers.

## A THEORY OF HARMONY

Now let us resume our story, at its second stage. The eardrum is attached to a system of three small bones, the ossicles, which in turn are attached to a membranous "oval window" opening on a snail-like structure, the cochlea. The cochlea is the critical organ for hearing, playing a role roughly analogous to the role the eye plays in sight. It is filled with fluid that is set in motion by the vibrations at the oval window. Immersed in that fluid is a long tapering membrane, the basilar membrane, that worms through the gyrations of the cochlear snail. Running parallel to the basilar membrane is the organ of Corti. The organ of Corti is where, finally, the message of the string—after many transformations—gets translated into nervous impulses. The details of all these transformations are complex, and fascinating to experts, but the big picture is simple and does not depend on those details. The big picture is that the frequency of the original vibration gets translated into firings of neurons that have the same frequency.

One important aspect of the translation is especially pretty, and Pythagorean in spirit. It led Georg von Békésy to a Nobel Prize in 1961. Because the basilar membrane tapers along its length, different parts of it prefer to oscillate at different rates. The thicker parts have more inertia, so they prefer to vibrate more slowly, at lower frequencies, whereas the thinner parts prefer to vibrate at higher frequencies. (This effect is responsible for the difference in the overall pitch between typical male and female voices. At puberty the male vocal cords thicken markedly, leading to lower frequencies of vibration and a deepened voice.) Thus when a sound, after its many tribulations, sets the surrounding fluid into motion, the response of the basilar membrane will be different at different places along its length. A low-frequency tone will put the thicker parts into vigorous motion, while a high-frequency tone will put the thinner parts into vigorous motion. In this way, information about frequency gets encoded into information about position!

If the cochlea is the eye of audition, the organ of Corti is its retina. The organ of Corti runs parallel to the basilar membrane, and close by. Its structure is complex in detail, but roughly speaking it consists of hair cells and neurons, one hair cell per neuron. The motion of the basilar membrane, coupled through intermediate fluid, exerts forces on the hair cells. The hair cells move in response, and their motion triggers electrical firing of the corresponding neurons. The frequency of the firing is the same as the frequency of stimulation, which in turn is the same as the frequency of the original tone. (For experts: The firing patterns are noisy, but they contain a strong component at the signal frequency.)

Because the organ of Corti abuts the basilar membrane, its neurons inherit the position-dependent frequency response of that membrane. This is very important for our perception of chords, because it means that when several tones sound simultaneously, their signals do not get completely scrambled. Different neurons respond preferentially to different tones! This is the physiological mechanism that allows us to do such a good job of discriminating different tones.

In other words, our inner ears follow the advice of Newton—and anticipate his analysis of light—by performing an excellent Analysis of the incoming sound into pure tones. (As we'll discuss later, our sensory ability to analyze the frequencies of signals in light, or in other words the color content of light, is based on different principles, and is much poorer.)

This sets the scene for the third stage of our story. In it, signals from the primary sensory neurons in the organ of Corti are combined and passed on to subsequent neural layers in the brain. Here our knowledge is considerably less precise. But it is only here that we can finally come to grips with our main question:

> *Why* do tones whose frequencies are in ratios of small whole numbers sound good together?

Let us consider what the brain is offered when two different sound frequencies play simultaneously. Then we have two sets of primary neurons responding strongly, each firing with the same frequency as the vibrations of the string that excites them. Those primary neurons fire their signals brainward, to "higher" levels of neurons, where their signals are combined and integrated.

Some of the neurons at the next level will receive inputs from both sets of firing primaries. If the frequencies of the primaries are in a ratio of small whole numbers, then their signals will be synchronized. (For this discussion, we will simplify the actual response, ignoring the noise and treating it as accurately periodic.) For example, if the tones form an octave, one set will be firing twice as fast as the other, and every firing of the slower one will have the same predictable relationship to the firing of the former. Thus the neurons sensitive to both will then get a repetitive pattern that is predictable and easy to interpret. From previous experience, or perhaps by inborn instinct, those secondary neurons—or the later neurons that interpret their behavior—will "understand" the signal. For it will be possible to anticipate future input (i.e., more repetitions) in a simple way, and simple predictions for future behavior will be borne out, over many vibrations, until the sound changes its character.

Note that the sound vibrations we can hear have frequencies ranging from a few tens to several thousand per second, so even brief sounds will produce many repetitions, except at the very low-frequency end. And at the low-frequency end our sense of harmony peters out, consistent with the line of thought we are pursuing.

Higher levels of neurons, which combine the combiners, need coherent input to get on with their job. So if our combiners are producing sensible messages, and in particular if their predictions satisfy the test of time, it is in the interest of the higher levels to reward them with some kind of positive feedback, or at least to leave them in peace. On the other hand, if the combiners are producing wrong predictions, the mistakes will propagate up to higher levels, ultimately producing discomfort and a desire to make it stop.

When will the combiners produce wrong predictions? That will happen when the primary signals are almost, but not quite, in synch. For then the vibrations will reinforce each other for a few cycles, and the combiners will extrapolate that pattern. They expect it to continue—but it doesn't! And indeed it is tones that are just slightly off—like C and C#, for example—that sound most painful when played together.

If this idea is right, then the basis of harmony is successful prediction in the early stages of perception. (This process of prediction need not, and usually does not, involve conscious attention.) Such success is experienced as pleasure, or beauty. Conversely, unsuccessful prediction is a source of pain, or ugliness. A corollary is that by expanding our experience, and learning, we can come to hear harmonies that were previously hidden to us, and to remove sources of pain.

Historically, in Western music, the palette of acceptable tone combinations has expanded over time. Individuals can also learn, by exposure, to enjoy tone combinations that at first seem unpleasant. Indeed, if we are built to enjoy *learning* to make successful predictions, then predictions that come too easily will not yield the greatest possible pleasure, which should also bring in novelty.

# PLATO I: STRUCTURE FROM SYMMETRY—PLATONIC SOLIDS

The Platonic solids carry an air of magic about them. They have been, and are, literally, objects to conjure with. They reach back deep into human prehistory, and live on as the generators of good or bad luck in some of the most elaborate of games, notably Dungeons & Dragons. Their mystique has inspired, besides, some of the most fruitful episodes in the development of mathematics and science. A worthy meditation on embodied beauty must dwell upon them.

Albrecht Dürer, in his *Melancholia I* (figure 4), alludes to the allure of regular solids, although the solid that appears is not quite a Platonic solid. (Technically, it is a truncated triangular trapezohedron. It can be constructed by stretching out the sides of an octahedron in a peculiar way.) Perhaps the philosopher is melancholy because she can't fathom why a baleful bat dropped that particular, not quite Platonic, solid into her study, rather than a straightforward example.

FIGURE 4. DÜRER'S *MELANCHOLIA I*. IT FEATURES A TRUNCATED PLATONIC SOLID, A VERY MAGIC SQUARE, AND MANY OTHER ESOTERIC SYMBOLS. TO ME, IT WELL DEPICTS THE FRUSTRATIONS I OFTEN ENCOUNTER WHEN USING PURE THOUGHT TO COMPREHEND REALITY. FORTUNATELY, IT'S NOT ALWAYS THIS WAY.

### *Regular Polygons*

To appreciate the Platonic solids, let us start with something simpler: their closest two-dimensional analogue, regular polygons. A regular polygon is a planar figure with all equal sides that meet at all equal angles. The simplest regular polygon, with three sides, is an equilateral triangle. Next we have squares, with four sides. Then there are regular pentagons (the chosen symbol of the Pythagoreans, and also the design of a famous military headquarters), hexagons (the unit of a bee's hive and, as we shall see, of graphene), heptagons (various coins), octagons (stop signs), nonagons. . . . The series continues indefinitely: For each whole number, starting with three, there is a unique regular polygon. In each case, the number of vertices equals the number of sides. We can also consider the circle as a limiting case of regular polygon, where the number of sides becomes infinite.

The regular polygons capture, in some intuitive sense, the notion of ideal regularity for planar "atoms." They will serve us as conceptual atoms, from which we build up richer and more complex ideas of order and symmetry.

## THE PLATONIC SOLIDS

As we move from planar to solid figures, searching for maximal regularity, we can generalize the regular polygons in various ways. A very natural choice, which turns out to be most fruitful, leads to the Platonic solids. We ask for solid bodies whose faces are regular polygons, all identical, that meet in identical fashion at every vertex. Then, instead of an infinite series of solutions, we find there are exactly five!

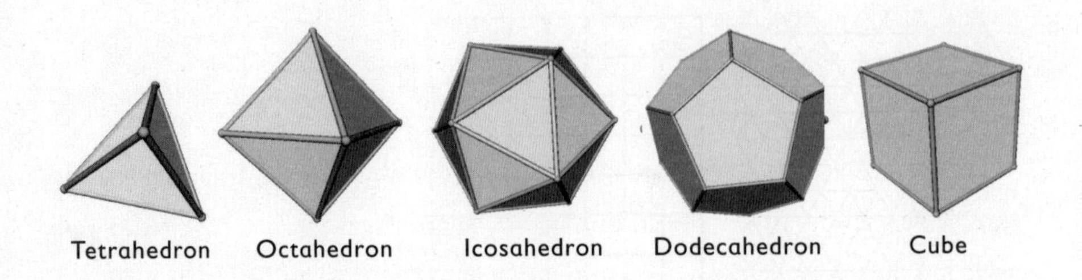


FIGURE 5. THE FIVE PLATONIC SOLIDS: OBJECTS TO CONJURE WITH.

These five Platonic solids are:

- The *tetrahedron*, with four triangular faces, four vertices, and three faces coming together at each vertex
- The *octahedron*, with eight triangular faces, six vertices, and four faces coming together at each vertex
- The *icosahedron*, with twenty triangular faces, twelve vertices, and five faces coming together at each vertex
- The *dodecahedron*, with twelve pentagonal faces, twenty vertices, and three faces coming together at each vertex
- The *cube*, with six square faces, eight vertices, and three faces coming together at each vertex

The existence of those five solids is easy to grasp, as one can imagine and construct models without great difficulty. But why are there just those five? (Or are there others?)

To get our head around that question, we notice that the vertices of the tetrahedron, octahedron, and icosahedron feature three, four, and five triangles coming together, and ask, What happens if we continue to six? Then we realize that six equilateral triangles sharing a common vertex *lie flat*. Repeating that flat building block will not allow us to complete a finite figure, bounding a solid volume. Instead, it leads to an infinite dissection of a plane, as shown in figure 6:

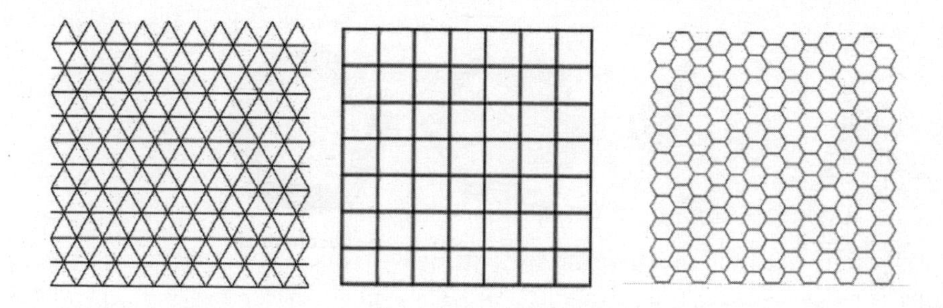

Platonic Prodigals

FIGURE 6. THE THREE INFINITE PLATONIC SURFACES. ONLY FINITE PORTIONS ARE SHOWN HERE. THESE THREE REGULAR DISSECTIONS OF A PLANE CAN AND SHOULD BE CONSIDERED RELATIVES OF THE TRADITIONAL PLATONIC SOLIDS—THEIR PRODIGAL SIBLINGS THAT WANDER OFF AND NEVER RETURN.

We find similar results if we put together four squares, or three hexagons. These three regular dissections of a plane are worthy supplements to the Platonic solids. We will find them embodied in the microcosm (figure 29, page 216).

If we try to put together more than six equilateral triangles, four squares, or three of any of the larger regular polygons, we run out of room—we simply can't accommodate the accumulated angles. And so the five Platonic solids are the only finite regular solids.

It is remarkable that a specific finite number—that is, five—emerges from considerations of geometric regularity and symmetry. Regularity and symmetry are natural and beautiful things to consider, but they have no obvious or direct connection to specific numbers. Plato interpreted this profound emergence in an astonishingly creative way, as we shall see.

## *Prehistory*

Famous people often get credit for the discoveries of others. This is the "Matthew Effect" identified by the sociologist Robert Merton, based on this observation from the Gospel of Matthew:

> For unto every one that hath shall be given, and he shall have abundance: but from him that hath not shall be taken even that which he hath.

So it is for the Platonic solids.

At the Ashmolean Museum of Oxford University you can see a display of five carved stones dating from 2000 BCE Scotland that appear to be realizations of the Platonic solids (though some scholars dispute this). They were most likely used in some sort of dice game. Let us imagine cave people huddled around the communal fire, rapt in paleolithic Dungeons & Dragons. But it was probably Plato's contemporary Theaetetus (417–369 BCE) who first *proved* mathematically that those five bodies are the only possible regular solids. It's not clear to what extent Theaetetus was inspired by Plato, or vice versa, or whether it was something in the Athenian air they both breathed. In any case, the Platonic solids got their name because Plato used them creatively, in work of imaginative genius, to construct a visionary theory of the physical world.

FIGURE 7. PRE-PLATONIC ANTICIPATIONS OF THE PLATONIC SOLIDS, PROBABLY USED IN DICE GAMES CIRCA 2000 BCE.

Going back much further, we now realize that some of the biosphere's simplest creatures, including viruses and diatoms (not pairs of atoms, but marine algae that often grow elaborate Platonic exoskeletons), not only "discovered" but have literally embodied the Platonic solids since long before humans walked the Earth. The herpesvirus, the virus that causes hepatitis B, the HIV virus, and many other nasties are shaped like icosahedra or dodecahedra. They encase their genetic material—either DNA or RNA—in protein exoskeletons, which determine their external form, as seen in plate D. The exoskeleton is color-coded in such a way that identical colors indicate identical building blocks. The dodecahedron's signature triply meeting pentagons leap to the eye. If we join the centers of the blue regions with straight lines, an icosahedron emerges.

More complex microscopic creatures, including the radiolaria lovingly portrayed by Ernst Haeckel in his marvelous book *Art Forms in Nature*, also embody the Platonic solids. In figure 8 it is the intricate silica exoskeletons of these single-cell organisms that we see. The radiolarians are an ancient life-form, represented in the earliest fossils. They continue to thrive in the oceans today. Each of the five Platonic solids is realized in a number of species. Several species names enshrine those shapes, including *Circoporus octahedrus*, *Circogonia icosahedra*, and *Circorrhegma dodecahedra*.

## *Euclid's Inspiration*

Euclid's *Elements* is, by a wide margin, the greatest textbook of all time. It brought system and rigor to geometry. From a larger perspective it established, by example, the method of Analysis and Synthesis in the domain of ideas.

Analysis and Synthesis is Isaac Newton's, and our, preferred formulation of "reductionism." Here is Newton:

> By this way of Analysis we may proceed from Compounds to Ingredients, and from Motions to the Forces producing them; and in general,

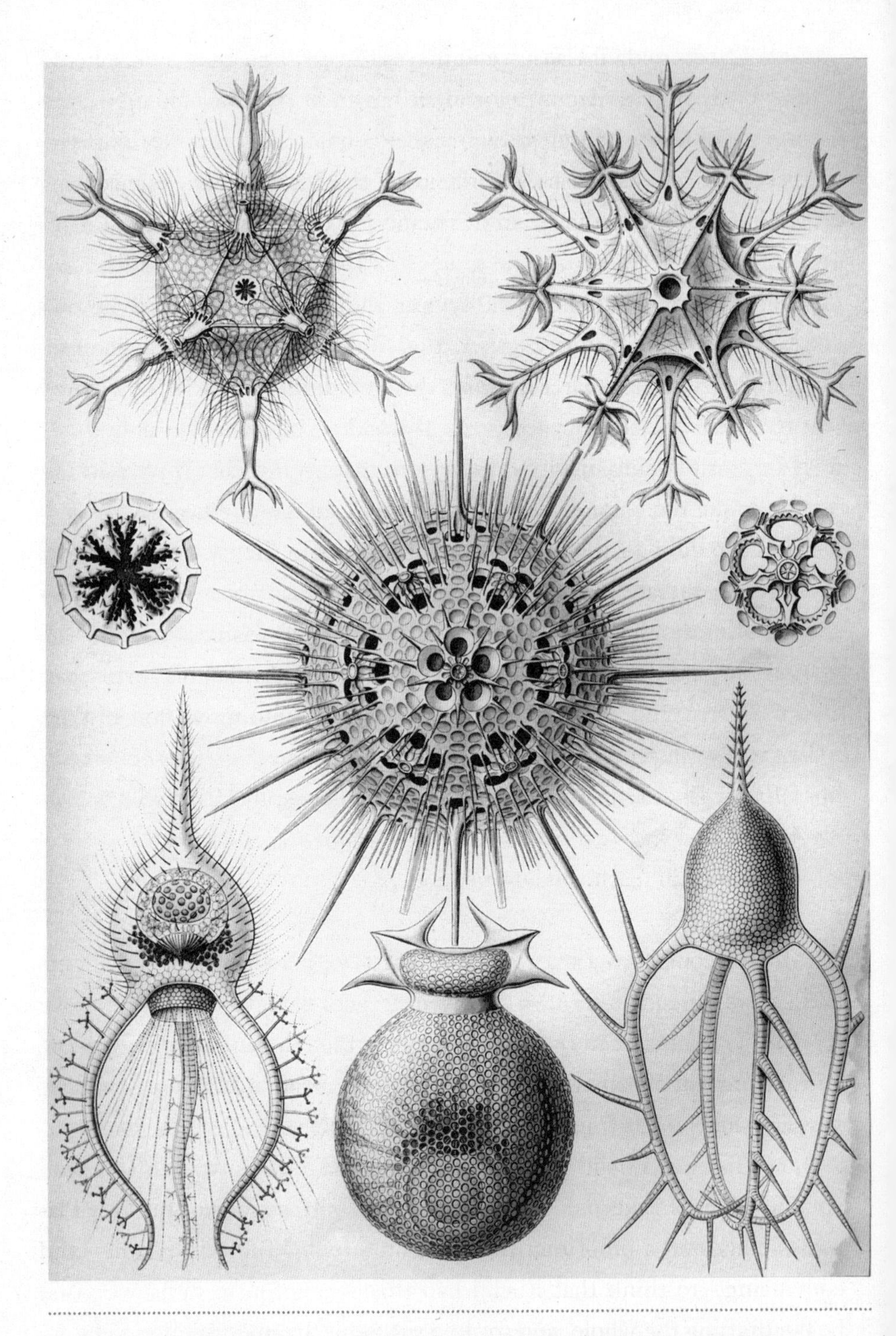

FIGURE 8. RADIOLARIA BECOME VISIBLE UNDER A MODEST MICROSCOPE. THEIR EXOSKELETONS OFTEN EXHIBIT THE SYMMETRY OF PLATONIC SOLIDS.

from Effects to their Causes, and from particular Causes to more general ones, till the Argument end in the most general. This is the Method of Analysis: And the Synthesis consists in assuming the Causes discover'd, and establish'd as Principles, and by them explaining the Phænomena proceeding from them, and proving the Explanations.

This strategy parallels Euclid's approach to geometry, where he proceeds from simple, intuitive *axioms* to deduce rich and surprising consequences. Newton's great *Principia,* the founding document of modern mathematical physics, also follows Euclid's expository style, building from axioms to major results step-by-step through logical construction.

It is important to emphasize that axioms (or laws of physics) don't tell you what to do with them. By stringing them together without purpose, it's easy to generate hosts of forgettable, worthless truths—like a play or a piece of music that wanders aimlessly, arriving nowhere. As those who have attempted to deploy artificial intelligence to do creative mathematics have discovered, identifying *goals* is often the hardest challenge. With a worthy goal in mind, it becomes easier to find the means to achieve it. My all-time favorite fortune cookie summed this up brilliantly:

The work will teach you how to do it.

Also, of course, as a matter of presentation, it's attractive to students and potential readers to have an inspiring goal in sight—and impressive for them to realize, at the start, that they can look forward to experiencing an amazing feat of construction that builds, by inexorable steps, from "obvious" axioms to far-from-obvious conclusions.

So: What was Euclid's goal in the *Elements*? The thirteenth and final volume of that masterpiece concludes with constructions of the five Platonic solids, and a proof that there are only five. I find it pleasant—and convincing—to think that Euclid had this conclusion in mind when he began drafting the whole, and worked toward it. In any case, it is a fitting, fulfilling conclusion.

## *Platonic Solids as Atoms*

The ancient Greeks recognized four building blocks, or elements, for the material world: fire, water, earth, and air. You might notice that four, the number of elements, is close to five, the number of regular solids. Plato certainly did! One finds, in his influential, visionary, inscrutable *Timaeus,* a theory of the elements based on the solids. Here it comes:

Each of the elements is built from a different variety of atom. The atoms take the form of Platonic solids. The atoms of fire are tetrahedra, the atoms of water are icosahedra, the atoms of earth are cubes, and the atoms of air are octahedra.

There is a certain plausibility to these assignments. They have explanatory power. The atoms of fire have sharp points, which explains why contact with fire is painful. The atoms of water are most smooth and well-rounded, so they can flow around one another smoothly. The atoms of earth can pack closely, and fill space without gaps. Air, being both hot and wet, features atoms intermediate between those of fire and water.

Now while five is close to four, it is not quite equal to it, so there cannot be a perfect match between regular solids, regarded as atoms, and elements. A merely brilliant thinker might have been discouraged by that difficulty, but Plato, a genius, was undaunted. He took it as a challenge and an opportunity. The remaining regular solid, the dodecahedron, he proposed, does figure in the Creator's construction, but not as an atom. No, the dodecahedron is no mere atom—rather, it is the shape of the Universe as a whole.

Aristotle, who was forever determined to one-up Plato, put forward a different, more conservative and intellectually consistent variation of that theory. Two of that influential philosopher's big ideas were: that the Moon, planets, and stars inhabit a celestial realm made from stuff different from what we find in the mundane world; and that "Nature abhors a vacuum," so that the celestial spaces could not be empty. Thus consistency required there to be a fifth element, or quintessence, different from earth, air, fire, and water, to fill the celestial realm. Dodecahedra, then, find their place as the atoms of quintessence, or ether.

It is difficult to agree, today, with the details of these theories, in either version. We haven't found it useful, in science, to analyze the world in terms of those four (or five) elements. Nor are modern atoms hard, solid bodies, much less realizations of the Platonic solids. Plato's theory of the elements, seen from today's perspective, is both crude and, in detail, hopelessly misguided.

## *Structure from Symmetry*

And yet, though it fails as a scientific theory, Plato's vision succeeds as prophecy and, I would claim, as a work of intellectual art. To appreciate those larger virtues, we have to step away from the details, and look at the bigger picture. The deepest, core intuition of Plato's vision of the physical world is that the physical world must, fundamentally, embody beautiful concepts. And this beauty must be of a very special kind: the beauty of mathematical regularity, of perfect symmetry. For Plato, as for Pythagoras, that intuition was at the same time a faith, a yearning, and a guiding principle. They sought to harmonize Mind with Matter by showing that Matter is built from the purest products of Mind.

It is important to emphasize that Plato pushed his ideas past the level of philosophical generalities to make specific claims about what matter is. His specific ideas, though wrong, do not fall into the ignominious category of "not even wrong." Plato even made some gestures in the direction of comparing his theory with reality, as we've seen. Fire stings because tetrahedra are sharply pointed, water flows because icosahedra easily slide past one another, and so forth. In his dialogue *Timaeus*, where all this is spelled out, you'll also find some fanciful explanations of what we'd call chemical reactions and properties of composite (i.e., non-elemental) materials, based on the geometry of atoms. But these throwaway efforts fall woefully short of anything we'd be tempted to consider serious experimental testing of a scientific theory, much less the exploitation of scientific insight for practical purposes.

Yet Plato's vision anticipates modern ideas now at the forefront of scientific thinking, in several ways.

Though the building blocks for matter that Plato proposed are not the ones we know today, the idea that there *are* just a few kinds of building blocks, existing in many identical copies, remains foundational.

Beyond that level of vague inspiration, the more specific strategy of Plato's vision, to derive *structure* from *symmetry*, is one that echoes through the ages. We are led to a small number of special structures from purely mathematical considerations—considerations of symmetry—and put them forward to Nature, as candidate elements for her design. The sort of mathematical symmetry that Plato invoked, to arrive at his list of building blocks, is quite different from the symmetry we use today. But the idea that there *is* symmetry at the root of Nature has come to dominate our understanding of physical reality. The far-fetched idea that *symmetry dictates structure*—that one can use stringent requirements of mathematical perfection to converge on a small list of possible realizations, and then use that list as the construction manual for our world-model—has become, at the unmapped frontiers of the unknown, our guiding star. It is an idea almost blasphemous in its audacity, for it claims that we can decode the Artisan's working methods, and suggests precisely how to do it. And, as we'll see, it has turned out to be profoundly correct.

Plato used the word "demiurge" to denote the Creator of the physical world as we know it. "Artisan" is a standard translation. The word is chosen carefully. It reflects Plato's belief that the physical world is not the ultimate reality. Rather, there is an eternal, timeless world of Ideas, which exist prior to and independent of any necessarily imperfect, physical embodiment of them. A restless, artistic Intelligence—the Artisan—molds his creations from Ideas, using them as templates.

The *Timaeus* is not easy to interpret, and it can be tempting to mistake obscurity for depth. That said, I find it interesting, and inspiring, that Plato does not stop at the Platonic solids, but considers how atoms in that form might in turn be constructed, as physical objects, from more primitive triangles. The details, of course, are "not even wrong," but the instincts to take your model seriously, speak in its language, and push its limits are profoundly right. The idea that atoms might have subcomponents anticipates the modern ambition to analyze ever deeper. And the idea that those subcomponents might not normally exist as separate objects, but could only be found within more complex objects, is a possibility realized by today's quarks and gluons, forever confined within nuclear interiors.

Above all, we find in Plato's speculations the idea that is central to our meditation: the idea that the world, in its deepest structure, embodies Beauty. It is the animating spirit of Plato's conjectures. He proposes that the most basic structures of the world—its atoms—are embodiments of pure concepts that can be discovered and articulated by unaided Mind.

## *Economy of Means*

Returning to viruses: Where did *they* learn their geometry?

This is a case of simplicity giving the appearance of sophistication, or more precisely of simple rules giving rise to apparently complex structures that on reflection become ideally simple. The point is that the DNA of viruses, which must instruct them in all facets of their existence, is very limited in size. To economize on the length of the construction manual, it helps to make your product from simple, identical parts, identically assembled. We've heard that tune: "simple, identical parts, identically assembled" fits the very definition of a Platonic solid! Because the part generates the whole, the virus does not need to "know" about dodecahedra, or icosahedra—but only about triangles, and a rule or two for latching them together. It is the more diverse, irregular, superficially haphazard bodies—like the human—that require more detailed assembly instructions. Symmetry emerges as the default structure when information and resources are limited.

## *Young Kepler and the Music of the Spheres*

Two millennia after Plato's work, young Johannes Kepler found his calling within it. Here too the number 5 was central. Kepler, an early and passionate convert to Copernicus's idea that put the Sun at the center of creation, sought to understand the structure of the Solar System. At that time, six planets were known: Mercury, Venus, Earth, Mars, Jupiter, and Saturn. Six, you will notice, is quite close to five. Coincidence? Kepler thought not. What could be more worthy of the Creator than to use, in designing Creation, the most perfect geometrical objects?

Copernicus, like Ptolemy, based his astronomy on circular motion. This was another miscarriage of beauty, endorsed (and largely engendered) by Plato and Aristotle. Only the most perfect figure—the circle—could be worthy of Creation. The planets were supposed to be carried about on celestial spheres. Copernicus and Ptolemy had different views about where those spheres should be centered (the Sun, or Earth), but both took their existence for granted, as did young Kepler. Thus Kepler considered that there were six great spheres centered upon the Sun. He asked: Why six? And why do they have the sizes they do?

The answers struck Kepler one day while he was lecturing to an introductory astronomy class. One can circumscribe a different Platonic solid around each of the first five spheres, and inscribe it within the next. Thus the five Platonic solids can mediate among six spheres! The system will only work, however, if the spheres have appropriate sizes. In this way, Kepler could predict the relative distances between the various planets and the Sun. Convinced he had discovered God's plan, Kepler announced his discovery in a rapturous book, *Mysterium Cosmographicum*, full of quotations like this:

> I feel carried away and possessed by an unutterable rapture over the divine spectacle of the heavenly harmony.

And this:

> God himself was too kind to remain idle, and began to play the game of signatures, signing his likeness into the world; therefore I chance to think that all nature and the graceful sky are symbolized in the art of geometry.

It is indeed a gorgeous system, as you can see from the splendidly realized model in figure 9.

Evidently Kepler had asked himself, and believed he had answered, our Question: The world *does* embody Beauty, very much along the lines Plato anticipated. He went on to discuss, in concrete detail, the precise nature of the music emitted by those rotating spheres—and wrote out the score!

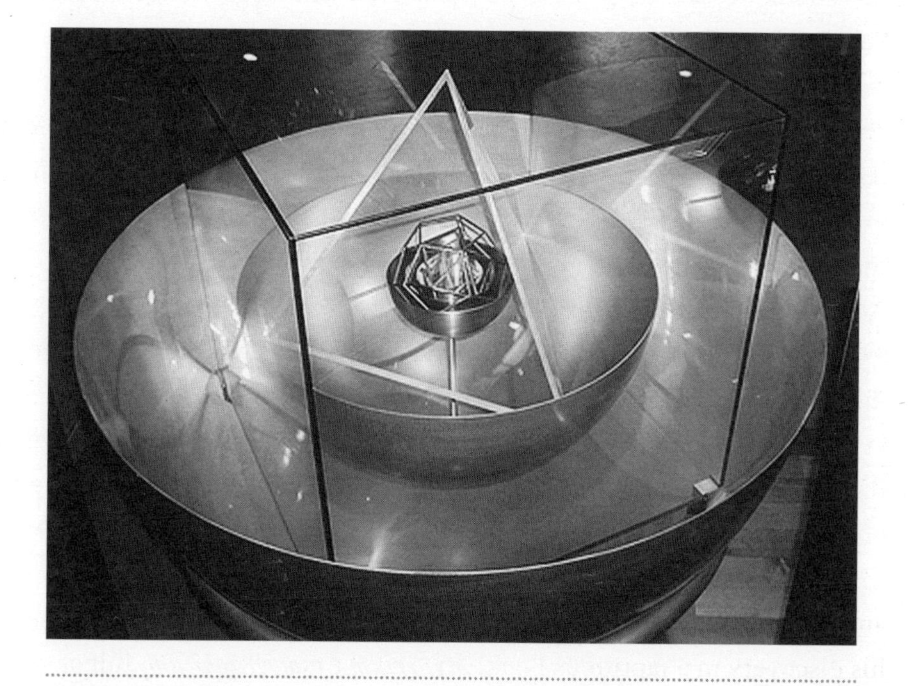

FIGURE 9. THE PLATONIC SOLIDS INSPIRED KEPLER TO PROPOSE A MODEL OF THE SIZE AND SHAPE OF THE SOLAR SYSTEM, EXHIBITED HERE. THE PLANETS ARE CARRIED ABOUT BY THE REVOLUTIONS OF CELESTIAL SPHERES WHOSE SPACING IS CONTROLLED BY PLATONIC SURFACES INTERPOSED BETWEEN THEM AS THEIR SCAFFOLDING.

Kepler's enthusiasm carried him through a life full of woes, both personal and professional. He lived close to the center of the turbulent vortex of war, religion, and politics that swept over middle Europe following the Reformation. His mother was tried as a witch. And his honest, toilsome work to describe the motion of the planets accurately resulted, through his own discoveries, in the overthrow of his youthful dream. For the planets describe not circles, but ellipses (Kepler's first law), and the Sun is not at the center of those ellipses (for experts: rather, it is at one focus). Eventually deeper beauties emerged from Kepler's mature and more accurate portrait of Nature, but they were quite different from the dreams of his youth, and he did not live to see them.

## DEEP TRUTHS

The great Danish physicist and philosopher Niels Bohr (1885–1962), a founding figure in quantum theory and author of the complementarity principle highlighted later in this book, was fond of a concept he called "deep truth." It exemplifies Ludwig Wittgenstein's proposal that all of philosophy can, and probably should, be conveyed in the form of jokes.

According to Bohr, ordinary propositions are exhausted by their literal meaning, and ordinarily the opposite of a truth is a falsehood. Deep propositions, however, have meaning that goes beneath their surface. You can recognize a deep truth by the feature that its opposite is also a deep truth. In this sense, the sober conclusion

> The world, alas, is not made according to mathematical principles in the way that Plato guessed.

. . . expresses a deep truth. As, of course, does its opposite:

> The world *is* made according to mathematical principles, as Plato guessed.

## *Dalí's* Last Supper

It seems fitting to conclude this sub-meditation with a modern work of art that plays on its themes.

Plate E, Salvador Dalí's masterpiece, *The Sacrament of the Last Supper*, contains many hidden geometrical themes. But of these themes, the strangest and most striking is the appearance of several large, but only partly realized, pentagons looming over the scene as a whole. It seems clear that they are meant to come together in a dodecahedron that embraces not only the participants in the supper, but also the viewer. And we are meant to recall Plato's conception that this shape frames the Universe.

# PLATO II: ESCAPING THE CAVE

Our Question, in asking after Beauty, hinges in part upon the relationship between physical reality and our perception of it. We have discussed this for hearing, and later we will discuss it for vision.

But there's another dimension to our Question, which is the relationship between physical reality and ultimate reality. Or, if you are (understandably) uncomfortable with the concept of ultimate reality, let's just say the big picture—how we connect the deep nature of physical reality to our hopes and dreams. What, if anything, does it all mean? Those issues are major elements in appreciating (or not) the world's beauty, as we pass beyond the level of raw perception.

Plato suggested some answers to those questions long ago. His answers were based on mystic intuition and dubious logic, rather than science. Nevertheless, they have inspired scientific work, and continue to do so. We'll have many occasions to look back to them. And their influence extends beyond science, to philosophy, art, and religion. Alfred North Whitehead famously wrote:

> The safest general characterization of the European philosophical tradition is that it consists of a series of footnotes to Plato.

So let us now visit Plato's Cave, where we find the mystic core of his worldview, captured in visionary imagery.

## ALLEGORY OF THE CAVE

Plato's allegory of the Cave occurs in his weightiest work, the *Republic*. He puts it, as he does many of his thoughts, into the mouth of Socrates, his revered teacher. Socrates describes the Cave to Glaucon, Plato's elder brother, who was likewise a student of Socrates. This setting, and this cast of characters, emphasize the central importance of the Cave in Plato's thinking.

Here is how he introduces it:

> **Socrates.** And now, I said, let me show in a figure how far our nature is enlightened or unenlightened: Behold! Human beings living in an underground den, which has a mouth open towards the light and reaching all along the den; here they have been from their childhood, and have their legs and necks chained so that they cannot move, and can only see before them, being prevented by the chains from turning round their heads. Above and behind them a fire is blazing at a distance, and between the fire and the prisoners there is a raised way; and you will see, if you look, a low wall built along the way, like the screen which marionette players have in front of them, over which they show the puppets.
>
> **Glaucon.** I see.
>
> **Socrates.** And do you see, I said, men passing along the wall carrying all sorts of vessels, and statues and figures of animals made of wood and stone and various materials, which appear over the wall? Some of them are talking, others silent.

**Glaucon.** You have shown me a strange image, and they are strange prisoners.

**Socrates.** Like ourselves.

The point is clear and simple: The prisoners see a projection of reality, not reality itself. Because that projection is all they know, they take it for granted. It is their world. But we should not feel superior to those benighted prisoners because our own situation is no different, according to Socrates (i.e., Plato). The words "Like ourselves" arrive with the force of a blow.

The story of the Cave does not *prove* that point, of course—it's only a story, after all. But it does persuade us to consider, as a logical possibility, that there's more to reality than our senses detect. And this deeply subversive story issues challenges: Do not accept limitations. Struggle to attempt different ways of viewing things. Doubt your perceptions. Be suspicious of authority.

The Platonic vision of a reality beyond the world of appearances is captured beautifully in plate F, a cosmic version of the Cave.

I should note that Plato, as a political thinker, was a utopian reactionary. He did not advocate those subversive ideas for general adoption. His prescription of freethinking was not meant as a recommendation for everyone, but only guidance for a small, select group of guardians who would be the philosophers in charge of running things. Presumably, they are his intended readers!

## *Vision of Eternity; Paradox of Stasis*

Plato's vision of a reality behind appearances unites two streams of thought. We've already dipped into one of them, the Pythagorean All Things Are Number. As we saw, several beautiful discoveries gave support to that credo. Plato's own atomic theory, which we discussed in the preceding chapter, was another attempt in the same spirit (lacking only in evidence or truth).

A second stream was more properly philosophical, in the modern sense. It is a piece of *metaphysics*, or "after physics." (The roots of this word are interesting. When Aristotle's works were collected, the books following his *Physics* were referred to as "after physics," simply because they were bound that way. The subject matter of those books was the first principles of things. In metaphysics such topics as being, space, time, knowledge, and identity are treated, not by experiment and observation, but by pure reasoning, like mathematics. That ambitious yet elusive intellectual quest has ever after been called metaphysics.)

Here is a typical piece of metaphysics, due to Parmenides, as described by Bertrand Russell, a master philosopher and mathematician of the twentieth century. It demonstrates why nothing can ever change (!):

> When you think, you think *of* something; when you use a name, it must be the name *of* something. Therefore both thought and language require objects outside themselves. And since you can think of a thing or speak of it at one time as well as another, whatever can be thought of or spoken of must exist at all times. Consequently there can be no change, since change consists in things coming into being or ceasing to be.

Despite that unassailable logic, it is not entirely easy to convince oneself, psychologically, that nothing ever changes. If change is an illusion, it's a pretty convincing one.

It certainly looks as though things move, for example. The first step in overcoming that illusion is to undermine naive faith in appearances. Parmenides's student, Zeno of Elea, was a master of subversion. He invented four paradoxes intended to show that naive ideas about motion are hopelessly confused.

The most famous of these is the paradox of Achilles and the tortoise. Achilles, the great hero of Homer's *Iliad*, was a warrior renowned for his speed afoot as well as his strength. We are asked to imagine a race between Achilles and an ordinary tortoise—let's say, to be concrete, a fifty-yard dash. The tortoise is allowed a ten-yard head start. One might expect that Achilles will win. "Wrong!" says Zeno. Zeno points out that in order to overtake the tortoise, Achilles must first catch up. And that's a big problem—in fact, an infinitely big problem. Suppose that at the start, the tortoise is at position *A*. Achilles will run to *A*, but by the time he arrives there, the tortoise will have moved ahead, to *A′*. Then Achilles will run to *A′*, but when he arrives there, the tortoise will have moved ahead, to *A″*. You see where this is going—no matter how many iterations we allow, Achilles never actually catches up.

It may be mind-bending to deny motion, as Parmenides recommends. But it is worse, argues Zeno, to accept it. That is not mind-bending, but mind-breaking.

Bertrand Russell looked back on Zeno in these words:

> Having invented four arguments, all immeasurably subtle and profound, the grossness of subsequent philosophers pronounced him to be a mere ingenious juggler, and his arguments to be one and all sophisms. After two thousand years of continual refutation, these sophisms were reinstated, and made the foundation of a mathematical renaissance.

Indeed a proper physical answer to Zeno only emerged with Newton's mechanics and the mathematics it embodies, as we'll take up a little later.

Today it seems possible, in the framework of quantum theory, to agree with Parmenides, and yet to do justice to appearances. Change might indeed be a mere appearance. I will justify that outrageous claim toward the end of this meditation.

But let us resume our narrative, and return to the historical order of things.

## *The Ideal*

In Plato's theory of Ideals, these two streams—Pythagorean intuitions of harmony and perfection, and Parmenides's changeless reality—flow together. (Plato's theory is generally called the theory of Ideas, but I think "Ideal" corresponds better to what Plato had in mind, so I'll use that word.)

The Ideals are the perfect objects, of which real objects are imperfect copies. Thus, for example, there is an Ideal Cat. Actual animals are cats to the extent that they share in the properties of that Cat. The Ideal Cat, of course, never dies, nor can it change in any way. This theory embodies Parmenides's metaphysics: There is a realm of Ideals, the deepest reality, which is eternal and unchanging, and provides the source of all we can name or speak of. And it builds upon Pythagoras: We come into close contact with the world of eternal, perfect Ideals when we deal in mathematical concepts like numbers, or Platonic solids.

There is a third, subterranean stream that also surely fed into the theory of Ideals. This is the stream of Orphic religion. That was the serious side, we might say, of Greek mythology. The details of Orphism, which featured secretive rituals, are obscured in the mists of history (that's the fate of secrets!), and they need not concern us here. But its centerpiece was the doctrine of immortality of souls, which had (and, of course, still has) sublime emotional appeal. Wikipedia describes it as

> Characterizing human souls as divine and immortal but doomed to live (for a period) in a "grievous circle" of successive bodily lives through metempsychosis, or the transmigration of souls.

These ideas fit into the theory of Ideals elegantly. Each of us, by our nature, participates in the world of Ideals. The part of us that participates is our soul, and is imperishable. While we live on Earth our attention is diverted by appearances, and if we do not transcend them, we are only dimly aware of the Ideals, and our soul slumbers. But through philosophy, mathematics, and a dose of mysticism (the mysterious ceremonies of Orphism) we can awaken it. There is a Cave—and there is a way out.

## *Liberation*

Plato describes the process of liberation:

> **Socrates.** And now look again, and see what will naturally follow if the prisoners are released. . . . The glare will distress him, and he will be unable to see the realities of which in his former state he had seen the shadows. . . . Will he not fancy that the shadows which he formerly saw are truer than the objects which are now shown to him?
>
> **Glaucon.** Far truer.
>
> . . . . . . . .
>
> **Socrates.** He will require to grow accustomed to the sight of the upper world. And first he will see the shadows best, next the reflections of men and other objects in the water, and then the objects themselves; then he will gaze upon the light of the moon and the stars and the spangled heaven; and he will see the sky and the stars by night better than the sun or the light of the sun by day?
>
> **Glaucon.** Certainly.

It is noteworthy that Plato (through Socrates) describes liberation as an active process, a process of learning and engagement. This is quite different from ideas that are more popular, though to me less inspiring, where salvation comes about through external grace, or through renunciation.

If liberation comes through engagement with a hidden reality, how are we to achieve it? Here there are two paths, inward and outward.

Along the inward path, we examine our concepts critically, and try to strip them of the dross of mere appearance, to reach their ideal (i.e., Ideal) meaning. This is the path of philosophy and metaphysics.

Along the outward path, we engage appearances critically, and try to strip them of complications, to find their hidden essence. This is the way of science and physics. As we've anticipated, and will discuss in depth, the outward path does in fact lead to liberation.

## *Undoing Projection: Looking Ahead*

In his central intuition, Plato was quite correct—indeed, more profoundly correct than he possibly could have known. Our naturally given view of the world is but a shadowy projection of the world as it truly is.

Our unaided senses take only paltry samples from the cornucopia of information the world puts on offer. With the help of microscopes, we discover a microcosmic universe full of tiny alien creatures, some our friends and some our enemies, and the yet more alien constituents of our material being, things that play by the weird rules of quantum mechanics. With the help of optical telescopes, we discover the vast size of the cosmos dwarfing our Earth, featuring vast, dark, (apparently) empty spaces sprinkled with billions of billions of varied suns and planets. With the help of radio receivers, we come to "see" invisible radiations that fill space, and to put them to use. And so on . . .

As for our senses, so for our minds. Without training and help, they cannot begin to do justice to the richness of reality we know, let alone what we don't yet know—the unknown unknowns. We go to school, read books, tap into the Internet, and use scratch pads, computer programs, and other tools to help us keep complicated ideas in order, solve the equations that govern the Universe, and visualize their consequences.

Those aids to sensation and imagination open the doors of perception, allowing us to escape from our Cave.

## THE TURN TO UNWORLDLINESS

But Plato, knowing nothing of that future, emphasized the inward path. Here he explains why:

**Socrates.** Accordingly, we must use the embroidered heaven as an example to illustrate our theories, just as one might use exquisite diagrams drawn by some fine artist such as Daedalus. An expert in geometry, faced with such designs, would admire their finish and craftsmanship; but he would not dream of studying them in all earnest, expecting to find all angles and lengths conforming exactly to the theoretical values.

**Glaucon.** That would of course be absurd.

**Socrates.** The genuine astronomer, then, will adopt the same outlook when studying the motions of the planets. He will admit that the sky and all it contains have been framed by their maker as perfectly as things can be. . . . He will not imagine that these visible, material changes go on for ever without the slightest alteration or irregularity, and waste his efforts trying to find perfect exactitude in them.

**Glaucon.** Now that you put it like that, I agree.

**Socrates.** So if we mean to study astronomy in a way which makes proper use of the soul's inborn intellect, we shall proceed as we do in geometry, working at mathematical problems, and not waste time observing the heavens.

We can summarize that one-sided dialogue in an inequality. It states, quite simply, that the Real does not live up to the Ideal. The Real is strictly less:

$$\text{Real} < \text{Ideal}$$

The Artisan, who creates the physical world from the world of Ideals, is an artist, and a good one. Yet ultimately the Artisan is a copyist whose creations reflect the messiness of the available materials. The Artisan paints with a broad brush and blurs details. The physical world is a flawed representation of the ultimate reality we should seek.

To put it another way: Plato recommends unworldliness. If your theories are beautiful, but do not exactly agree with observations—well, then, so much the worse for the observations.

## *Two Kinds of Astronomy*

Why did Plato, in seeking ultimate truth, turn inward, away from the physical world? Part of the reason, no doubt, was that he loved his theories too much, and could not bring himself to contemplate their possible failure. That all-too-human attitude is still with us—it is standard in politics, common in social sciences, and not unknown even in physics.

But part of the reason emerged from the study of Nature, in astronomy, the subject his dialogue alludes to.

Keeping an accurate calendar was important for the societies of the ancient world, whose economic base was agricultural, and especially so for those that relied on irrigation systems. It was also, not coincidentally, important for religious purposes, because rituals were timed to get godly assistance at planting and at harvest. All this required astronomy. So too did the art of divining the future through astrology. The ancient Babylonians became extremely adept at predicting the timing of astronomical events, including the variation of the Sun's position at dawn and on setting, equinoxes, solstices, and eclipses of the Moon and Sun. Their method was simple, in principle, and almost theory-independent. They accumulated centuries of accurate observations, noted regularities (periodicities) in the behaviors, and extrapolated those regularities into the future. In other words, they assumed that future cycles of behavior in the celestial realm would reproduce past behavior, as it had been observed to do repeatedly in the past. "Big data" is all the rage today, but the basic concept goes back a long way, for it is none other than the method of ancient Babylonian astronomy.

The Babylonian work was maturing as Plato wrote, and most likely he had no more than vague knowledge of it. In any case, their "bottom-up," data-heavy, theory-light approach was completely at odds with his goals and methods.

To Plato, as we've seen, what seemed overwhelmingly important is the human soul—its ascent to wisdom, purity, and a transcendent Ideal. Thus in building an account of planetary motion, what is most important is that the theory should be beautiful, not that it should be completely accurate. The primary goal is to identify the Ideals that inspired the Artisan. The compromises that coarse building materials forced upon Her are a secondary concern.

The dominant, as well as the simplest, periodicities in astronomy are the regular cycles of day and night and of the seasons, which associated the apparent motion of stars across the sky and with the Sun's apparent trajectory. Today, we understand that those cycles are associated with the daily rotation of the Earth around its axis, and its yearly revolution around the Sun. Because both these motions are fairly close to being circular motions at constant speed, the observed phenomena could be described (to an excellent approximation) by an extremely beautiful theory, as follows:

The most perfect geometric figure is a circle. For uniquely among closed figures, a circle has the same appearance everywhere along its extent. Any other figure exhibits differences among its different parts, and so not all of those parts can be the best possible, and therefore neither can the whole. Similarly, the most perfect motion in a circle is motion at constant speed. Also, motion in a circle at constant velocity is as unchanging as motion can possibly be, because it takes the same form at every moment. From these "top-down" considerations, we deduce that the Ideal of motion is motion in a circle at constant velocity. And when we look to the sky, we discover that by combining two such perfect motions, we can match the observed motion of the Sun and stars, pretty nearly.

This is, at first sight, a stunning success. It continues in the spirit of Pythagoras's discoveries, identifying hidden numerical and geometric relationships at play in the physical world. It surpasses those discoveries in grandeur and nobility, because the Sun and stars emanate directly from the Artisan, while mere human artisans are responsible for musical instruments.

Unfortunately, as we attempt to go beyond that initial success, things get very messy very fast. The apparent motions of the planets, and the Moon, prove much more difficult to describe. The top-down approach demands that we account for the appearances in terms of Ideal (that is, uniform circular) motion. Mathematical astronomers responded to that challenge by putting the (hypothetical) circular paths of the planets into circular motion. That still didn't quite work, so they put the circles of motion of the circular paths of the planets into circular motion. . . . With enough of these cycles upon cycles, artfully arranged, it is possible to reproduce appearances. But in those complicated, manifestly artificial systems, the initial promise of purity and beauty was lost. One could have either beauty or truth, but not both at once.

Plato insisted on beauty, and was ready to compromise—or we might better say, to surrender—accuracy. That disdain for facts, beneath its veneer of pride, betrays deep lack of confidence, and a kind of exhaustion. It gives up on the ambition to have it all, marrying beauty *and* accuracy, Real *and* Ideal. Plato gestured in that unworldly direction, and his disciples took it much further. The attraction of unworldliness in those dark times is understandable, considering that the world had war, poverty, disease, and the collapse of classical Greek civilization on offer.

Plato's successor and rival, Aristotle, was in some ways a truer student of Nature. He and his students collected biological specimens, made many acute observations, and recorded their results honestly and in detail. Unfortunately, by focusing from the start on very complex objects and problems, they missed the clarifying simplicity of geometry and astronomy. They did not seek, nor could they hope to find, the mathematically Ideal within those gnarly branches of the Real. They emphasized description and organization, and did not strive for beauty or perfection. When the Aristoteleans turned to physics and astronomy, they limited their ambitions in the same way. Where later (and earlier) scientists would demand precise equations, they were content with broad verbal descriptions.

## OBJECTIVE SUBJECTIVITY: PROJECTIVE GEOMETRY

Centuries later, in the Renaissance, a newly confident culture rediscovered Plato. It took up his quest for the Ideal, while abandoning his otherworldliness.

Artists and artisans—human ones—led the way. The challenge they took up is very basic: How can a two-dimensional painting represent the geometry of objects in three-dimensional space? This was a concrete, practical problem. At a time when photography was unknown, and significant private fortunes were accumulating, patrons wanted to have pictures that would give a permanent record of how they and their possessions actually looked.

At first sight, it might seem that nothing could be further from Plato's desire to get to a deeper reality beyond the superficial appearance of things. The artistic science of perspective is all about getting the superficial appearance right!

Yet there is a sense in which mastering the appearance of things brings us closer to their essence. For by understanding how the same scene can appear different, depending on the viewpoint from which it is perceived, we learn to separate the accidents of viewpoint from the properties of the thing itself. By treating subjectivity objectively, we master it.

So much for generalities. The first steps in this work already reveal some delightful surprises. Let us simplify the problem to its bare essence, taking a cross-section of canvas and landscape, so that both appear as lines, in figure 10:

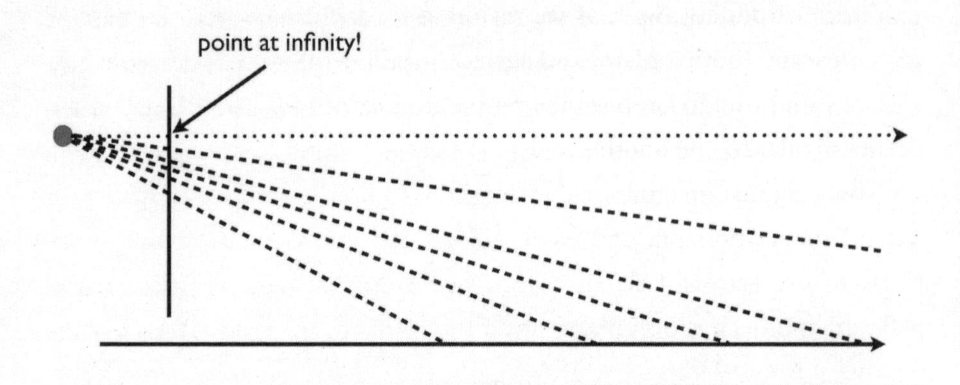


FIGURE 10. THE POINTS ALONG THE HORIZONTAL LINE (A FLOOR) PROJECT ONTO A LINE SEGMENT ALONG THE VERTICAL LINE (A CANVAS). THE INFINITE LIMIT OF THE HORIZONTAL LINE, WHICH IS NEVER ATTAINED IN REALITY, NEVERTHELESS PROJECTS TO A REAL, FINITE "POINT AT INFINITY" ON THE CANVAS.

The points on our maximally simplified landscape—a flat horizontal plane, reducing to a line in cross-section—project light in straight lines to the viewer. They are the dashed lines in the figure. By following those lines to where they intersect the canvas (whose cross-section is the vertical solid line) we determine where the different landscape points should appear in the painting.

As you can see, points that are farther away get projected higher up, vertically, on the canvas. But as we consider more and more distant landscape points, the rate at which their images climb the canvas decreases. The connecting light rays approach a horizontal limit, depicted by the dotted line in the figure. That limiting line does not correspond to an actual point on the landscape—yet it intersects the canvas at a specific point.

Right before our eyes, a conceptual miracle has occurred: We have captured infinity! As we view the landscape, there is a horizon. The horizon is not a physical thing, but an idealization. It represents the boundary of our vision, and lies at infinite distance. Yet the image the horizon casts on our canvas is unquestionably real. It is a unique, specific point—the point at infinity.

Further wonders await, as we restore both the canvas and the base of our landscape (both a plain, and a plane) to their full two dimensions.

Let's suppose, to keep things simple, that the canvas and the plain are perpendicular to one another.

Now we must imagine many straight lines on the landscape. Each will extend to the horizon, and each will project its associated point at infinity to the canvas. One discovers, however, that parallel lines on the plain all approach the same point on the horizon. In figure 11, this leaps to your eye.

FIGURE 11. PARALLELS MEET AT THE HORIZON, IN A COMMON "VANISHING POINT." ONCE YOU ARE ALERT TO THIS PHENOMENON, YOU WILL SEE IT ALL AROUND YOU.

We call that point the *vanishing point* of the family of parallels. In the language appropriate to the canvas, we can say that *parallel lines meet in the point at infinity.*

Here mystical poetry emerges as a straightforward description of artistic reality.

Different families of parallel lines define different vanishing points, which together define the horizon. Projected back to the canvas, the horizon generates a horizontal line, capturing the horizon as a collection of points at infinity. The conceptual horizon, in other words, projects onto canvas as the tangible line at infinity.

Discoveries such as these both excited and empowered the pioneering Renaissance artist/scientist/engineer Brunelleschi. He developed these insights into a powerful technique for producing realistic drawings. In a famous experiment, he used projective geometry to make an accurate representation of how the Baptistery of St. John, in Florence, should appear, as seen from an entrance to the nearby cathedral, then under construction. As shown in figure 12, he arranged so that a viewer could compare the drawing, seen reflected in a mirror, to the actual Baptistery,

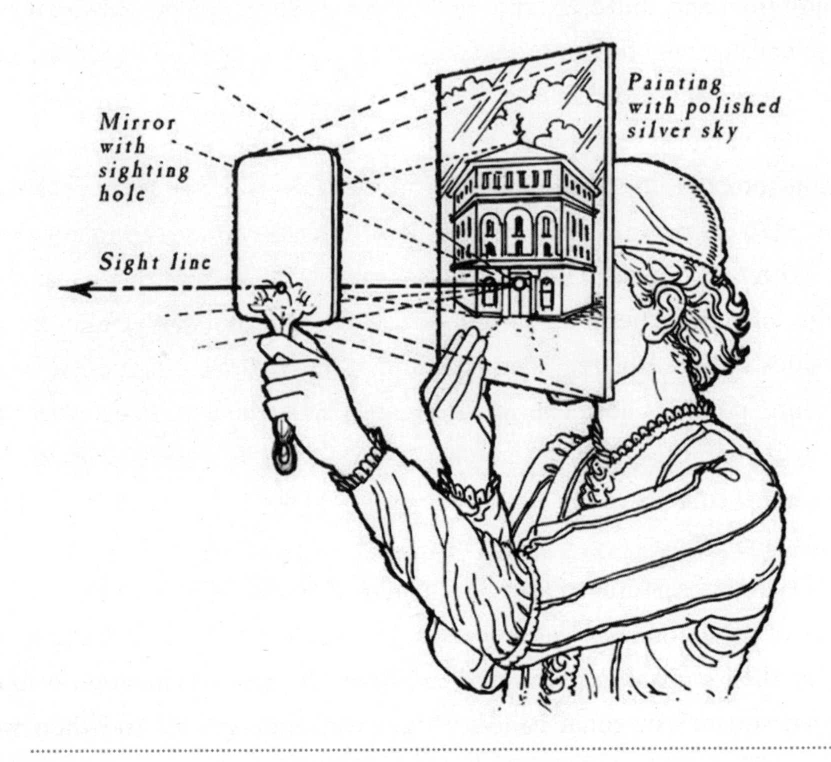


FIGURE 12. BRUNELLESCHI'S DEVICE FOR COMPARING HIS DRAWINGS, BASED ON THE NEW SCIENCE OF PERSPECTIVE, TO REALITY.

revealed when the mirror is removed. (A small hole in the drawing permitted viewing.)

This ingenious demonstration made a huge impression on contemporary artists, who took up Brunelleschi's techniques with enthusiasm and developed them energetically. Before long, exuberant joy in perspective infused masterpieces such as *Giving of the Keys to Saint Peter* (plate G), by Pietro Perugino. Here perspective is an active player, lending a special sense of order, harmony, and authority to this founding event for the Catholic Church. This fresco is in the Sistine Chapel.

There's no better way to understand the joy of the artists who discovered and experimented with perspective than to share in one of their simpler creations. In plate H, I've indicated the process by which you can make an accurate perspective drawing of a floor tiled by squares, viewed from in front and above, extending off to an infinite horizon. All you need is a pencil, a straightedge, and an eraser. ("Straightedge" is the term of art for a ruler without distance markings. Of course, a ruler with distance markings will also serve—just ignore the markings!)

The process of construction is indicated in the top portion of the figure. We draw a line, indicated in black, which will be the horizon. We start with one square tile, indicated in blue at the bottom. It isn't drawn square, of course, because we're viewing the floor obliquely. The opposite sides of the "square," when continued, meet at the horizon, at their vanishing points. These continuations are also in blue. So that's what we start with: one tile, and the horizon. The challenge is then to draw all the other equal squares in the tiling as they would appear (in perspective) to an actual viewer.

The key observation is that the diagonals of the squares also form a family of parallel lines. That family of parallels will also meet at the horizon, at their vanishing point. We can draw the red continuation of our original square's diagonal to locate that vanishing point. And then we continue back from the vanishing point, with the orange lines, to get diagonals for the neighboring squares! Having located those diagonals, we know that the intersections of the orange and blue lines are vertices of the neighboring squares. The yellow lines, through those vertices and the appropriate vanishing points, therefore contain the sides of those squares. And now we can keep going—the intersections of the yellow "side" lines with the orange "diagonal" lines are the vertices of new squares. . . . You can keep going as long as you like, until you lose patience, or your pencil wears down—or your squares shrink down to atomic dimensions.

To complete the construction, you can just erase the diagonals, and (optionally) make all the lines the same color, to arrive at the bottom figure. The foreshortening in this figure is extreme—as though we're viewing the floor from an ant's-eye view, close by and at a small height—to emphasize just how different equal figures, such as these squares, can appear. Of course, you can pick up the book and view the figure from different perspectives—you'll see different apparent sizes of the squares, but always the same pattern of intersections.

I've done this construction a dozen times and more, and still each time I feel a thrill when that tiled floor emerges. It's a small, but genuine, act of creation.

I feel it's the sort of thing any Artisan would enjoy.

I've found that appreciating these basic ideas of perspective opens my eyes. To put it more accurately, those ideas bring the message of my eyes into more intimate contact with my conscious mind. Especially in an urban environment, I often discover many sets of (physically) parallel lines, going off toward different vanishing points. When I'm alert to those things, my experience is fuller and more vivid. I hope you will find this happens for you too. Through disciplined imagination, we transcend the Cave of ignorant sensation.

## *Matters of Perspective: Relativity, Symmetry, Invariance, Complementarity*

Many of the central ideas of modern fundamental physics are unfamiliar to most people. They can seem abstract and forbidding if they are introduced abruptly, in the strange contexts that are their natural habitat. For this reason, those of us who try to bring those ideas to a broad audience often work in metaphors and analogies. It's challenging to find metaphors that are both faithful to the original ideas and readily accessible, and even more challenging to do that in a way that does justice to their beauty. I've struggled with that problem many times over the years. Here, I'm happy to present a solution that's given me a real feeling of satisfaction.

*Projective geometry*, that artistic innovation of the Renaissance, contains not merely metaphors, but genuine *models*, in a gallery of big, cunning, and awesomely pregnant ideas:

- *Relativity* is the idea that the same subject can be represented, faithfully and without loss, in many different ways. Relativity, in this sense, is the very essence of projective geometry. We can paint the same scene from many different perspectives. The dispositions of paint on canvas will be different, but all will represent the same information about the subject, just differently encoded.
- *Symmetry* is an idea closely related to relativity, but with attention directed to the subject, rather than the viewer. For example, if we rotate the subject of our painting, then from any fixed perspective it will look different. But its projective description—that is, the totality of views offered by all possible perspectives—remains the same (because painters can relocate their easels to compensate). We summarize this situation by saying that rotation of the object is a *symmetry* of its projective description. We can change the object, by rotating it, without changing its projective

description. As we shall emphasize again in later encounters, Change Without Change is the essence of symmetry.

- *Invariance* is the counterpoint to relativity. Many aspects of the subject are differently represented as we change perspectives, but some features are common to all those representations. For example, straight lines in the subject will always appear as straight lines from any perspective (though their orientation and position on the canvas will vary); and if three straight lines intersect in the subject, their representatives will meet at a point, from any perspective. Features that are common to all representations are said to be *invariant.* Invariant quantities are profoundly important, because they define features of the subject that are valid from any perspective.
- *Complementarity* is an intensification of relativity. It is one of the deep principles of quantum theory, but its importance, as an insight into the nature of things, goes beyond physics. (Complementarity is, I think, a genuine metaphysical insight—a rare bird indeed.)

  At its simplest level, complementarity says that there can be many different views of your subject that are equally valid, in principle, but that to observe (or paint, or describe) the subject you must choose a particular one.

  If that were all there is to it, complementarity would be a minor gloss on relativity. The novelty that arises in quantum theory is that it is generally *impossible* for two quantum portraitists to portray, from different perspectives, the same subject at the same time. For in the quantum world we must take into account that observation is an active process, wherein we interact with our subject.

  Let us try to see an electron, for example. To do that we must irradiate the electron with light (or X-rays). But light transmits energy and momentum to the electron, and can therefore disturb its position, which is the very thing we hoped to determine!

  By taking appropriate precautions, and fiddling with the electron just so, we can arrange our measurements to capture some aspects of our subject correctly. But others must be sacrificed, as they are destroyed in the process of observation. With different arrangements and precautions, we can choose our division between the observed and the sacrificed in different ways, but we cannot avoid choosing. In painting the quantum world, we must choose one among all possible perspectives, and work to achieve it. If another painter is also at work, fiddling with our electron in his own way for his own purposes, he will cloud our vision, and ruin our portrait (and we will ruin his).

The strong form of complementarity, which takes it beyond relativity, is this: There are many equally valid views of your subject—perspectives, in the general sense of that word—but they are mutually exclusive. In the quantum world, we can realize only one perspective at a time. Quantum cubism is a no-go.

These big ideas—relativity, symmetry, invariance, complementarity—form the heart of modern physics. They should be, though they are not yet, central to modern philosophy and religion. In all those contexts they sometimes appear in strange and abstract forms that can seem bewildering. It is a wonderful possibility, in case of bewilderment, to think back to projective geometry, where we can visit them embodied in tangible, artistically beautiful images.

# NEWTON I: METHOD AND MADNESS

The classic Scientific Revolution was not a single historical event, but an intense period extending roughly over the years 1550–1700, marked by dramatic progress in many fields, but above all in physics, mathematics, and astronomy. The energetic curiosity and invention of Renaissance artist-engineers such as Filippo Brunelleschi and Leonardo da Vinci foretold its spirit, but Copernicus's *De Revolutionibus Orbium Coelestium* (*On the Revolutions of the Celestial Spheres*) is usually taken to be its seminal document. Copernicus put forward serious arguments, based on mathematical analysis of astronomical observations, that Earth was neither the center of the Universe nor immobile, but a rotating satellite of the Sun. That conclusion seemed an outrage to common sense, not to mention Church doctrines in cosmology, which had been heavily influenced by Plato and Aristotle. But the math would not go away. The radical thinkers who chose to build out from its precision, rather than to sequester its influence, eventually triumphed. Breakthrough work by Galileo, Kepler, and René Descartes culminated in the synthesis of Isaac Newton—an intellectual singularity, on which this part of our meditation centers.

## ANALYSIS AND SYNTHESIS

Beyond its wealth of specific discoveries, the Scientific Revolution was a revolution in ambition and, profoundly, in *taste.* The new thinkers were not satisfied with a bird's-eye overview of reality, in the style of Aristotle. They demanded that the ant's perspective, as well as the bird's, be fully respected. Nor would they renounce any detail of it to accommodate some higher mind's-eye view, in the style of Plato. They demanded observation, measurement, and precise description, using geometry, equations, and systematic mathematics whenever possible.

Newton captured the essence of the new vision here:

> As in Mathematicks, so in Natural Philosophy, the Investigation of difficult Things by the Method of Analysis, ought ever to precede the Method of Composition. . . . By this way of Analysis we may proceed from Compounds to Ingredients, and from Motions to the Forces producing them; and in general, from Effects to their Causes. . . . And the Synthesis consists in assuming the Causes discover'd, and establish'd as Principles, and by them explaining the Phænomena proceeding from them, and proving the Explanations.

Let us elaborate this powerful statement of method more fully, and enrich its context.

### *Demanding Precision*

An ant must be concerned with getting local topography exactly right, while a bird flies fast through a mostly open sky. An ant gazing at the sky will walk into roadblocks, or fall into pits, while a bird that focuses on surface details eventually flies into a cliff face. Similarly there is tension between the goals of *precision* and *ambition*—between speaking only truth on the one hand, and having a lot to say on the other.

Earlier, we discussed Plato's choice to abandon precision, while going all in on ambition. For him that was a conscious decision, motivated by his hope to discover, through intellectual and spiritual exercises, a better world, of which ours is an imperfect copy. Pythagoras had discovered wonderful but subjective, and therefore imprecise, laws of musical harmony. Astronomy appeared to encode precise, but not quite accurate, laws, as we've discussed. Only the laws of mathematics itself—our window to the Ideal—could be, in Plato's view, both precise and unfailingly correct.

This tension between Real and Ideal reaches the level of Orwellian doublethink in the writings of Newton's predecessor Johannes Kepler. We've already mentioned Kepler's youthful infatuation with a model of the Solar System based on the Platonic solids. Though equally (that is, completely) wrong, Kepler's conception reaches a higher level, scientifically, than Plato's speculations in *Timaeus*. For Kepler, unlike Plato, tries to be both precise and specific. Mercury's sphere supports a circumscribed octahedron, which is inscribed within Venus's sphere. Then we have an icosahedron, a dodecahedron, a tetrahedron, and a cube interpolating, respectively, Venus-Earth, Earth-Mars, Mars-Jupiter, and at last Jupiter-Saturn. That scheme makes concrete numerical predictions about the relative sizes of planetary orbits, which Kepler compared to their observed values. The agreement, while not exact, was close enough to convince Kepler he was on the right track. Encouraged, he then courageously set out to sharpen his model and to compare it with better data, so as to bring out more clearly the Music of the Spheres.

Thus Kepler's model set him on his storied career in astronomy. His arduous labors of calculation led him to discover regularities among the orbits of the planets—his famous three laws of planetary motion—that really do hold precisely. Kepler's laws of planetary motion came to play a central role in Newton's celestial mechanics, as we'll discuss in the chapter "Newton III."

Kepler was delighted with, and justly proud of, those discoveries. Yet they fatally undermined the foundation of his own beautiful system of celestial spheres supported by Platonic solids. In his efforts to do justice to Tycho Brahe's exquisitely accurate observations, Kepler discovered that the orbit of Mars is not circular at all, but follows an ellipse. So much for celestial spheres!

Kepler's own work had destroyed his model's conceptual foundation, and its approximate agreement with observation did not survive in more accurate work. Yet Kepler never abandoned his ideal system. He prepared a later, much expanded edition of the *Mysterium* in 1621. There the accurate laws appear in footnotes, undermining the text like a sober cross-examination that belies the words of a witness prone to fantasy. Symbol or model? Ambition or precision? In refusing to choose, Kepler fell back into the Platonic temptation, to put his theoretical Ideal above the reality it contradicted.

With Newton, the break is decisive. Theories that do not describe reality are, to Newton, mere hypotheses, beyond the pale:

> Whatever is not deduced from the phenomena must be called a Hypothesis; and Hypotheses, whether metaphysical or physical, or based on occult qualities, or mechanical, have no place in experimental philosophy.

And the description theories provide must be precise. The historian and philosopher of science Alexandre Koyré identified this *raising of standards* as Newton's most revolutionary achievement, forming the capstone of the Scientific Revolution:

> To abolish the world of the "more or less," the world of qualities and sense perception, the world of appreciation of our daily life, and to

> replace it by the (Archimedean) universe of precision, of exact measures, of strict determination.

Those high standards of realism and precision are not easy to meet! Plato declared that they were mutually exclusive, and in practice even Kepler was content to satisfy one or the other. Newton, in his work on light and on mechanics, showed that they *could* be met—and thereby provided models of the sort of theoretical excellence we, his followers, aspire to today. To meet those standards, one must rein in premature ambition, as Newton recognized:

> To explain all nature is too difficult a task for any one man or even for any one age. . . .' Tis much better to do a little with certainty & leave the rest for others that come after you than to explain all things by conjecture without making sure of any thing.

## *Nurturing Ambition*

And yet Newton himself was supremely ambitious. His curiosity extended in many directions, and in his vast notebooks one finds hypotheses galore about all sorts of things. Reading Newton is an exhilarating but an exhausting experience because the ideas—clever ones—come thick and fast. He makes extensive observations on fermentation, muscular contraction, and the transformations of matter chronicled in ancient alchemy and modern chemistry.

To reconcile his ambition with his demands for rigor, Newton used two basic techniques. One was a method of intellectual work, the other a trick of presentation.

I like to think of his *method* as a process of selection—a kind of Darwinian struggle for survival in the world of ideas. Newton always tried to put his guesses to work by deriving consequences he could compare with observations. Some survived this trial, or left surviving descendants, while others went extinct.

There are many ideas in the notebooks that fizzled out and never went public. Newton's famous statement

> I do not know what I may appear to the world, but to myself I seem to have been only like a boy playing on the sea-shore, and diverting myself in now and then finding a smoother pebble or a prettier shell than ordinary, whilst the great ocean of truth lay all undiscovered before me.

. . . is often interpreted as indicating his becoming modesty. I don't see it that way. Newton was not a modest man, but he was an honest one. He, of all people, was aware of how much he'd left on the table.

Others of his guesses survived, but did not sufficiently thrive to reach Newton's avowed standards. Those, he smuggled into public view by a trick.

Newton's *trick* is charming in its transparency. It is the technique of putting a question mark at the end of statements. For then they are not assertions, or Hypotheses, but only Queries. Newton's last scientific work is, in fact, a set of thirty-one *Queries* he attached to later editions of his *Opticks.*

The early Queries are short, leading questions, usually phrased negatively. Here, for example, is the first:

> Do not Bodies act upon Light at a distance, and by their action bend its Rays; and is not this action strongest at the least distance?

This Query, like many of the others, was really a suggestion for research. And like many of the others, it proved quite fruitful. We can read it as an anticipation of the bending of light by the Sun and by distant galaxies—major discoveries of *twentieth-century* physics.

Although he doesn't seem himself to have worked it out in detail, it is not difficult to apply Newton's law of universal gravity to light, regarding light—as Newton usually did—as composed of material particles. The orbits of light particles would, in this concept, follow the same kinds of orbits as planets that have the same velocity. (Gravitational force is proportional to mass, and force in general is equal to mass times acceleration. And so in calculating acceleration due to gravity, the mass cancels out.) Newton knew of Ole Rømer's astronomical determination of the speed of light, and refers to it in the *Opticks*, where he mentions that light takes somewhere between seven and eight minutes to travel from the Sun to Earth. So Newton was in a position to estimate the bending of light by the Sun due to gravity. It is a very small effect, well beyond the technology of Newton's time to measure. Einstein calculated the bending of light by the Sun, first essentially as Newton might have, and then using his new theory of general relativity, which gives twice as large an answer, in 1915. His prediction was tested by an international expedition during the solar eclipse of 1919, when the positions of surrounding stars were seen to be distorted. The expedition's success, marking a return of common European values after the fiasco of World War I, was a sensation, and established Einstein as a world celebrity.

Much larger masses and distances are in play when the light of distant galaxies passes close to other foreground galaxies, leading to the spectacular phenomenon of gravitational lensing. Images of distant galaxies are sometimes distorted by the passage of their light through the gravitational fields of intervening matter, similar to how the image of a straw is distorted when the straw is viewed through water. In figure 13, for example, the arcs are distorted images of a very distant galaxy population extending five to ten times farther than the lensing cluster.

Newton would surely have enjoyed this cosmic vindication of his first Query!

FIGURE 13. THE GRAVITATIONAL FIELDS OF BODIES CAUSE LIGHT TO BEND, CREATING COSMIC LENSES. HERE YOU CAN SEE SOME HUGELY DISTORTED IMAGES OF GALAXIES, MADE TO APPEAR AS WISPY ARCS.

## *Looking Everywhere*

The later Queries open increasingly wide-ranging discussions, until in Query 31 the question mark is left implicit. Here we reach Newton's grandest hypothesis of all, and his last words on light and Nature:

> For so far as we can know by natural Philosophy what is the first Cause, what Power he has over us, and what Benefits we receive from him, so far our Duty toward him, as well as that toward one another, will appear to us by the Light of Nature. And no doubt, if the Worship of false Gods had not blinded the Heathen, their moral Philosophy would have gone farther than to the four Cardinal Virtues; and instead of teaching the Transmigration of Souls, and to worship the Sun and Moon, and dead Heroes, they would have taught us to worship our true Author and Benefactor, as their Ancestors did

> under the Government of Noah and his Sons before they corrupted themselves.

To some it may seem strange that the greatest hero of the Scientific Revolution would venture into these questions of theology and ethics. But Newton saw the world whole.

John Maynard Keynes, a polymath most famous for his work in economics, made a pioneering study of Newton's voluminous unpublished papers. Keynes summarized his impressions in "Newton, the Man," a remarkable essay that I highly recommend (see "Recommended Reading"). According to Keynes,

> He regarded the Universe as a cryptogram set up by the Almighty.

For Newton, Nature was not the only source of answers to the riddles of existence:

> Philosophy both speculative & active is not only to be found in the volume of nature but also in the sacred scriptures, as in Genesis, Job, Psalms, Isaiah & others. In the knowledge of this philosophy God made Solomon ye greatest philosopher in the world.

He believed that the ancients possessed vast knowledge that they encoded in esoteric texts and symbology, including specifically the prophetic visions of Ezekiel and Revelation, the measurements of the Temple of Solomon, and the symbol-laden works of alchemists. Newton wrote millions of words of dense commentary on these subjects, including one he published, the polished *The Chronology of Ancient Kingdoms Amended,* a work of more than eighty thousand words, whose unreadable genius foreshadows *Finnegans Wake.* Newton also spent many years at Cambridge doing intense experimental work in a special laboratory of his own construction, aiming both to clarify and to augment the transformations of the alchemists.

It should be emphasized that as Isaac Newton worked at biblical scholarship or at alchemy, he remained Isaac Newton. Keynes wrote:

> All his unpublished works on esoteric and theological matters are marked by careful learning, accurate method and extreme sobriety of statement. . . . They were nearly all composed during the same twenty-five years of his mathematical studies.

And here I will add a Query of my own: Is it not unnatural to separate our understanding of the world into parts that we do not seek to reconcile?

It is that Query to which this book, for me, responds.

## JOTTINGS OF BIOGRAPHY

Isaac Newton's achievements present a challenge to eugenicists, and to theorists of child-rearing. His father, also named Isaac, was an uneducated, illiterate yet prosperous yeoman farmer, described as "a wild and extravagant man." His mother, Hannah Ayscough, was a poor relation of country aristocracy. He was a posthumous child, born prematurely on Christmas Day 1642, so small that his mother said he could "fit into a quartpot." His mother remarried when Newton was three years old, leaving him (at the demand of her new husband) in the care of his grandmother until Hannah was again widowed and rejoined her son in 1659. Newton's origins were, in short, both humble and troubled.

And so there is a spark of divine grace in the wide-ranging curiosity, creativity, and sense of intellectual adventure the young Newton brought to this world.

As a boy he tracked the shadows of the Sun, constructing carefully calibrated sundials and recording the seasonal variations of the Sun's rise and fall, becoming the trusted local arbiter of time in that clockless countryside. He also built and flew elaborate kites, and once alarmed the neighbors by attaching lanterns to several and flying them at night (inspiring an early UFO alert!).

Young Isaac was expected to be a farmer, but hated farming and did it poorly. On the other hand he had been an excellent scholar at the local grammar school, and Henry Stokes, the master there, somehow persuaded both Hannah and the university that Isaac should attend Cambridge. He was admitted as a subsizar, meaning that he received financial assistance in return for acting as a servant to wealthier undergraduates.

In the years 1665–66 bubonic plague broke out in England. Cambridge University shut down, sending the twenty-two-year-old undergraduate back to his family farm at Woolsthorpe. During that time Newton had breakthrough insights in mathematics (infinite series and calculus), mechanics (idea of universal gravity), and optics (theory of colors). As he describes it:

> All this was in the two plague years 1665 and 1666, for in those days I was in my prime of age for invention, and minded mathematics and philosophy more than at any time since.

Perhaps nothing captures Newton's self-transcending spirit than an experiment he performed around that time, to clarify the relationship between the external world and the inner sensation of vision. Here is his description, including both text and drawing (figure 14):

> I tooke a bodkine gh & put it betwixt my eye & [the] bone as neare to [the] backside of my eye as I could: & pressing my eye [with the] end of it (soe as to make [the] curvature a, bcdef in my eye) there appeared severall white darke & coloured circles r, s, t, &c. Which circles were plainest when I continued to rub my eye [with the] point of [the] bodkine, but if I held my eye & [the] bodkin still, though I continued to presse my eye [with] it yet [the] circles would grow faint &

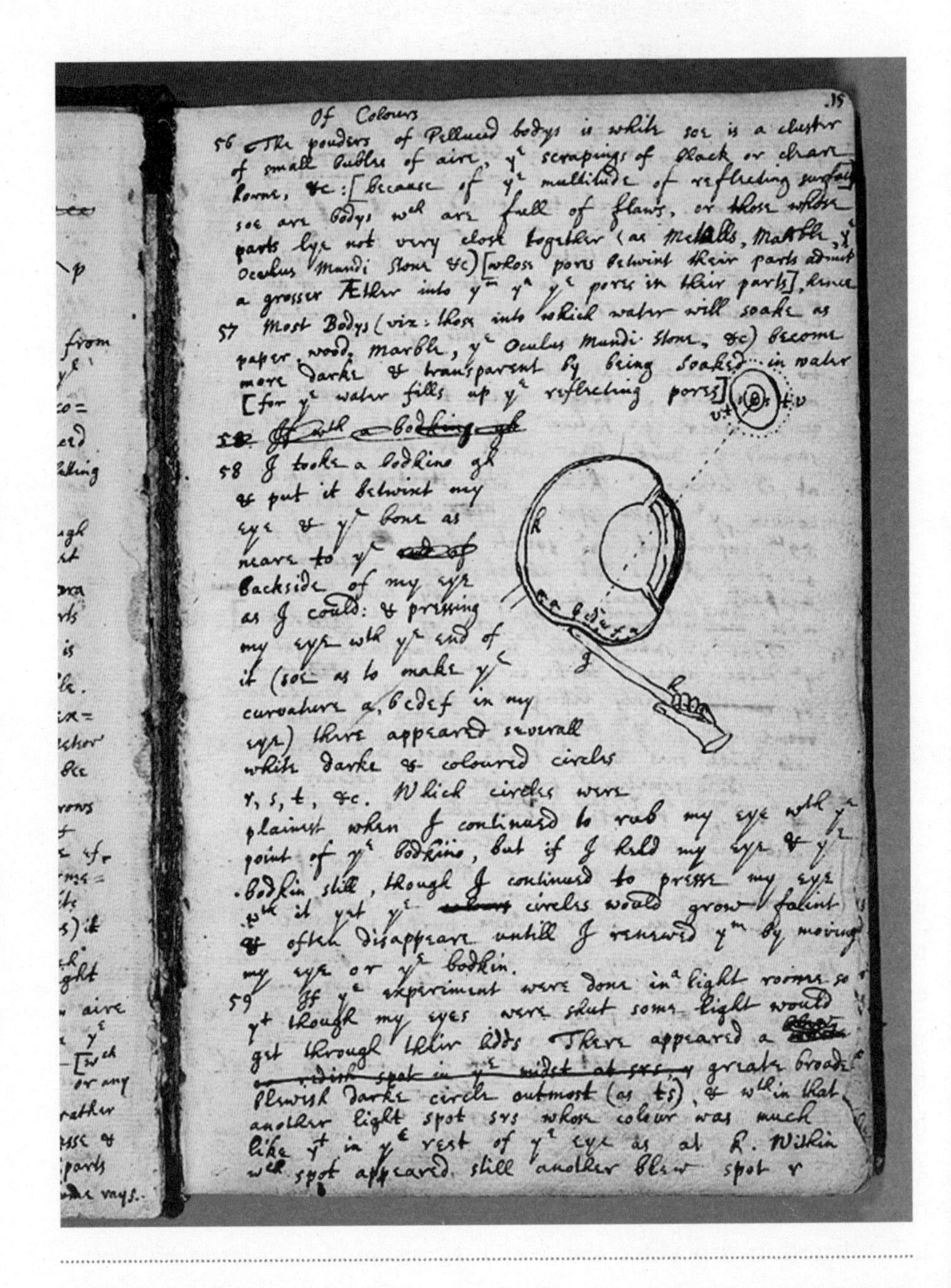

FIGURE 14. NEWTON'S COOL ACCOUNT OF AN ALARMING EXPERIMENT HE PERFORMED ON HIS OWN EYE TO BETTER UNDERSTAND THE PERCEPTION OF LIGHT AND WHETHER IT MIGHT HAVE A MECHANICAL CAUSE.

FIGURE 15. ISAAC NEWTON IN HIS PRIME.

often disappeare untill I removed [them] by moving my eye or [the] bodkin.

Newton worked with furious intensity continually through mid-1693, when, after more than twenty-five years of concentration rarely if ever matched in human history, he suffered what we'd now call a psychotic episode. He was unable to sleep for days at a time, imagined his friends were conspiring against him (and wrote them vitriolic letters), and suffered from tremors, amnesia, and general confusion. Newton described himself as "extremely troubled by the embroilment I am in, have neither ate nor slept well in the last twelve months, nor have my former consistency of mind." These symptoms persisted for several months, then gradually abated. It is possible that mercury poisoning, a result of his work with that most alchemical of materials, played a part in his illness.

In 1694 he left Cambridge to take up a position at the Royal Mint in London. His concerned friends had arranged what they thought would be a sinecure for him there. Newton became a much more "normal" person and, for the next twenty-five years, a very effective civil servant, but his days of furious seeking were over.

The haunting image in figure 15 is the only portrait of Newton which, to me, seems to convey his spirit and power. He went gray at a very early age.

# NEWTON II: COLOR

*Colors are the smiles of nature.*

• *Leigh Hunt*

*From what has been said it is also evident, that the Whiteness of the Sun's Light is compounded all the Colours wherewith the several sorts of Rays whereof that Light consists, when by their several Refrangibilities they are separated from one another, do tinge Paper or any other white Body whereon they fall. For those Colours . . . are unchangeable, and whenever all those Rays with those their Colours are mix'd again, they reproduce the same white Light as before.*

• *Isaac Newton*

The first of those quotes needs no explanation—humans enjoy color, as they enjoy smiles, spontaneously. The second, whose explanation will occupy this chapter, begins a deeper consideration of color, which will occupy much of our meditation, and greatly illuminate our Question.

*The purest and most thoughtful minds are those which love color the most.*

• *John Ruskin,* The Stones of Venice

That's us—so let's get to it.

## PURIFYING LIGHT

White has long been a color signifying purity. In ancient Egypt the priests and priestesses of Isis dressed only in white linen—as did mummies, readied for their afterlife! It is the traditional color for weddings, a union of the pure of heart. In Christian symbolism it is the color of the Lamb, and of the heavenly host, and of Christ triumphant, as seen in plate I.

This association of white with purity feels right. White is the color of the primary source of natural illumination, our Sun, when it is highest in the sky. White is the color of the brightest surfaces, such as snowfall, which best reflect the Sun.

But scientific analysis tells a different story.

When a beam of sunlight passes through a glass prism, a rainbow—or as we say a spectrum—of colors emerges. A similar effect, involving passage of sunlight through water droplets, is responsible for natural rainbows.

Prior to Newton's work, most believed that the colors emerging from prisms or raindrops resulted from the degradation of white light as it traveled through them. Colors were commonly thought to be mixtures of black (darkness) and white, in varying ratios. Depending on how far the light passes through the prism, it is more or less degraded, and so appears as a different color. That idea has the appeal of simplicity: Why introduce many ingredients, when two (or one) will do?

Newton, on the other hand, proposed that white light—including, specifically, the white light that comes from the Sun—is a mixture of many basic ingredients. According to his idea, the prism doesn't degrade the white light. Rather, the prism separates sunlight into more basic ingredients—ingredients that were present beforehand.

A simple but profound experiment, which Newton cited as the *experimentum crucis* (crucial experiment) for his proposal, makes the case visibly, as seen in plate J. The spectral colors, into which white light is analyzed by a prism, can be reassembled into white light using a second prism. When only part of the spectrum is used, the combined light is not white, but a mixture of the colors that originally came through. When natural sunlight is used as the original source, and the blue end of the spectrum is lopped off, then green dominates the perceived color. When only a narrow band of colors is let through—as, here, red—only light of that color emerges.

The point is that with a second prism, you can reverse the separation and get back to white light, indistinguishable in its properties from the sunlight you started with. As our picture suggests, you can also choose to combine only part of the spectrum. Then beams of intermediate colors, rather than white, emerge. In short, the prism performs an Analysis of the incoming white light.

It is easy to interpret this experiment, if we assume that sunlight consists of photons. (Although the term did not come into use until centuries later, to avoid confusion I'll refer to atoms of light as photons.)

The photons could come in different sorts—with different shapes, say, or different masses—that are differently affected by the prism's glass. Then the prism, by bending the trajectories of the different kinds of atoms by different amounts, would separate and effectively sort them. It would act like a modern vending machine that separates different kinds of coins. The different sorts of photons would also affect our eyes differently, producing the sensation of different colors.

Newton did not commit himself to that, or to any specific model. That would have been a Hypothesis! But he kept such thoughts in mind, as guides to further experimental investigation.

How far can the idea of sorting light be taken? We can block off all but a small portion of the spectrum to obtain beams of pure spectral color. The components of these filtered beams, whatever they are, were all bent by the same amount as they passed through the prism. Has that process sorted the light into identical, fundamental parts? Or could further processing reveal additional structure, and call for further purification?

Newton subjected his purified beams, the beams of spectral colors, to a wide variety of indignities. He reflected them off various surfaces, and passed them through lenses and prisms of many different transparent (or partially transparent) materials, not only common glass. He found that those processes respect the spectral sorting performed by the prism.

Spectral yellow, when reflected, remains yellow; spectral blue remains blue, and so forth. The light will often be absorbed by objects we see as colored. For example, an object we see as blue might absorb all the spectral colors except for those close to blue, which it reflects—and that's why it will appear blue. But one never finds spectral yellow reflected as spectral blue, or as anything other than spectral yellow.

The same rule applies to passage *through* materials (refraction). Spectral colors retain their integrity. The different colors will generally be bent by different amounts, of course—that's how the prism separated them in the first place—but a given material will refract rays of a given spectral color in a definite way.

Through experiments like these, Newton established that light beams derived from spectral colors are pure substances, with fixed, reproducible properties. White does not appear in the spectrum. Beams of white light can always be analyzed into component spectral colors, and will always be revealed to be a mixture. Ironically, despite its symbolism, regarded simply as light, white is never pure.

(Here, to be accurate, I should mention an interesting complication. It's not quite true that spectral colors can't be analyzed further. A further twofold separation is possible, into different *polarizations*. It will be natural to discuss this complication later, in connection with Maxwell's work. Though possible, it is not entirely easy to separate light of a single spectral color into its two polarized components, and so for many purposes we can neglect the distinction. There is a similar situation for chemical elements of substance, which can be mixtures of isotopes that are difficult, but not impossible, to separate.)

Though I've never heard it described that way, I think it's fitting to regard what Newton is doing here, and throughout the *Opticks*, as establishing the chemistry of light. Analysis, or purification, is the first step in chemistry.

## *The Chemistry of Light*

Having purified light, we are prepared to pursue its chemistry further.

Our analysis so far has been consistent with the guiding idea that light is made from photons, and that the different kinds of photons are differently bent by glass, so that they can be separated by passage through a prism. Each spectral color, then, is a purified selection, containing only photons of one kind. In this way, we've identified the elements of light.

Let's compare and contrast the chemistry of light with the more familiar, though later to develop, and vastly more complicated, chemistry of substance, starting with their periodic tables:

- The periodic table of light has just one row—the rainbow of spectral colors. The spectral colors are its elements. The periodic table of substance has several rows, with the elements arranged in columns featuring broadly similar, but distinct chemical properties. There are also two long, odd stretches—the lanthanides (rare earths) and actinides—where not much changes, chemically.
- The periodic table of light can be realized in a tangible, physical form. Indeed, upon putting a beam of sunlight, or the light produced by any glowing hot body, through a prism, and projecting the output on a screen, you've basically got it. The periodic table

of substance, by contrast, is an intellectual construct. There is no corresponding object in Nature.

- The periodic table of light is continuous, while the periodic table of substance is discrete.
- The elements of light interact only very weakly with one another. Indeed, crossing light beams pass through one another freely, without reacting. (They do not shed sparks, for example, nor leave behind a residue of light molecules.) In those ways, every element of light resembles the "noble" or "inert gas" elements of substance chemistry.

From a larger perspective, it is natural to consider both sorts of chemistry together, as the science of atoms and their interactions, including both atoms of light and those of substance. In that larger framework, the atoms of light no longer appear as inert. Though they do not readily combine with one another, they do combine with atoms of substance, according to definite rules. We'll have a lot more to say about this below, and go deep when we get to "Quantum Beauty I: Music of the Spheres."

A major goal of alchemy was to produce the Philosopher's Stone, which would transform one sort of atom into another, for example, lead into much more valuable gold. For atoms of light there is a Philosopher's Stone—motion! If we are moving toward a beam of spectrally pure light, it will appear to us as spectral light of a different color. The colors shift in the direction from red toward blue, and we say they are blueshifted. Similarly, motion away from the beam, or motion of the beam away from us, produces a redshift. The size of the shifts are proportional to the velocities of relative motion, and are very small unless those velocities are close to the speed of light. The shifts were much too small for Newton to observe. For most practical purposes, we can neglect them. But the redshift of light from distant galaxies—in particular, the change that redshift makes in the location of dark and bright lines in the spectrum—encodes how rapidly those galaxies are moving away from us, and allows us to map the expansion of the Universe.

The idea that light is made from particles or, as we've been calling them, photons, has had a checkered history. As we've discussed, Newton was sympathetic to the idea, but not married to it. (He flirted with it, but did not commit himself to it exclusively.) Such was his authority, however, that particle theories of light dominated science until well into the nineteenth century, when wave theories took over. With the advent of Maxwell's electromagnetic explanation of light, which we'll discuss at length later, the triumph of wave theory seemed complete. But in the twentieth century, with the advent of quantum mechanics, the particle theory made a comeback, with the atoms of light officially christened photons. Now we understand that particle and wave offer complementary perspectives on the reality of light. Newton's practice of keeping many alternatives in play, while refusing to put forward any one Hypothesis exclusively, anticipates modern complementarity.

## *Profiting from Analysis*

Newton put his fundamental understanding of color to good practical use, by improving telescopes. Before his work, telescopes used two lenses, typically at opposite ends of a long tube, basically to gather light from, and then to focus (in magnified form) images of, distant objects. Because different colors of light follow different trajectories through a lens, not all the colors can be focused accurately at the same time, and one gets a blurry image. This is the problem of *chromatic aberration*. Newton suggested that one could use a concave, reflecting mirror to gather light, instead of a lens, and designed telescopes based on that idea. His reflecting telescopes reduce chromatic aberration, and are also simpler to make. Essentially all modern telescopes are reflecting telescopes.

The analysis of light has been a profoundly fruitful source of scientific discovery. Among many possible examples, here I'll describe one that's simple to describe, hugely important, and has a touch of poetry. (We'll encounter others later.)

The overall impression one gets, looking at the spectrum produced by sunlight, is of a continuous gradation of intensity. But if one uses very high-quality glass in the prism, and separates the light more accurately, one discovers an abundance of fine detail. Joseph von Fraunhofer, who pioneered this field in the early 1800s, discovered no less than 574 dark lines within the apparent continuum. The reason for these lines was not understood until the 1850s, when Robert Bunsen and Gustav Kirchhoff showed how similar lines can be produced on Earth. When one puts a volume of cold gas in front of a hot source of light, the gas will absorb some of the light. The gas is generally quite selective in its absorption, taking out only the components of light within narrow spectral bands. When the light is analyzed, the colors that have been absorbed are missing, leaving dark lines in the spectrum.

Different kinds of gases (e.g., gases based on different chemical elements) absorb different spectral colors. So if one has a gas of unknown composition, one can deduce what it's made of by seeing what light it absorbs! In the language of our generalized chemistry, the message of Fraunhofer's dark lines, as interpreted by Bunsen and Kirchhoff, is that a given atom of substance will combine with—that is, absorb—only specific elements of light—that is, spectral colors—while ignoring others. There is also a converse effect, that heated gas will emit light in preferential colors, creating bright lines in the spectrum. Altogether, these dark and bright lines are like fingerprints identifying the responsible substances.

So by analyzing the light from a star, and comparing its bright and dark lines to those observed in laboratory gases, astronomers can determine what the star is made of (and many other details about conditions in its atmosphere, where the light originates). This quickly became, and remains to this day, the bread and butter of physical astronomy. Most fundamentally, it teaches us that stars are made of the same materials, and are subject to the same physical laws, as those we observe on Earth.

Some puzzling observations by Norman Lockyer and Pierre Janssen of the solar corona once seemed to challenge that great conclusion, but in the end reinforced it. In 1868, during an eclipse of the Sun, they observed a bright line that no earthly gas had been observed to produce. A new, intrinsically celestial element, "corunium," was held to be responsible. But in 1895 two Swedish chemists, Per Cleve and Nils Abraham Langlet, and independently William Ramsay, discovered that the same line is produced by a gas emanating from uranium ore. In this way, family relations between heaven and earth were restored. The new element was (re)named *helium*, after Helios, the Greek god of the Sun.

# NEWTON III: DYNAMIC BEAUTY

The basic laws of Newtonian mechanics are dynamical laws, that is, laws about how things change. Dynamical laws are different from the rules of geometry, or the sorts of laws discussed by Pythagoras and Plato, which describe particular objects or relationships.

Dynamical laws invite us to expand our search for beauty. We should consider not only the world of what *is*, but also—and primarily—the much larger, imaginative world of *what can happen*. The world of Newtonian mechanics is a world of *possibilities*.

That expanded search strikes gold at Newton's Mountain (figure 17, page 104). But a little preparation is in order before we make our visit.

## EARTH VERSUS COSMOS

Newton's immediate predecessors left natural philosophy with a great unfinished task.

Galileo's *Sidereus Nuncius* (*Starry Messenger*) contains dozens of his sketches of the Moon, as he viewed it through the first astronomical tele-

FIGURE 16. SOME OF GALILEO'S STRIKING DRAWINGS OF THE TELESCOPIC MOON.

scope, a 20x magnifier of his own construction. The patterns of light and shadow clearly reveal that the Moon has a rugged topography (figure 16).

As Galileo brought the celestial spheres down to earth, Copernicus put the Earth in motion, as one planet among many, and Kepler found precise regularities in planetary motion. Although their details need not concern us, I'll record Kepler's three laws here, so that I can bring out two crucial points:

1. The orbit of a planet is an ellipse, with the Sun at one focus.
2. The line connecting a planet to the Sun sweeps out equal areas in equal times.
3. The square of the period (i.e., the length of a planet's "year") is proportional to the cube of the long axis of the ellipse.

The first point is that these are *not* dynamical laws. They describe accomplished relationships, not rules of change. The second point is that they are rules about the motion of planets. They have nothing to say about the motion we observe closer to home, in earthly experience. They are alien reports from a different conceptual universe—even though Earth is a planet itself!

The great unfinished task, then, was to reconcile Earth and cosmos. What are the common laws that govern both those visibly similar realms?

## NEWTON'S MOUNTAIN

There are many geometrical diagrams in Newton's *Principia*, and several numerical tables, but only one drawing (figure 17). To me, it is the most beautiful drawing in all of scientific literature.

Obviously, viewed simply as a piece of draftsmanship, this drawing is a modest achievement. What makes it beautiful is the ideas it invites us to imagine. It is an invitation to a thought experiment, which suggests that objects falling to Earth and celestial bodies orbiting in space are doing the same sort of thing, thus the possibility of a universal force of gravity.

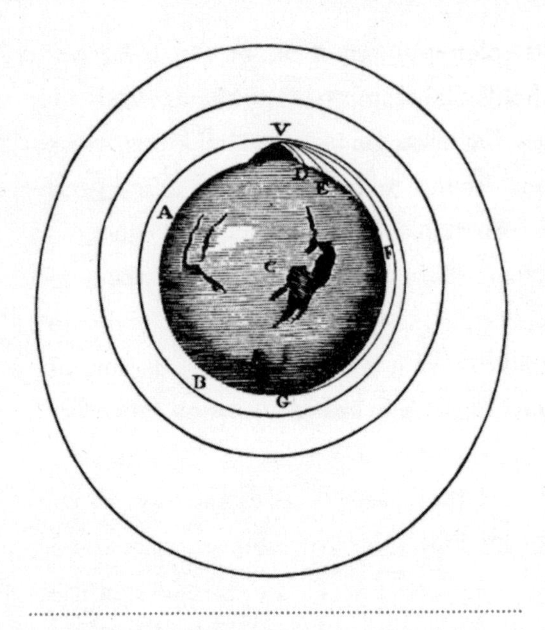


FIGURE 17. NEWTON'S MOUNTAIN—A GREAT THOUGHT EXPERIMENT.

You stand on a mountaintop and throw a stone horizontally—that is, parallel to the Earth's surface. If you launch it with a small velocity, it will progress only a short distance while descending to the surface. If you throw harder, it will progress farther. In practice, no mortal can throw a stone hard enough that it progresses over a significant fraction of the Earth's circumference. Never mind—this is a thought experiment, and you're encouraged to substitute mental for physical power. Throw harder. With your mind's eye you see, as in the drawing, that the end point of the trajectories start creeping up on the starting point.

And then, as you throw still harder, you must duck, or the stone will hit you in the back of the head! Having taken that precaution, you can watch the stone repeat its performance, for it has achieved a circular orbit. (Air resistance? Please, it's a thought experiment!) You can put your thought-self atop imaginary mountains, and apply the same logic, to see the possibility of bodies orbiting Earth at any distance, under the influence of its gravity.

You can imagine a really tall mountain, and a really big stone . . . and, when that stone is in orbit, call it the Moon.

I have spoken of the body in the figure as Earth, but it is drawn as an idealization, with vaguely defined surface features and a mountain of unrealistic proportions. The point here is precisely that the body *need not* be Earth. The same thought experiment can be applied to the Sun, and explain how the Sun's gravity holds planets in their orbits. Or to Jupiter, explaining how Jupiter's gravity holds Galilean moons in their orbits.

The idea of *universal* gravity, acting between bodies of all kinds, takes this thought experiment far beyond where we (or Newton) have a right to extrapolate the sort of everyday experience—a stone falling to Earth—that launched us on this voyage of imagination. Thought experiments do not prove anything! But they may suggest fruitful directions to follow up with more careful investigation. If the imagined outcome of a thought experiment seems logical, that's good. If it's beautiful, that's better. If you're getting more out than you put in, that's better still. Newton's Mountain has it all.

There is a famous legend, apparently stemming from some offhand recollections by Newton in his old age, that he was led to think about the possibility of universal gravity by observing an apple fall to the ground at his home in Woolsthorpe. In his written account there is no apple, but only this:

> I began to think of gravity extending to the Orb of the Moon, and having found out how to estimate the force with which a globe revolving within a sphere presses the surface of the sphere: from Kepler's Rule of the periodical times of the Planets . . . I deduced that the forces which keep the Planets in their Orbs must be reciprocally as the squares of their distances from the centers about which they revolve: and thereby compared the force requisite to keep the Moon in her Orb with the force of gravity at the surface of the Earth, and found they answer pretty nearly.

Whether or not the fall of an apple aroused his thinking, in itself that phenomenon could not supply much food for thought. I think it is likely that something close to Newton's Mountain was the realization that made universal gravity, for Newton, a compelling vision.

I also think there is a plausible path that starts from the apple thought as a flash of insight, and leads to the Mountain thought as its elaboration. The idea is simple, but very pretty. If we think of Earth's influence "extending to the Orb of the Moon" through gravity as an explanation of the Moon's orbit, we are assuming a connection between two kinds of motion that seem very different. Gravity as observed on Earth—by watching apples, say—is a process of *falling* toward the center of the Earth. The motion of the Moon *orbiting* the Earth is, on the face of it, something else entirely. The whole point of the Mountain thought experiment, however, is to show that orbiting is a process of constantly falling—but toward (from the stone's point of view) a moving target! And you can see in the drawing that at each point in the circular orbit the velocity of the stone is parallel to the surface of Earth (i.e., it is locally "horizontal"), while the curvature of the inward orbit bends toward the surface. Having seen, in our view from the Mountaintop, that orbiting is a form of falling, we can connect the Moon to the apple.

## *Time as a Dimension*

Even the best thought experiments prove nothing. A journey opens before us, leading from the vision of Newton's Mountain to the sort of precise mathematical theory that he aspired to. It is a journey through a new dimension: Time, newly imagined.

The curves in the Newton's Mountain drawing are trajectories—each is a collection of points occupied by a body (our stone) at successive times. They are not, of course, themselves bodies in space, nor physical objects in any obvious direct sense. Yet trajectories do define geometrical objects, and trajectories are—as we shall see—foundational for understanding the physics of motion. To appreciate them properly, let's give them a home.

A single trajectory carries some information about the motion of a single body, but from the curve alone we can't infer when the different parts were traversed. We could put time labels along the points of the curve,

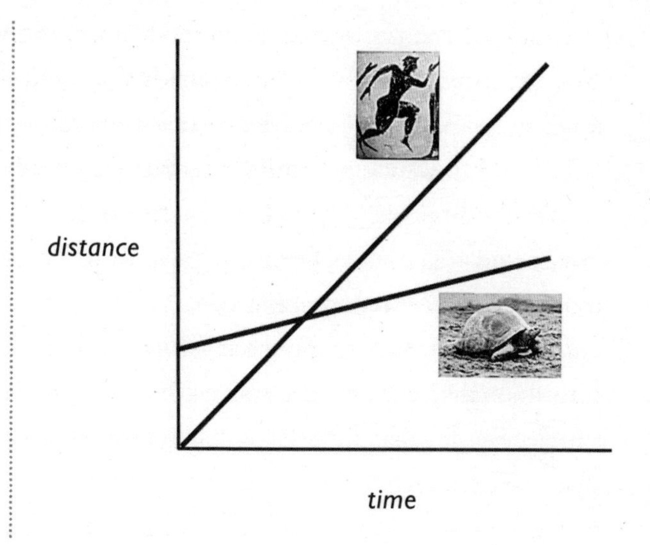


FIGURE 18. AS TIME PROGRESSES (TO THE RIGHT) BOTH ACHILLES AND THE TORTOISE ADVANCE ALONG THE RACECOURSE (HERE DEPICTED UPWARD). ACHILLES' TRAJECTORY IS STEEPER, BECAUSE HE COVERS A GREATER DISTANCE IN A GIVEN INTERVAL OF TIME. HERE TIME HAS BECOME A FULLY FLEDGED DIMENSION, ON THE SAME FOOTING AS DISTANCE (I.E., SPACE).

restoring the missing information. But that becomes awkward if we want to consider several trajectories at once, because any one time corresponds to a hodgepodge of points, one on each trajectory, with their pattern shifting as time flows. A better procedure is to consider time as another dimension. The trajectories have a natural home in the more expanded conceptual universe that results: space-time.

To bring out the essential nature of this profound reimagining, let's revisit a situation simpler than Newton's Mountain, namely Zeno's paradoxical race between Achilles and a tortoise. First, notice that here the trajectories in space are just two partially overlapping straight lines—not very informative! By ascending the space-time view, we can reenvision the race between Achilles and the tortoise in a way that does their race—and, by extension, motion in general—better justice.

If we want to synchronize our descriptions of our two trajectories, it makes sense to introduce time as a separate quantity—a new dimension—and put points marking both positions at each time. Figure 18 lays it out.

In this figure, the logical structure of Zeno's argument is laid bare, and the paradox dissolves. In space-time the two trajectory-curves, one steeper than the other, can hardly help but cross! (You can have a bit of fun tracking the time when Achilles reaches the tortoise's starting point, then the time when Achilles reaches the point where the tortoise has gone, to when Achilles reaches the tortoise's starting point, . . . thus recreating, and defusing, Zeno's original formulation.)

We get back to trajectories, in the original sense, by *projecting* the space-time trajectories horizontally onto the distance axis, thereby suppressing all information about time.

The trajectories in Newton's Mountain are already drawn in two-dimensional space, so their space-time versions will come to live in three dimensions. In that three-dimensional space-time, the circular orbits unfold into helices.

You can also, in an act of mathematical imagination, work things the other way: Take ordinary two- (or three-) dimensional space, and pretend that it's space-time! In that way ordinary geometric curves get reinterpreted as dynamical trajectories. Or, to put it another way, we consider them as *motions* of a point through space. Newton developed that basic thought in great depth. For him, it was the conceptual essence of what today we call calculus. Newton, who invented the subject, called it the *method of fluxions.* According to this method, curves (and other geometric objects) are regarded not as completed objects, but as entities built up over time, through smooth changes in their infinitesimal components.

## *Analysis of Motion*

Figure 19 is a key diagram from the *Principia*, showing how motion is analyzed. Kepler had derived mathematical laws describing the motion of the planets, but he did not have a derivation of those laws from deeper physical principles. In this figure, using his characteristic method of Analysis—breaking things into small parts—Newton unveils the inner meaning of Kepler's laws.

The orbit is analyzed into a very large number of steps, each extending over a small interval of time. Because these are not physical steps, but

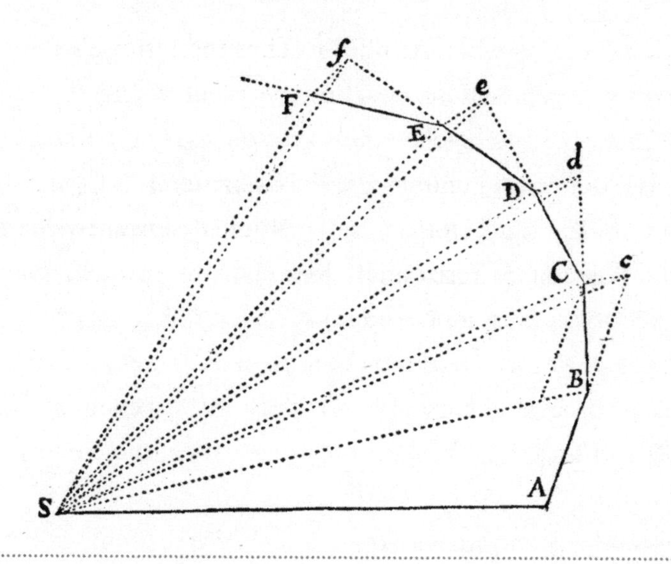


FIGURE 19. NEWTON'S ANALYSIS OF MOTION. DEVIATIONS FROM STRAIGHT-LINE MOTION ARE DUE TO FORCES.

mathematical idealizations, we can make them as small as we want. Over a sufficiently small interval, the orbit can be approximated by a straight line, and the object's speed is roughly constant. One of Newton's laws of motion is that a body subject to no forces will continue in its state of motion—that is, it will continue to move in the same direction, at the same speed. In the diagram, we see dotted-line extensions of the orbital segments, indicating the path that would be followed were the force suddenly turned off. But the actual orbit differs from those extrapolations, precisely because there is a force.

By a close mathematical examination of the problem, we can determine what sort of force is required to support a given orbit. Newton did that for the orbits of planets, using the regularities Kepler had discovered (the three laws we recorded on page 103). Through this analysis, Newton deduced that the force is directed toward the Sun, and falls off according to the square of the distance to the Sun.

We should not fail to notice that this analysis is, at its core, a mathematical implementation of the basic concepts we saw in Newton's Mountain, his visionary thought experiment.

## *Newton's Anagram*

The analysis of motion into infinitesimal parts, with forces driving any deviation from "natural" motion (i.e., motion at constant velocity), is the essence of Newtonian mechanics. Reluctant to share his secrets, but eager to ensure his priority, Newton published this anagram:

6a cc d æ 13e ff 7i 3l 9n 4o 4q rr
4s 8t 12u x

Its solution* is a phrase in Latin

Data æquatione quotcunque fluentes quantitates involvente, fluxiones invenire; et vice versa

Vladimir Arnold, a distinguished twentieth-century mathematician and a profound student of Newton's work, helpfully translates this as

It is useful to solve differential equations.

Herewith a more expansive translation that summarizes our discussion:

Analysis of motion by considering its smallest parts is a good thing. It allows you to determine the forces from the trajectories, or the trajectories from the forces.

## *The System of the World*

Having inferred a universal law of gravity from Kepler's laws of planetary motion, Newton used it to predict an impressive range of new

*There is an extra *t*—Newton didn't want this to be too easy.

consequences. This was the Synthesis, flowing from his Analysis. A partial list:

- The general nature of the force of gravity we feel on Earth. Given the motion of the Moon, its magnitude. How it varies with position on Earth.
- The motions of the moons of Jupiter and Saturn, and of our Moon
- The motion of comets
- The cause of the tides (which is gravitation from the Moon and Sun) and their main features
- The shape of the Earth: slightly oblate
- A slow wobble in the direction of the Earth's axis, approximately 1 degree every 72 years. The effects of this "precession of the equinoxes" had been observed by the ancient Greek astronomers, but previously neither they nor anyone else had come close to explaining it.

Each of these applications was quantitative, and several of them could be checked with great precision. All of them could be refined by better observations and more extensive calculations, with no change in the principles.

Newton called the third book of the *Principia*, where he carried out his Synthesis, *The System of the World.* There had never been anything like it before. He had taken on some of the grandest problems of cosmology, and solved them according to mathematical principles, with unheard of, and potentially unlimited, precision.

Real *and* Ideal.

## DYNAMICAL BEAUTY

Newton's dynamical laws reveal beauty in the physical world, but it is a rather different kind of beauty from what Pythagoras and Plato anticipated. Dynamical beauty is less obvious, and requires more imagination to appreciate. It is a beauty of laws, rather than of objects or sensations.

We can see the difference by comparing Kepler's model Solar System, based on the Platonic solids, with Newton's System of the World. In Kepler's model, the Solar System itself is a beautiful object, realizing perfect symmetry. Its elements are spheres, spaced by Plato's five Ideal solids. In Newton's System, the actual orbits of the planets reflect God's initial conditions, perhaps somewhat eroded over time. (More on this below.) God may, and probably did, have other considerations than mathematical mysticism in mind, so beauty in the actual orbits is not to be expected—and is not found. What *is* beautiful is not the particular orbits, but general principles that underlie all possible orbits, and the totality of all orbits. It is the beauty of Newton's Mountain, enhanced in its precise elaboration.

## *Reduction Is Expansionary*

Newton's method of Analysis and Synthesis also goes by another name: reductionism. People say that a complex object or subject has been "reduced" to something simpler when it has been shown, or made plausible, that the more complex thing can be analyzed into simpler parts, and its behavior synthesized from the behavior of those parts.

Reductionism has a bad name, not least because "reductionism" *is* a bad name. The word's plain implication is that when you have understood something, by the method of Analysis and Synthesis, you have somehow reduced it. Your rich and complex object is "no more" than the sum of its parts. For that matter—and here, close to home, is where it gets disturbing—perhaps you yourself, and those you love, are "no more" than collections of molecules just doing their thing, which is behaving according to mathematical rules.

The poets and artists of the Romantic era, responding to the triumphs of Newtonian "reductionist" science, expressed their disquiet with its implicit "no more than"-ism. John Keats, most lyrical of lyric poets, wrote:

Do not all charms fly
At the mere touch of cold philosophy?
There was an awful rainbow once in heaven:
We know her woof, her texture; she is given
In the dull catalogue of common things.
Philosophy will clip an Angel's wings,
Conquer all mysteries by rule and line,
Empty the haunted air, and gnomed mine—
Unweave a rainbow . . .

William Blake protested against reductionism's blinkered vision (plate K). In this depiction of Isaac Newton at work, Blake's conflicted feelings for his subject are on display. His Newton is a figure of extraordinary concentration and purpose, not to mention superhuman anatomy. On the other hand, he is shown looking down, lost in abstractions, having literally turned his back on the strange, colorful landscape. Yet Blake admitted (as did Keats) that mathematical order governs the world (plate L). In Blake's complex mythology Urizen, depicted here, is a dualistic Father figure, who both brings life and constrains it. One can hardly fail to notice a certain resemblance to the preceding drawing. Is Newton Urizen's interpreter, or his incarnation?

In countering an emotional reaction, a good image can be more powerful than edifying rhetoric. Here, truly, "a picture is worth a thousand words." Please ignore for a moment the caption as you take in plate M, a strikingly beautiful work of abstract art.

OK, now read the caption (if you haven't already). Does knowing that the image can be "reduced" to strict mathematics detract from its beauty? For me, and I trust for you, the revelation that simple mathematics can encode this structure *adds* to its beauty. It still looks the same, of course, as an image. But now you can also, with your mind's eye, see it from another perspective, as an embodiment of concepts. It is both Real, and Ideal.

Conversely, the beauty of the image enhances the beauty of the mathematics. Going over the logic of the generating program, without having seen what it is capable of producing, is a mildly entertaining exercise. Once you've seen the output, that same process becomes a spiritual quest, reaching for the sublime.

The Real is more compelling for being Ideal, and the Ideal is more compelling for being Real.

As with fractal images, so more generally: Understanding does not detract from experience; rather, it adds alternative perspectives. In the spirit of complementarity, we can enjoy every alternative, sequentially if not all at once.

By the way, it's a safe wager that Keats did not master the scientific theory of rainbows. If he had, we'd have his poems praising its beauty! For the same John Keats also wrote this:

> When old age shall this generation waste,
> Thou shalt remain, in midst of other woe
> Than ours, a friend to man, to whom thou say'st,
> "Beauty is truth, truth beauty,"—that is all
> Ye know on earth, and all ye need to know.

## *Getting Started*

There is another aspect to the dynamical view of the world, which led Newton to God, and which poses challenges that are yet unmet.

The dynamical laws are laws of motion. They relate the state of the world at one time to its state at other times. If we know the state at one time, we can predict the future, or extrapolate to the past. In Newton's mechanics, specifically, once we know the positions, velocities, and masses of all particles at one time, and the forces that act among them, we can infer their positions and velocities (and masses—those don't change) at all other times, by pure calculation. Those quantities specify the state of the world because, in Newtonian mechanics, they provide a complete description of matter.

There are severe practical difficulties that stand in the way of actually performing those calculations, as every student of the weather can testify. There are lots and lots of particles, and it's completely unrealistic to get all their coordinates and measure all their velocities. Even if you could, and even if you knew the force laws exactly, the required computation boggles any conceivable mind. To top it off, a central result of chaos theory is that small errors anywhere along the line—in the initial conditions, in the force laws, or in the numerical computations—tend to grow into much larger errors as time progresses.

Practical difficulties aside, the fundamental point is this: You need a starting point! The dynamical equations are not self-sufficient. In the standard jargon, we say that they require *initial conditions.* In order to begin computing the world's behavior using the dynamical equations, you must first specify the state of the world at one time, as input.

(Of course, if you're interested in something smaller than the whole world, and you can effectively isolate it from the rest, you only need to know the state of your subsystem. For simplicity, I'll continue to speak of "the world.")

The description of the world divides into two parts:

- Dynamical equations
- Initial conditions

From the regularity and order of the Solar System—with all its planets revolving around the Sun very nearly in circles, all in common plane, all in the same direction—Newton surmised, in the "General Scholium" that concludes the *Principia,* that the initial conditions had been mindfully arranged:

> This most elegant system of the sun, planets, and comets could not have arisen without the design and dominion of an intelligent and powerful being.

Today we have more mundane, physically based ideas about the origin of the Solar System, but the deeper issue remains. Although Newtonian mechanics, as a fundamental theory, has been superseded, this feature of it has survived. We still have dynamical equations, and they still need initial conditions. Our description of the world divides into two parts, dynamics and initial conditions. We have an excellent theory for the former, but only empirical observations and incomplete, more or less plausible speculations regarding the latter.

If we take the Universe in space-*time*, unfolded to yield a "God's-eye" view of reality, to be fundamental, we are led to a modern form of Parmenides's changeless One. The great twentieth-century mathematician and physicist Hermann Weyl, whose books were a big part of my education, put it this way, in what I think is among the most beautiful, as well as most profound, passages in all of literature:

> The objective world simply *is,* it does not *happen.* Only to the gaze of my consciousness, crawling along the lifeline of my body, does a section of this world come to life as a fleeting image in space which continuously changes in time.

If Parmenides and Weyl are right, and space-time as a whole is the primary reality, then we should aspire to a fundamental description of its totality. In such a description, there would be no place left for initial conditions.

# MAXWELL I: GOD'S ESTHETICS

Truly modern physics begins in 1864, with James Clerk Maxwell's paper "A Dynamical Theory of the Electrodynamic Field." In that paper you can find, for the first time, equations that still appear in today's Core Theory.

Those equations—Maxwell's equations—changed many things.

Maxwell's equations changed space from a receptacle into a material medium—a sort of cosmic ocean. No longer a mere Void, space is filled with fluids that run the world.

Maxwell's equations gave us an entirely new understanding of what light is, and predicted the existence of unsuspected forms of radiation, which are new kinds of "light." They led directly to radio, and inspired several other major technologies.

Maxwell's equations also mark a great advance toward the answer to our Question, for they display beautiful ideas deeply embodied in the world. Their beauty derives from many sources: from the way they were discovered, from their shape, and from their power to inspire other good ideas.

- Beauty as a tool: For Maxwell imagination and play, guided by a sense of mathematical beauty, were prime tools of discovery—and he showed that those tools work!
- Beauty as an experience: The Maxwell equations can be written pictorially, in terms of flows. When so presented, they depict a sort of dance. I often visualize them that way, as a dance of concepts through space and time, which is a joy. Even at first sight the Maxwell equations give an impression of beauty and balance. Like the impact of more conventional works of art, that impression is easier to experience than to explain. Paradoxically, there's a word to describe beauty that can't be described in words—"ineffable." Having experienced the ineffable beauty of Maxwell's equations, one would be disappointed if they were wrong. As Einstein said in a similar context, when asked whether his general theory of relativity might be proved wrong, "Then I would feel sorry for the good Lord."
- Beauty and symmetry: Deeper appreciation of the Maxwell equations, acquired over several decades following their discovery, led to a complementary, more intellectually precise perspective on their beauty. They are a very *symmetric* system of equations, in a precise mathematical sense of that word, as we'll discuss. The lessons drawn from the Maxwell equations—that *equations* can display symmetry, and that Nature loves to use such equations—guides us to the Core Theory, and perhaps beyond.

Let us open our minds to their spirit!

## ATOMS AND THE VOID?

Newton's physics left space empty, but he was not happy about it. His law of universal gravity postulates attractive forces that act immediately, with no delay in time, between bodies that are separated in space. Furthermore, the magnitude of the force depends on the separation of the bodies, falling off as the square of the distance. But if space, in the absence of bodies, is simply *nothing,* how is the force transmitted? How does it jump the gap? And why should the magnitude of the force depend upon exactly how much "nothing" intervenes?

To Newton, his own theory begged those questions. But he found no answers. It was not for lack of trying—Newton filled many pages of his private notebooks with alternative ideas about gravity, but none could compete with the law that he himself, in private correspondence, called an absurdity:

> That one body may act upon another at a distance through a vacuum without the mediation of anything else, by and through which their action and force may be conveyed from one to another, is to me so great an absurdity that, I believe, no man who has in philosophic matters a competent faculty of thinking could ever fall into it.

Newton was also drawn, with misgivings, into using the Void in his work on light. His particles of light move in straight lines through space, in the absence of material, very much in the spirit of ancient atomism. According to that doctrine, as expressed poetically by Lucretius: "By convention there is sweetness, by convention bitterness, by convention color, in reality only atoms and the Void."

Yet at the very end of the *Principia* we find this expression of faith, or yearning, that seems to belong to a different book:

> And now we might add something concerning a certain most subtle Spirit, which pervades and lies hid in all gross bodies; by the force and action of which Spirit, the particles of bodies mutually attract one another at near distances, and cohere, if contiguous; and electric bodies operate to greater distances, as well repelling as attracting the neighbouring corpuscles; and light is emitted, reflected, refracted, inflected, and heats bodies; and all sensation is excited, and the

> members of animal bodies move at the command of the will, namely, by the vibrations of this Spirit, mutually propagated along the solid filaments of the nerves, from the outward organs of sense to the brain, and from the brain into the muscles. But these are things that cannot be explain'd in few words, nor are we furnish'd with that sufficiency of experiments which is required to an accurate determination and demonstration of the laws by which this electric and elastic Spirit operates.

In the decades that followed, physics rooted in the Void went from triumph to scientific triumph. More precise observations of lunar motion, the tides, and the motion of comets agreed perfectly with more precise calculations from Newton's laws. Amazingly, both electric forces (between charged bodies) and magnetic forces (between magnetic poles) were measured to follow the same pattern as gravitation: they act through empty space, and fall off as the square of the distance. (Thus the force is four times weaker at twice the distance, nine times weaker at three times the distance, and so forth.)

Newton's followers soon abandoned his scruples. They became more "Newtonian" than Newton ever was. Newton's own abhorrence of Void was put down to philosophical or, at bottom, theological prejudice, and passed over in embarrassed silence. The new orthodoxy aimed to describe all the forces of physics, and ultimately chemistry, in the style of Newton's gravity, as forces acting at a distance, with strength dependent on the distance. Mathematical physicists forged elaborate mathematical tools to win consequences from those sorts of laws. A few more force laws, and the story would be complete.

## *Void Avoided*

Michael Faraday was born in England as the third child in a poor family of unorthodox Christians. His father was a blacksmith. He had little formal schooling. During the seven teenage years that Faraday was apprenticed to a London bookbinder, he became fascinated by some of the books that passed his way, especially those on self-improvement and on science. By attending public lectures of the popular chemist Humphry Davy, and working up meticulous notes, Faraday attracted Davy's attention, and got hired as his assistant. Soon he was making discoveries on his own . . . and the rest is history.

Faraday never got far in mathematics. He knew some algebra and a bit of trigonometry, no more. Unprepared to grasp the existing ("Newtonian") mathematical theories of electricity and magnetism, he developed his own concepts and images. Here is Maxwell's description of the result:

> Faraday, in his mind's eye, saw lines of force traversing all space where the mathematicians saw centers of force attracting at a distance; Faraday saw a medium where they saw nothing but distance; Faraday sought the seat of the phenomena in real actions going on in the medium, they were satisfied they had found it in a power of action at a distance.

The key concept here is *lines of force.* Its meaning is plainer in imagery than in words—see figure 20.

Iron filings, moving freely on a thin sheet of paper held over a bar magnet, trace out a striking pattern. The iron filings line up under the influence of the bar magnet and align into a space-filling system of curved lines. Those are Faraday's lines of (magnetic) force.

The Void-based theory of forces at a distance has no trouble accounting for this demonstration: Each shred of iron feels forces from the two poles of the magnet, acting through empty space, and aligns accordingly. The lines of force are an emergent, almost accidental by-product of deeper, simpler underlying principles.

But Faraday's interpretation was different, and more visceral. According to Faraday, the filings merely *reveal* the state of a space-filling medium, *which is there whether or not there are filings, or for that matter a magnet,*

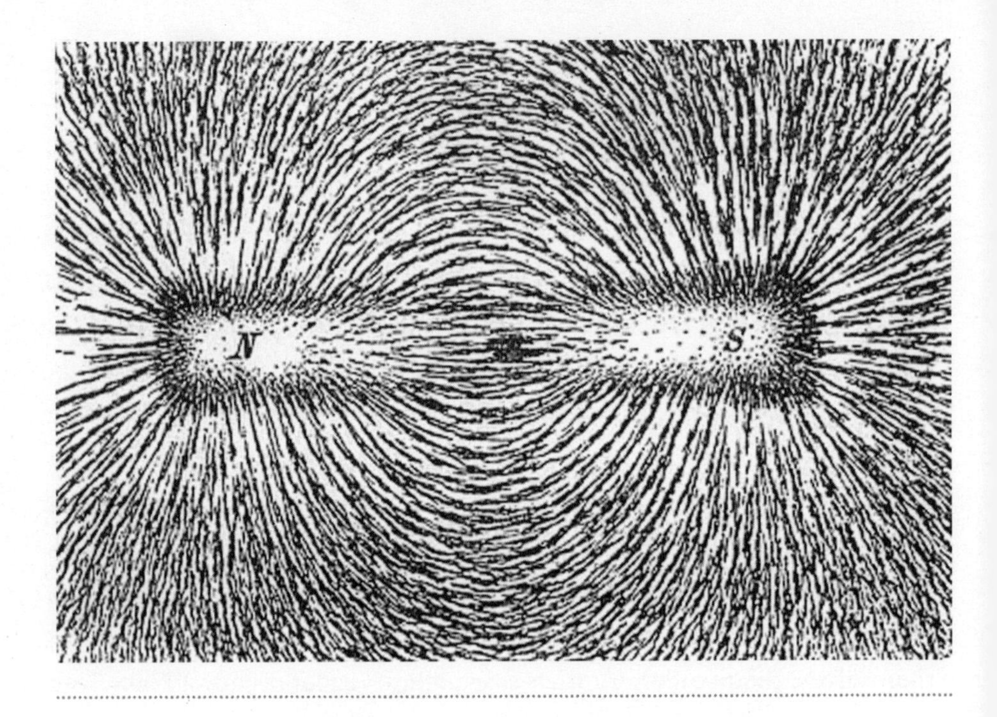


FIGURE 20. FARADAY'S LINES OF FORCE MADE VISIBLE.

*present.* The magnet stirs up that medium, or as we shall say (following Faraday and Maxwell) fluid, and the filings sense the fluid's state of excitation, through the pressure of its pushes and pulls.

We can make analogies with a more familiar fluid: our atmosphere, near Earth's surface. That atmosphere surrounds us, filling space. If the atmosphere is in motion we say there are winds. Winds are invisible in themselves, but are revealed through the forces on more readily visible, tangible bodies, such as weather vanes, birds, or clouds. If we imagine stirring up air with a fan, and using a system of weather vanes to track the pattern, the weather vanes' many alignments would define atmospheric "lines of force" very much in the spirit of Faraday's iron filings. In this case, of course, the weather vanes align with the local direction of airflow, or wind.

Pursuing that analogy further, we can imagine supplementing our weather vanes with speed-sensing devices (anemometers), to sample both the direction and the speed of the wind. We can do this at any point in space, and also at different times. In this way we define a *field* of velocities, which fills both space and time.

The field of wind velocities encodes the state of excitation of a fluid, namely air.

Faraday suggested that a similar logic applies to magnetism, and also to electricity. According to Faraday, an electrically charged test body acts as a combination weather vane and anemometer, sampling the state of the electric fluid. The test body feels a force due to the state of excitation of the electric fluid—the "electric wind," so to speak—at a particular place and time. After dividing the force the test body experiences by the test body's charge, we get a quantity that is independent of what kind of test body we use to measure it. We call this ratio the value of the electric field.

Now here, to avoid later confusion, I must briefly digress to describe, and then resolve, an annoying ambiguity that physicists have inflicted on themselves, their students, and the outside world for decades. Namely, it is standard practice to use the term "electric field" for two things that are quite distinct. One is the field of *values* of force divided by charge. As we just discussed, this is analogous to wind speed. Unfortunately, it is also standard to use "electric field" when referring to the underlying medium (i.e., the electric fluid itself), as opposed to its state of excitation. It is as if one used the same word for wind and for air. In this book, I will use terms like "electric fluid" and "magnetic fluid" (and later "gluon fluid"...) for the fluids whenever the distinction is important. That decision will lead me to use some slightly unconventional expressions, such as "quantum fluid theory" where you would elsewhere find "quantum field theory." I think the gain in clarity is worth the price in apparent eccentricity. (End of digression.)

Faraday's approach led him to several remarkable discoveries, one of which—the most remarkable—we'll discuss momentarily. Nevertheless, most of his contemporaries were unimpressed with Faraday's *theoretical* ideas. Those must have seemed not so much revolutionary as counter-revolutionary. Prior to Newton's celestial mechanics, the most influential system had been that of Descartes, who ascribed the motion of planets to the influence of space-filling vortices, which sweep them along. Newton had replaced those vague notions with simple, mathematically precise laws of motion and gravity, which worked spectacularly well. The same basic principles—action at a distance, inverse square falloff—were giving a good account of electric and magnetic forces too. To trade that solid framework, supported by concrete calculations and quantitative measurements, for the discredited dreams of a self-educated visionary? Hardly sound scientific strategy!

But Maxwell had a different perspective on Faraday's speculations. On pages 162–164, I will describe Maxwell himself, as an individual personality. (Full disclosure: He's my favorite physicist.) For now, I'll just say that he faced problems in science, and life in general, in a playful spirit. I think he saw Faraday's new fluids as wonderful toys, and was happy to be patient as he played with them.

## THE ROAD TO MAXWELL'S EQUATIONS

Maxwell's first major paper on electricity and magnetism (electromagnetism), written in 1856, predated "Dynamical Theory" by almost ten years. It was titled "On Faraday's Lines of Force." He says:

> My design is to shew how, by a strict application of the ideas and methods of Faraday, the connexion of the very different orders of phenomena he has discovered may be clearly placed before the mathematical mind.

Over the course of seventy-five pages of substantial work, Maxwell developed Faraday's imaginative visions into precise geometrical concepts, and then into mathematical equations.

His second big paper followed in 1861, and was titled "On Physical Lines of Force." His previous work, and that of Faraday, had packaged the observed facts of electromagnetism into a new form. They were interpreted as laws governing the excitations of a space-filling medium, the electromagnetic fluid (consisting, naturally, of the electric and magnetic fluids). Now Maxwell was ready to venture a mechanical model of the fluids themselves. Here it is, as he drew it (figure 21).

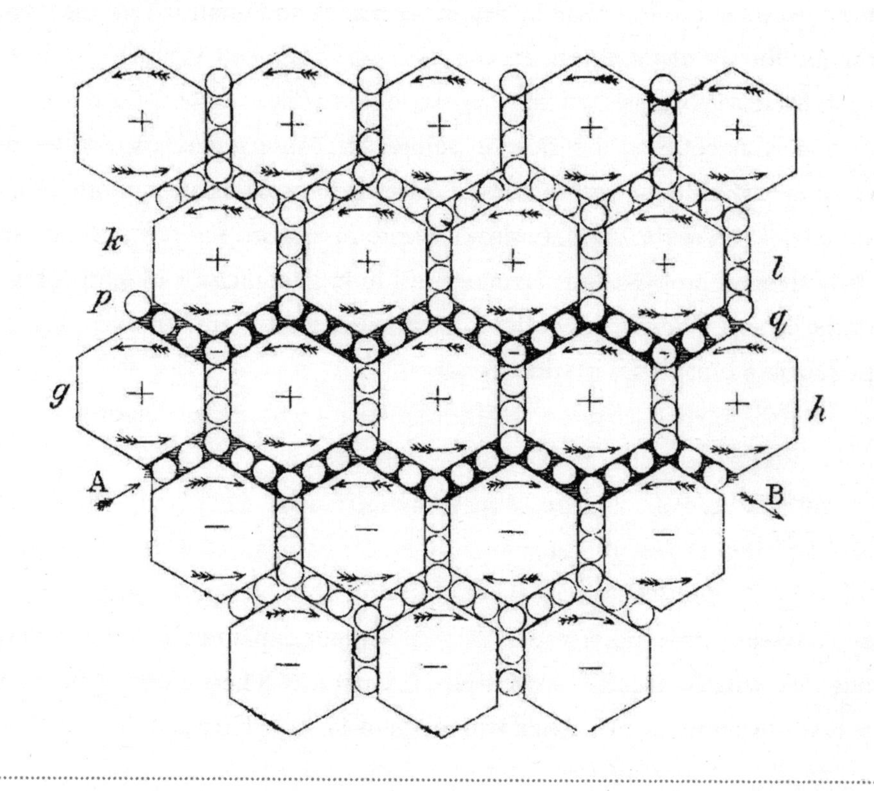


FIGURE 21. MAXWELL'S MECHANICAL MODEL OF SPACE FILLED WITH A MATERIAL MEDIUM WHOSE MOTIONS CAUSE ELECTROMAGNETIC FIELDS AND FORCES.

Maxwell's model includes magnetic vortex atoms (the hexagons) lubricated by electrically conducting spheres. Magnetic fields describe the speed and direction of magnetic vortex rotations, and electric fields describe the velocity field, or "wind," associated with flow of the spheres. Though entirely fictitious, the model gave a faithful representation of the known laws of electricity and magnetism, and suggested new ones.

It can be great fun and mind-stretching to play with Maxwell's model, but that's a difficult hobby not to everyone's taste, so I won't provide a manual here. In any case, the details of the model are more clever than believable, as Maxwell both recognized and freely admitted.

But the great virtue of a working model—or the design of one—is that it forces you to be both specific and consistent. Writing equations, or computer programs, involves the same sort of discipline. You must balance ambition *and* precision.

In Maxwell's model, when magnetic vortex atoms spin, they become oblate—flattened at the poles, fattened at the equator, like Newton's spinning Earth!—and push around the electric conducting spheres. Conversely, flows of the conducting spheres exert forces on the vortex atoms, and can set them spinning. Excitation of either fluid can lead to excitation of the other. Thus the model predicts that magnetic fields induce electric fields, and vice versa, in specific ways.

In this way, in addition to the known phenomena of electromagnetism, Maxwell's model predicted something new.

Faraday had discovered, experimentally, that when magnetic fields change in time, they produce electric fields. Faraday's law of induction, as it is called, underlies the design of electric motors and generators, and was an enormous stimulus to technology. It was also a vindication of Faraday's intuition that fields, in themselves, are independent elements of physical reality. For here is a law that's almost impossible to state without referring to them! Maxwell's model, built to accommodate Faraday's law, brought a dual effect, with the role of electric and magnetic fields reversed. Maxwell's law, as I call it, says that electric fields that change in time produce magnetic fields.

The combination of these two effects leads, as Maxwell saw, to dramatic new possibilities. From time-varying magnetic fields, Faraday gives us electric fields, which also vary in time. From time-varying electric fields, Maxwell gives us magnetic fields, which also vary in time. And the beat goes on:

. . . → Faraday → Maxwell → Faraday → Maxwell → . . .

Thus an excitation in electric and magnetic fields can take on a life of its own, with the fields dancing as a pair, each inspiring the other.

Maxwell could calculate from his model how fast such excitations will move through space. He found that they travel at a speed that matches the measured speed of light. And it is here we find the most extensive use of italics, by far, in all of Maxwell's work:

> The velocity of transverse undulations in our hypothetical medium . . . agrees so exactly with the velocity of light . . . that we can scarcely avoid the inference that *light consists in the transverse undulations of the same medium which is the cause of electric and magnetic phenomena.*

To Maxwell, it was obvious that this agreement of speeds was no coincidence: His electromagnetic disturbances *are* light, and light is, at bottom, nothing but a disturbance in electric and magnetic fields.

As I read this passage I find myself imagining what Maxwell felt, and then reliving some episodes from my own career when things suddenly came together, and then recalling these lines of Keats:

> Then felt I like some watcher of the skies
> When a new planet swims into his ken;
> Or like stout Cortez when with eagle eyes
> He star'd at the Pacific—and all his men
> Look'd at each other with a wild surmise—
> Silent, upon a peak in Darien.

It's a passage I like to revisit!

Maxwell's inference represented, if correct, a fantastic unification of electromagnetism and optics. Above all, it provided a startling new vision of light itself. It "reduced" light to electromagnetism—an expansive reduction, if ever there was one!

But Maxwell's wild surmise, for so long as it remained mixed up in a cockamamie model, was like a sparkle of gold amid a messy melt—a promise of beauty, yet undelivered. A wild surmise. The next step would be to remove the dross.

## *Spider Men*

Before we move on to Maxwell's equations themselves, I'd like to share a fantasy that occurred to me while preparing this chapter.

Imagine that a race of spiders arose, so intelligent that they begin to construct spider physics. What would that be like?

Spiders have poor vision, so they would not begin from the starting point that our visual perception suggests to us: a world of unconnected objects free to move within a receptacle, space. Instead, a spider's sensory universe is based on touch. More specifically, spiders sense vibrations of the strands of their webs, and from those vibrations they infer the exis-

FIGURE 22. INTELLIGENT SPIDERS WOULD HAVE A HEAD START ON FIELD THEORY. COMPARE THIS SPIDERY CONSTRUCTION WITH FIGURES 20 (PAGE 122) AND 21 (PAGE 125)!

tence of objects that cause them (especially, potential meals). For intelligent spiders, to conceive of lines of force would require no great leap of imagination. Force-conducting, space-filling webs are how they make their living. Their world is a world of connections and vibrations.

They'd know in their bones, so to speak, that forces are transmitted by space-filling media, and travel through them at finite velocity. They'd avoid Void, instinctively. They'd all be Faradays, and they'd soon conceive of a World Wide Web (figure 22).

## THE MAXWELL EQUATIONS

In "A Dynamical Theory of the Electrodynamic Field" Maxwell starts afresh. "On Physical Lines of Force" had resembled a gigantic Query, spinning out the consequences of a speculative Hypothesis, looking for encouraging signs in Nature. "Dynamical Theory" follows instead the style of the *Principia*, reasoning from observed facts to a system of basic equations.

Whereas Newton relied on Kepler's laws of planetary motion, Maxwell assembled four laws discovered by several earlier investigators—two Gauss laws, Ampere's law, and Faraday's law of induction. (These are described below, and also in "Terms of Art.") He expressed those laws in Faraday's language of electric and magnetic fluids, which he'd made precise and mathematical in his earlier work.

Maxwell also added his own law, dual to Faraday's. That addition was *not* based on experimental evidence. As we discussed, Maxwell was originally led to postulate that new law by working out the consequences of his cockamamie model. In the new treatment, he showed that the new law was necessary in order to make the old ones consistent!

Plate N spells out the Maxwell equations. It is an important aspect of their beauty that they can be represented in pictures! This system of four equations, combining four previously known laws together with a new addition, is now universally known as the Maxwell equations. (The four equations encode five laws, since one of the equations sums two physical effects.) They remain to this day our best fundamental description of electromagnetism and light.

Here, I can't resist spelling out the actual content of the Maxwell equations. After all this buildup, you might be curious to see *exactly* what the fuss is all about!

I've tried to do this in a way that is reasonably brief, accurate, and comprehensible. But there is some tension among those goals, and as a result you may find this patch hard going. I recommend that you approach it as you might approach an unfamiliar work of art—as an opportunity, not a burden. You might want to read it through lightly at first, and ponder the pictures, to get a general impression. Then you can decide whether you want to make a closer reading. And I hope you do—after all, the Maxwell equations are a *great* work of art. You can do that at leisure because nothing in our later meditations will refer to the details. You can also consult "Terms of Art," which surveys the same territory from slightly different perspectives. I've also indicated, in the endnotes, some excellent, free Web sites where you can explore the Maxwell equations interactively.

For each of the five laws that go into the four Maxwell equations, I will first give an informal version, then a more precise version, in words and in pictures. To follow along, please refer to plate N because we'll be reading it through, line by line.

First let me explain the labels on the drawings: $\vec{E}$ stands for electric field, $\vec{B}$ for magnetic field, and $\dot{\vec{E}}$, $\dot{\vec{B}}$ for the rate of change of those quantities in time. $Q$ stands for electric charge, and $\vec{I}$ stands for electric current. (The little arrows are reminders that these quantities have directions as well as magnitudes.)

Now to the laws:

- The *electric Gauss's law* equates the flux of electric field leaving a volume to the amount of electric charge inside that volume. It tells you that electric charges are the seeds of electric lines of force (or the seeds of their destruction). They are where electric lines of force can begin or end.

The definition of flux is most easily understood by reference to flow of a fluid. The electric field at a point is a quantity with magnitude and direction, as we've discussed. The field of velocities in a flowing fluid has the same character. Now given a volume and the velocity field, we can calculate how rapidly fluid is leaving the volume. That, by definition, is the flux of fluid leaving the volume. If we perform the same mathematical operations on the electric field that we performed on a velocity field to calculate its flux, then we obtain (by definition) the flux of electric field.

- The *magnetic Gauss's law* states that the flux of magnetic field leaving any volume is zero. The magnetic Gauss's law is, of course, very much like the electric Gauss's law, but with the additional simplification that there can be no magnetic charge! It tells you magnetic fields have no seeds—magnetic lines of force can never end, but must either continue forever, or close on themselves.
- *Faraday's law* is especially interesting because it brings in time. It forges a relationship between electric fields and the *rate of change* of magnetic fields. It says that when magnetic fields change with time, they cause electric fields to swirl around them.

To formulate Faraday's law precisely, we consider a curve that forms the boundary of a surface. Faraday's law equates the circulation of the electric field around the curve to (minus) the rate of change of magnetic flux through the surface. Circulation, like flux, is most easily understood by reference to the velocity field in fluid flow. We imagine expanding the curve into a narrow pipe, and compute the amount of fluid moved around the pipe per unit time. That is the circulation of the fluid flow. If we perform the same mathematical operations on the electric field that we performed on the velocity field to calculate its circulation, then we obtain (by definition) the circulation of electric field.

Finally, to be completely precise, we have to resolve the ambiguities of direction: In defining the circulation, which way do we go around the curve? In defining the flux, which way do we go through the surface? To get a definite relationship, we need to correlate those choices. The standard way is the so-called right-hand rule: If we go around the curve following the fingers of a right hand, then we take the flux in the direction of the thumb.

- *Ampere's law* forges a relationship between magnetic fields and electric currents. It says that electric currents cause magnetic fields to swirl around them.

  To formulate Ampere's law precisely, we consider a curve that forms the boundary of a surface. Ampere's law equates the circulation of magnetic field around the curve to the flux of electric current through the surface.

It's worthy of notice that the same concepts of flux and circulation recur several times in these laws. Flux and circulation are very basic ways of getting a mental grip on fields. They encode the sprouting and looping of lines of force, respectively. Their prominence in physical laws is a gift from Matter to Mind.

Now, when Maxwell assembled those four laws together, he found . . . a contradiction! (But Maxwell's fifth law fixes it.) To see it, consider plate O.

The problem occurs when you try to apply Ampere's law to situations where the electric current is interrupted. In the drawing of plate O we have electric current flowing into and out of a pair of plates, with a gap in between. (Experts may recognize this as the setup of a capacitor.) According to Ampere, the magnetic circulation around a loop is equal to the flux of current through any surface it bounds. But here we get different answers for that flux, depending on which surface we pick! If we take a disk within the gap (blue), we get zero. If we take a hemisphere that intersects the wire (yellow), we get the whole current.

Oops.

To fix this contradiction, we need something new. Thanks to his earlier work with his model, Maxwell was ready with:

- *Maxwell's law*, a kind of converse to Faraday's law, interchanging the role of electric and magnetic fields. It says that when electric fields change with time, they cause magnetic fields to swirl around them.

The disk within the gap intercepts no flux of current, but it does span a changing electric field. The yellow hemisphere predicts magnetic circulation due to Ampere's law, while the blue disk predicts magnetic circulation due to Maxwell's law—but they both predict the same result! Contradiction removed. After the addition of Maxwell's law, the entire system of Maxwell's equations becomes consistent.

In this treatment—a cleaned-up version of "A Dynamical Theory of the Electrodynamic Field"—Maxwell's law achieves a new status. It has lost its ties with mechanical models, vortex atoms, and lubricating spheres. Here we see it as a logical necessity, required for the consistency of other laws that had been derived from experiments.

## *Maxwell's Rapture*

Maxwell was a believing Christian, and he took his faith very seriously. Reflecting on his work establishing the world-fluids of electricity and magnetism, he was well gratified:

> The vast interplanetary and interstellar regions will no longer be regarded as waste places in the universe, which the Creator has not seen fit to fill with the symbols of the manifold order of His kingdom. We

> shall find them to be already full of this wonderful medium; so full, that no human power can remove it from the smallest portion of space, or produce the slightest flaw in its infinite continuity.

## *"Wiser Than We"*

James Clerk Maxwell died in 1879, at the age of forty-eight. At that time, his theory of electromagnetic fields was regarded as interesting, but not compelling. Serious work on rival action-at-a-distance theories continued. The most dramatic prediction of Maxwell's theory, that electric and magnetic fields could take on a life of their own and propagate in self-renewing waves, had not been verified.

It was Heinrich Hertz who first designed, and then starting in 1886 carried out, experiments that could test Maxwell's idea. In retrospect, we can say that Hertz created the first generation of radio transmitters and receivers.

The (apparently) magical ability to communicate across great distances, through empty space, through radio, was born of envisioning that empty space is *not* Void. It is filled with fluids, and pregnant with possibility.

Heinrich Hertz died in 1894, at the age of thirty-six. But before he died, he wrote this beautiful tribute to the Maxwell equations that gets to the heart of our Question

> One cannot escape the feeling that these mathematical formulae have an independent existence and an intelligence of their own, that they are wiser than we are, wiser even than their discoverers, that we get more out of them than was originally put into them.

We've looked at, and admired, the wisdom of the Maxwell equations themselves. Here Hertz is saying the sort of thing about equations that is often said of great works of art—that their meaning extends far beyond the creator's intention.

What is the "more" to which Hertz refers?

Three things, at least:

- Power
- Generative beauty
- An inspiring new idea: *symmetry of equations*

## *Power*

Maxwell surmised, from his equations, that light is an electromagnetic wave. But visible light is just the tip of a much larger iceberg, invisible to us, and almost entirely unknown in Maxwell's time. One can have electromagnetic waves with any specified wavelength,* and Newton's spectrum of visible light is only a sliver within that continuum, as we see in plate P. The solutions of Maxwell's equations describe much more than visible light. There are solutions where the oscillations between electric and magnetic fields take place over different distances (wavelengths). The visible spectrum corresponds to a narrow range of wavelengths within the infinite continuum of pure electromagnetic waves.

I've already mentioned Hertz's pioneering work, which gave us radio waves and blossomed into radio technology. Radio waves are "light" with much longer wavelength, and lower frequency, than visible light. In other words, in radio waves the oscillations between electric and magnetic fields take place more gradually in space, and more slowly in time. Going from radio toward shorter wavelengths, we meet microwaves, infrared, the visible, ultraviolet, X-rays, and gamma rays. Each of those many forms of "light" has evolved from a purely theoretical construction—in other

*We'll be discussing these things in the following chapter.

words, a dream—into a fountain of modern technology. They're all in Maxwell's equations. Now that's power!

## *Generative Beauty*

When you solve them, you often find that Maxwell's equations reveal beautiful, surprising structures.

In plate Q, for example, is the shadow cast by a razor blade, or anything with a sharp, straight edge, if you illuminate it with purified light. Magnifying the shadow cast by purified light, we discover a rich and beautiful pattern.

Geometric reasoning, based on the crude idea that light travels in strictly straight lines, would have told you the shadow is a sharp division between darkness and light. But when we calculate wavy disturbances in electric and magnetic fields, we discover that there is a lot more structure. Light penetrates darkness (i.e., the geometric shadow region) and darkness penetrates light. You can calculate the pattern precisely, using Maxwell's equations. And nowadays, with bright monochromatic lasers available, you can compare their predictions directly with reality. Looking at this image, you just have to say: Isn't that pretty?

## *Symmetry of Equations*

Study of Maxwell's equations brought out an essentially new idea that had not really played a big role in science before. That is, the idea that *equations*, like objects, can have symmetry, and that the equations Nature likes to use in her fundamental laws have enormous amounts of symmetry. Maxwell himself was unaware of that idea. So this is most definitely an example of getting more out than was put in!

What does it mean to say that equations have symmetry? While the word "symmetry" has various, often vague meanings in everyday life, in mathematics and physics "symmetry" has come to mean something quite precise. In those contexts, symmetry means Change Without Change. That definition may sound mystical, or even paradoxical, but it means something quite concrete.

Let's first consider how that strange definition of symmetry applies to objects. We say an object is symmetric if we can make transformations on it that might have changed it, but in fact do not. So, for instance, a circle is very symmetric because you can rotate a circle around its center, and though every point on it moves, overall it remains the same circle; whereas if you took some more lopsided shape and rotated it, you'd always get something different. A regular hexagon has less symmetry because you have to rotate through 60 degrees (a sixth of the way around) to get back to the same shape, and an equilateral triangle still less because you have to rotate through 120 degrees (a third of the way around). A general, lopsided shape won't have any symmetry at all.

One can also work in the opposite direction. We can start with symmetry, and get to objects. For example, we might ask for curves that are unchanged by rotations around some point, and then discover that circles are the unique embodiment of that symmetry.

The same idea can be applied to equations. Here's a simple equation:

$$X = Y$$

. . . which you can see is neatly balanced between X and Y. You'd be tempted to say it is symmetric. And indeed it is, according to the mathematical definition. For if you change X into Y and Y into X, you get a different equation, namely,

$$Y = X$$

This new equation differs in form, but it has exactly the same content as the old one. So we've got a Change Without Change: symmetry.

On the other hand, when we interchange X and Y, the equation $X = Y + 2$

changes into $Y = X + 2$, which does not mean the same thing at all. So that equation is *not* symmetric.

Symmetry is a property that certain equations, or systems of equations, have while others don't.

Maxwell's equations, it turns out, have an enormous amount of symmetry. There are many kinds of transformations you can make on Maxwell's equations that change their form but not their overall content. The interesting symmetries of Maxwell's equations are considerably more complicated than the toy example we've just considered, but the principle is the same.

As for objects, so also for equations, we can explore in the opposite direction. Instead of setting up equations, and then finding the symmetries they allow,

$$\text{equations} \Rightarrow \text{symmetry}$$

. . . we can start with symmetries, and ask for the equations that allow them:

$$\text{symmetry} \Rightarrow \text{equations}$$

Remarkably, that path is a road back to Maxwell's equations! In other words, Maxwell's equations are essentially the only equations to have the symmetry that they do. They are the circles defined by their own highly evolved rotations. In this way Maxwell's equations embody a perfect correspondence:

$$\text{equations} \Leftrightarrow \text{symmetry}$$

It is not a great stretch to see this relationship as an instance of our desired

$$\text{Real} \Leftrightarrow \text{Ideal}$$

In modern physics we have taken this lesson to heart. We have learned to work from symmetry toward truth. Instead of using experiments to infer equations, and then finding (to our delight and astonishment) that the equations have a lot of symmetry, we propose equations with enormous symmetry and then check to see whether Nature uses them. It has been an amazingly successful strategy.

This chapter's themes of connection, symmetry, and light come together in the art of the *mandala.* Mandalas are symbolic representations of the Universe. They are used as tools for meditation and trance. They typically display large-scale symmetry among connected, intricate parts, and are often colorful. Plate R provides, I think, a fitting conclusion.

# MAXWELL II: THE DOORS OF PERCEPTION

*If the doors of perception were cleansed every thing would appear to man as it is, infinite.*

*For man has closed himself up, till he sees all things thro' narrow chinks of his cavern.*

• *William Blake,* The Marriage of Heaven and Hell

In this chapter we focus on a special aspect of our Question: the beautiful idea, that by better understanding our experience of the world, we can expand that experience.

William Blake's dreamlike multimedia book *The Marriage of Heaven and Hell* yearns to unite, as he puts it, "what the religious call Good and Evil" (see plate S). According to Blake, "Good is the passive that obeys Reason. Evil is the active springing from Energy. Good is Heaven. Evil is Hell." The goals of our meditation, to harmonize Ideal and Real, and to see things whole, speak to the same yearning.

Blake refers to a cavern, and that reference calls to mind Plato's Cave. The prisoners in Plato's Cave experience the world in black and white, and miss out altogether on the beauty of color. While our case is less extreme, we too experience only a small part of what light has to offer.

We will compare the full reality of vision's subject matter—light—with the projection of reality that human vision captures. This was a subject that Maxwell loved, and greatly clarified.

In that context, we will substantiate Blake's visionary intuition by answering the two questions it raises:

- *Are there Infinities to which we're shuttered?* Yes. The world of physical colors is a space of doubly infinite dimension, of which we perceive only a three-dimensional projection.
- *Can we unveil them?* Yes. The question is not *whether that is possible,* for it certainly is, but *how to do it,* in a practical way.

Our exploration of color perception will also prove to be wonderful preparation for understanding Nature's deep design, in later chapters.

## *Two Kinds of Yellow*

Yellow is one of the colors that appears in rainbows, and in the spectrum produced by a prism from sunlight. Spectral yellow is one of Newton's pure colors. So are spectral red, green, and blue.

But there is also another, quite different form of light that appears yellow. We can combine spectral red and spectral green to produce a *non-spectral* but convincing color we perceive as yellow (see plate T). The yellow produced in this way is quite different from spectral yellow, as a physical entity, though they are perceived as identical.

Similarly, you need not add all the colors in the spectrum of sunlight, in just the same proportions as in sunlight, to get a white that *looks* like sunlight. As you can see in plate T, you can get a convincing perceptual white by mixing just three spectral colors, as here red, green, and blue. If you passed that "white" beam through a prism, you wouldn't get a continuous rainbow, but just three lines. As a physical entity this beam is quite different from sunlight, but human vision perceives it as identical.

Please note that the results you get by mixing several colored *light beams*, as illustrated in plate T, differ radically from the results you get by mixing *pigments* of the same colors, as you do in mixing paints or smudging crayon marks. When you combine colored light beams, you are simply adding up the light they contain. With pigments, it is quite different. We typically view pigments, say as they appear in a painting, in reflected sunlight (or some artificial near substitute). The color we see in the reflected light depends on what spectral colors the pigments have taken out, or absorbed, from the reflection. The color we see, of course, represents the mixture of light that hasn't been absorbed. Now when, in painting or drawing, you combine the effect of two pigments, you are adding up their powers of *absorption*. The processes of adding colors, as in beams, or adding absorption of colors, as in pigments, are very different. For example, you'll rather easily get black—the absence of reflection—by combining enough different pigments, but never by adding different beams. So it shouldn't be surprising that, in general, there are very different rules for combining colors of beams, as opposed to combining colors of pigments. Adding beams is conceptually simpler and physically more fundamental than adding pigments, and that's what I'll be discussing in what follows.

## COLOR TOPS AND COLOR BOXES

The basic observation that different mixtures of spectral colors can look the same naturally leads to a broader question: What mixtures, exactly, look the same? What kind of space is the space of perceived colors?

Before, during, and after his epochal theoretical work establishing the electromagnetic nature of light, Maxwell did a great deal of experimental work on exactly those questions. His results in this field are, within its narrower scope, equally fundamental. They have led to important technologies and promise more, as we shall discuss.

In the picture of a youthful Maxwell in figure 23, you may notice that

FIGURE 23. MAXWELL HOLDING AN EARLY COLOR TOP.

he is holding a circular object with a peculiar design. What you're seeing is a spinning top he designed to help elucidate color perception. You'll notice that the photograph is in black and white. Color photography hadn't yet been invented—Maxwell would do that, a little later!

The color top appears to be a toy, and at one level it is, but it is also much more. A simple but profound idea turns the color top into a powerful instrument for elucidating color perception.

Although we have the impression that our vision reveals the instantaneous state of the world, giving us a continuous, seamless reflection of events as they happen in time, the reality is different. Our vision is more nearly a series of snapshots, each with an exposure time of about one twenty-fifth of a second. Our brain fills in between those snapshots, to give the illusion of continuity. Movies and television exploit this fact: If the images they present are updated fast enough, we don't sense that they are a sequence of stills, or a rapid series of pixel updates. The color top takes advantage of that same effect, the persistence of vision.

In Maxwell's color tops, he arranged colored papers in two circular strips, as in plate U. Thanks to the persistence of vision, when we spin these two-banded disks rapidly around their center, we will perceive the color mixtures that the color segments within each band generate, as if those colors were produced by beams. That's the genius of Maxwell's color top: When we view the top, our eyes add up the reflected *beams*. Color tops following Maxwell's design allow us to map out, in a completely systematic and quantitative way, which combinations of colors look the same.

Of course, we should also check that different people report the same matches. This is basically true, though there are small variations among normal individuals. We must also allow exceptions for several kinds of color blindness, and possibly a population of people with superior powers of color discrimination. We'll discuss those departures from the norm later. For the most part, though, most people agree. Now, whether different people have the same *subjective experience* of, say, red, is an inexhaustible topic for murky philosophical debates. What we can say for sure is that my *mapping*, or projection, of physical light into perceived colors matches yours, very nearly. We both see many mixtures of spectral colors as yellow, and many others as magenta. Most important, we agree on which mixtures make which. Human discourse about colors would be quite confusing, if that weren't the case!

The central result to emerge from these studies is that by using just *three* colors in the inner strip, we can match *any* color in the outer one.

Thus, for example, we can use spectral red, green, and blue, in appropriate proportions, to get orange, mauve, chartreuse, puce, azure, burnt peony, or any other color you care to name. The three base colors need not be red, green, and blue (RGB)—almost any three will do, including mixtures, as long as they are independent. (If one of your base colors can be obtained as a mixture of the other two, it doesn't enable new possibilities.) On the other hand, we do need three base colors. If you restrict yourself to two base colors—no matter what they are—then most colors can't be matched by mixing them.

To put it another way, we can specify any perceived color by saying how much red, how much green, and how much blue it takes to match it. This is completely analogous to how we can specify a place in space by saying how far it is in the north-south, east-west, and vertical directions. Ordinary space is a three-dimensional continuum, *and so is the space of perceived colors.*

Referring back to plate T, we can see that our central result states that by adjusting the relative intensities of the different beams, we can make *any* perceptual color, not just white, appear in the middle, where the light from all three beams overlaps.

In later work, Maxwell contrived to combine light beams directly, using devices he called color boxes. The strategic idea is simple: extract colors from a prismatic rainbow, at places and in proportions under your control, then recombine them, using mirrors and lenses. The tactical details, given the limited technology of the time, were very tricky. The only available light source was sunlight, and the available detector was a human eye, for example. Maxwell's color boxes were huge—six feet long, and more—housing mirrors, prisms, and lenses. Unwieldly though they were, the color boxes enabled much more precise work than the color tops could.

Maxwell's idea to process color by separation, manipulation, and recombination was ahead of its time. Modern technologies should empower us to attempt much more ambitious color manipulations, as I'll discuss below.

### *Using It*

The fact that one can synthesize all perceived colors by mixing three is widely exploited in modern color photography, television, and computer graphics. For example, in color photography three kinds of color-sensitive dyes are used. In computer displays, there are three kinds of color light sources. When you see "millions of colors" given as an option, it refers to millions of different ways of adjusting the relative intensity of those sources. In other words, one samples *millions* of different *points*, but all within a *three*-dimensional *space.*

For artists, the possibility to obtain the same perceptual color in many different ways opens up creative possibilities. You can add local texture, while keeping the overall (averaged) color intact. This is basically another sort of color top that exploits the persistence of vision in space, rather than time. The spatial averaging is less crude, and so supports a richer palette of variations. The Impressionists, especially, built masterpieces around such possibilities, as in the painting shown in plate V, Claude Monet's *Grainstack* (*Sunset*), of his Haystacks series.

In applying the different pigments *separately*, in distinct (though very close) parts of the canvas, rather than *overlaying* them, the Impressionists followed a strategy similar to that used in Maxwell's color tops, carried over from time to space. In both cases, the light from different regions combines according to the rules of beams—for one does not mix the pigments, but rather their reflected light.

## INFINITIES LOST

Maxwell gave us new conceptions of what light is, and of what our perception of light is. And they are very different things! As Blake anticipated, they are *infinitely* different.

By comparing the totality of what's out there to the information we capture, we can formulate what's been lost quite precisely. And then we can think intelligently about how to recover some of it.

## *The Raw Material: Electromagnetic Waves*

I advertised the emergence of light from Maxwell's equations in the previous chapter. Now I'd like to explore it a bit more deeply. As payoff, we'll get a firm grip on the lost infinities.

Maxwell described his fundamental account of light this way:

> What, then, is light according to the electromagnetic theory? It consists of alternate and opposite rapidly recurring transverse magnetic disturbances, accompanied with electric displacements, the direction of the electric displacement being at the right angles to the magnetic disturbance, and both at right angles to the direction of the ray.

Plate W unpacks that description.

The electric and magnetic fields at any point have both magnitude and direction, so we can draw them as colored arrows emanating from the point. Now if we did that at every point in space, we'd get quite a mess of overlapping arrows, so in the picture only the fields along one line are shown.

If you imagine this whole pattern moving in the direction of the black arrow, you'll see that at every point the electric fields (in red) change and the magnetic fields (in blue) change. As we discussed in the preceding chapter, changing electric fields produce magnetic fields, and changing magnetic fields produce electric fields. You can see how if things are just right, the moving disturbance might be self-generating—that is, that the change in the electric fields makes the magnetic fields change in such a way as to make the electric fields that made the magnetic fields, and the

PLATE A. Taiji figure by He Shuifa, also used in the frontispiece to this book.

PLATE B. Pythagoras at work,
from *The School of Athens* by Raphael.

Plate C. "So simple"—Pythagoras's theorem at a glance.

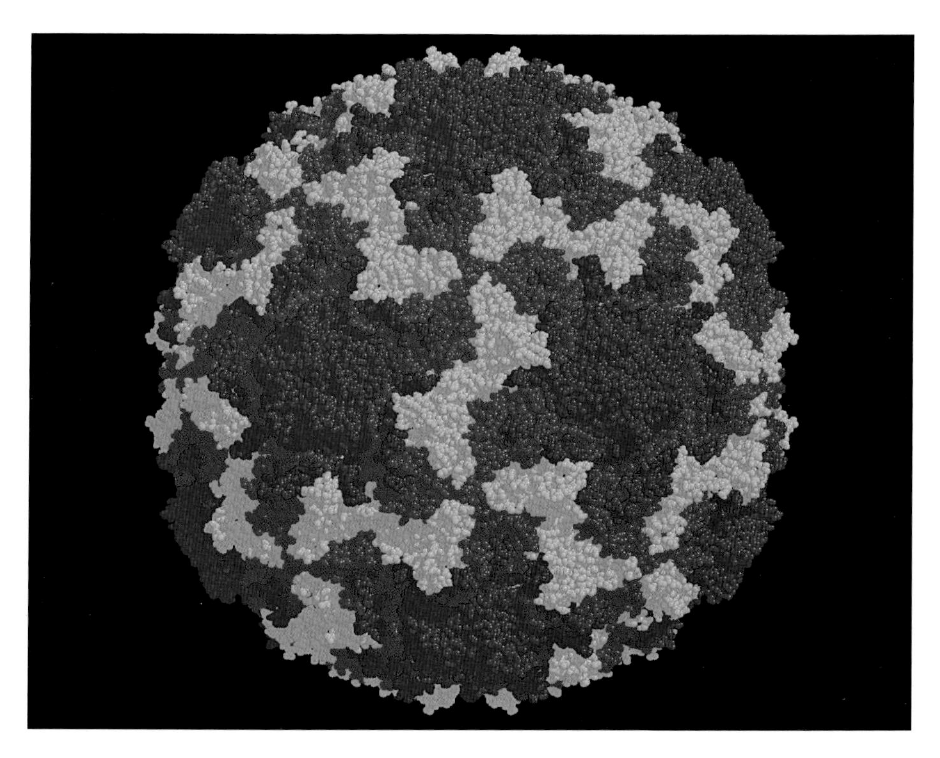

Plate D. A typical viral exoskeleton, exhibiting the structure of a dodecahedron—and also of an icosahedron!

Plate E. In Dalí's *Last Supper*, the sacrament unfolds within a dodecahedron.

Plate F. Plato urges us to look beyond appearances if we wish to discover the deep structure of reality.

PLATE G. Perugino's *Giving of the Keys to Saint Peter:* the joy of perspective.

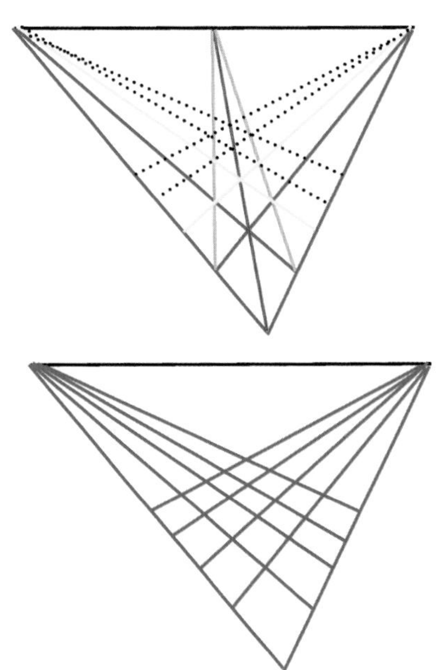

PLATE H. The beautiful geometric construction at the heart of perspective drawing.

Plate I. In Western Christian iconography, white is a powerful symbol both of purity and of power. Fra Angelico's *Transfiguration* is a sublime example.

Plate J. Spectral colors, which emerge when white light is analyzed by a prism, can be reassembled into white light using a second prism.

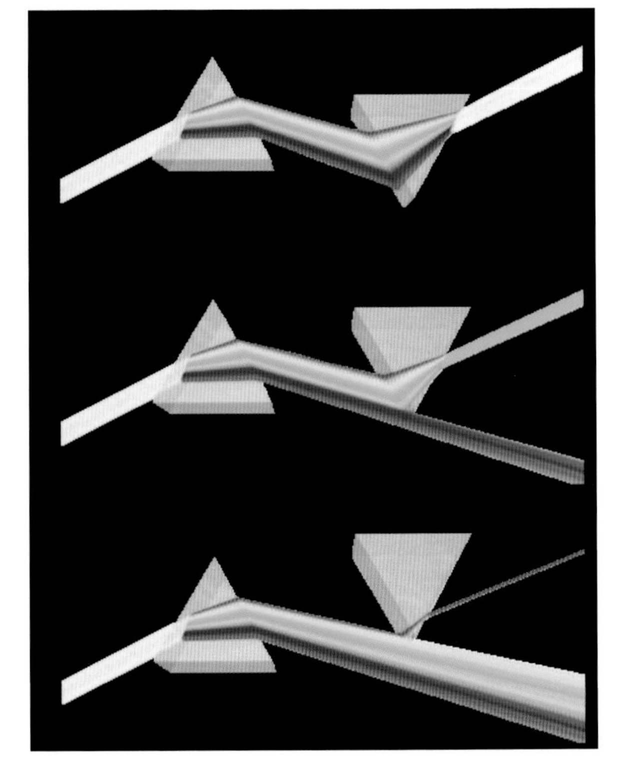

PLATE K. William Blake's depiction of Isaac Newton at work.

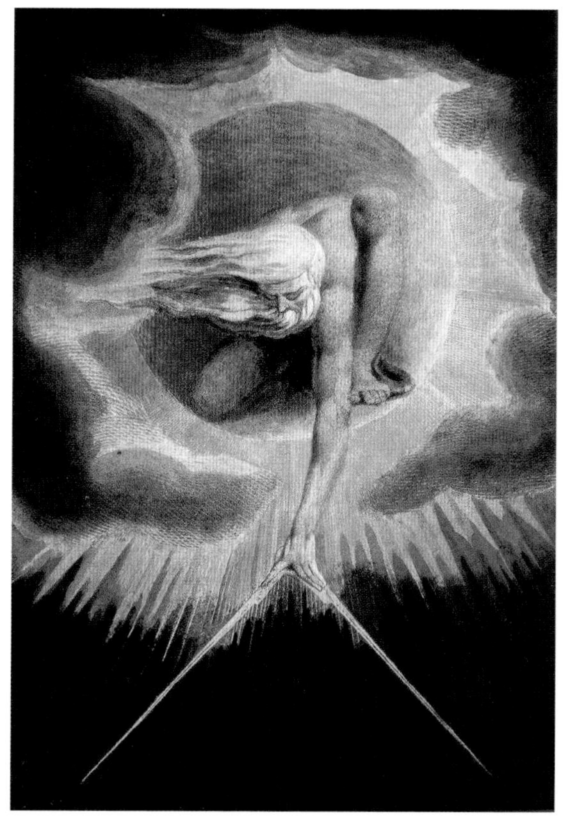

PLATE L. Blake's depiction of Urizen, creator and lawgiver.

PLATE M. Regular fractals are constructed according to simple, strict mathematical rules. A short computer program generated this intricate image.

## MAXWELL'S EQUATIONS

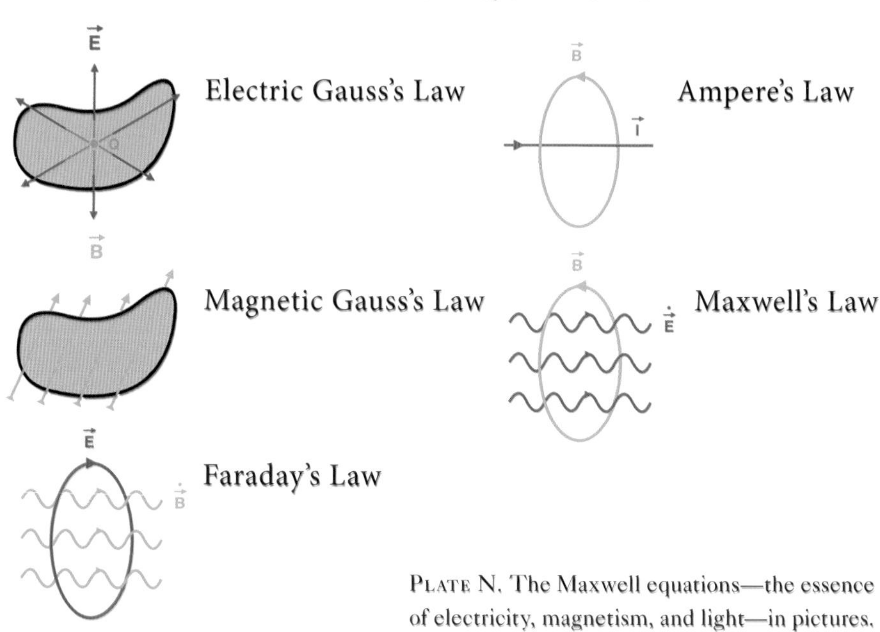


PLATE N. The Maxwell equations—the essence of electricity, magnetism, and light—in pictures.

## MAXWELL'S CONTRADICTION

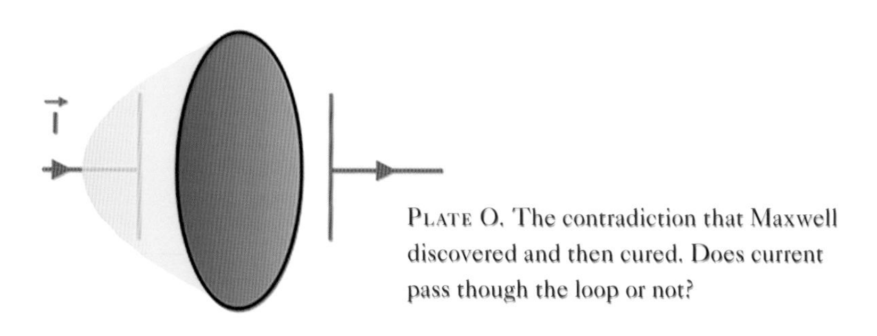


PLATE O. The contradiction that Maxwell discovered and then cured. Does current pass though the loop or not?

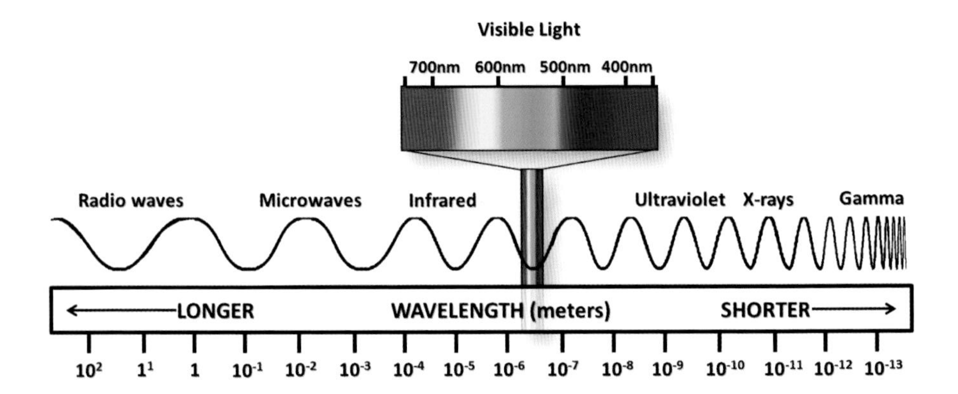


PLATE P. The solutions of Maxwell's equations describe much more than visible light. In modern technology, we exploit many other kinds of "light."

Plate Q. The shadow cast by a sharp edge—for example, a razor's edge.

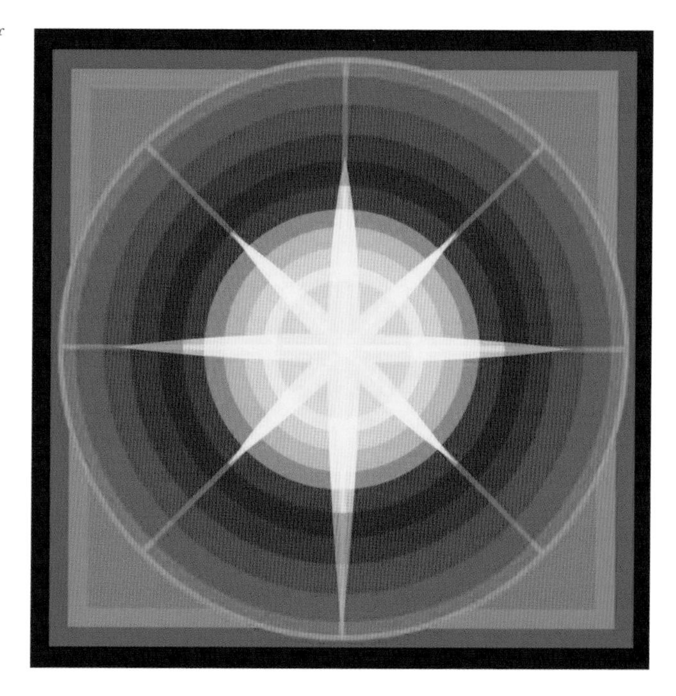

Plate R. *The Birth of the Son of God*, a digital painting by R. Gopakumar.

PLATE S. Title page from William Blake's dreamlike multimedia book *The Marriage of Heaven and Hell.*

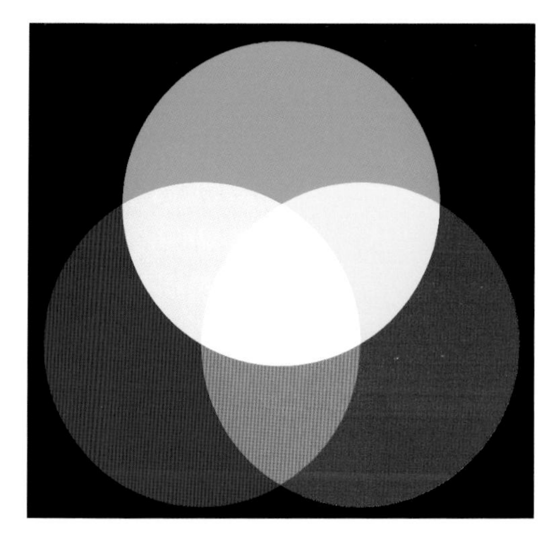

PLATE T. By combining beams of spectral red, green, and blue, we produce a variety of perceptual colors, including yellow and white. The perceptual white produced in this way is quite different from the white of sunlight.

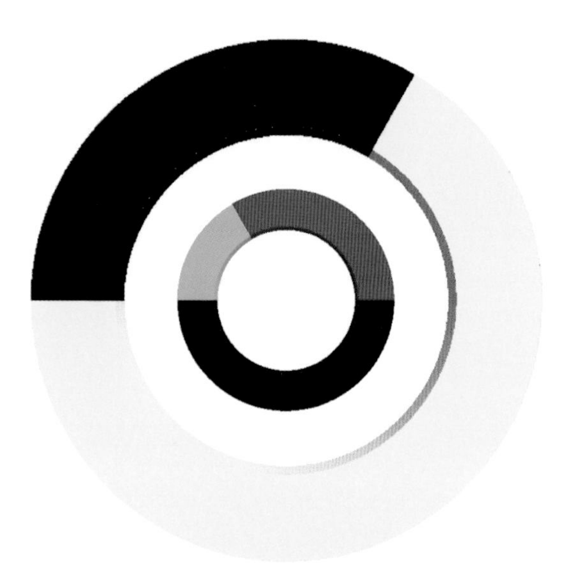

PLATE U. When this disk is mounted on stiff cardboard and spun rapidly around its center, it generates a color mixture of red and green from one strip and yellow from the other that we can easily compare. We can adjust the overall brightness of each strip by including black sectors, which contribute no reflected light.

PLATE V. Impressionist painters took advantage of the possibility of creating a given perceptual color from varied mixtures of others. Shown here is Monet's *Grainstack* (*Sunset*) from his Haystacks series.

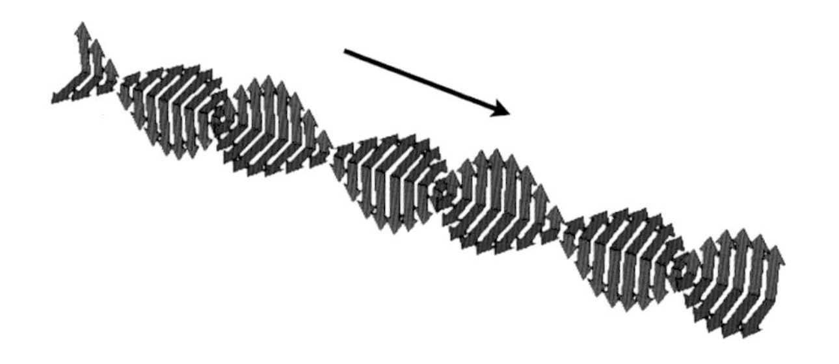

PLATE W. A snapshot of the electromagnetic reality of light, according to Maxwell's (correct) theory. Electric fields are shown as red arrows, magnetic fields as blue arrows. As time progresses, this complex of disturbances moves along its defining line toward the southeast—at the speed of light!

What you see

What Fido sees

PLATE X. By doing some image processing on a normal image, to project from three to two color dimensions, we can get a rough impression of what dogs—and color-blind people—miss out on.

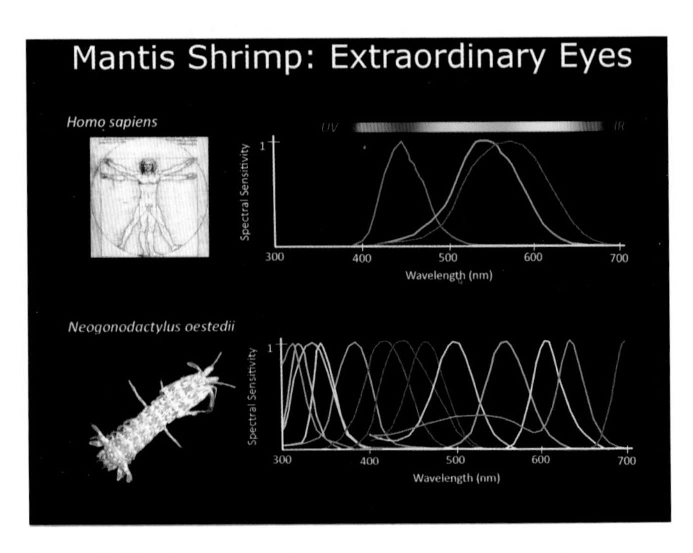


Plate Y. The human visual system is based on three color receptors, while mantis shrimp have many more. These renditions of the respective spectral sensitivity curves give a sense of the shrimps' superior color resources.

Plate Z. The most visually advanced species of mantis shrimp is also very colorful, as you can see in this picture. Of course, the picture conveys only how a mantis shrimp looks to *us*.

whole process takes on a life of its own. It's a feat worthy of Baron von Münchausen, who—according to Baron von Münchausen—can pull himself up by his own bootstraps. But for electromagnetism it's not a tall tale, nor magic realism, but *realistic magic.*

At any one point, as time progresses, the electric field arrow goes up and down, like the surface of water in a wave. In general, we call moving, self-generating electromagnetic disturbances electromagnetic waves.

Our picture shows a particularly simple electromagnetic wave, where the pattern of electric and magnetic disturbances repeats after a certain distance (and, technically, follows a sine function). I'll call this a *pure* wave, for reasons that will become clear momentarily. In that case we call the spacing between the repeats the wavelength of the wave. The pattern will also repeat in time, at a rate we call the frequency of the wave.

A *very* important property of electromagnetic waves is that you can multiply and add them. That is, if you have an electromagnetic wave solution of Maxwell's equations, and multiply the electric and magnetic fields in it by a common factor, the result will still be a solution of Maxwell's equations. So if you make all the fields in a solution twice as big, for example, you get another kind of disturbance that's still a solution. That amounts to adding your original solution to itself. You can also add one solution to another, and the result is a solution. These *mathematical* possibilities correspond to the *physical* possibilities of turning the brightness of a light beam up or down (multiplication) or combining one beam with another (addition).

We know, from experience, that turning up the brightness and adding are things you can do with light beams. So we'd be in trouble, trying to identify light as a form of electromagnetic waves, if we couldn't do them with electromagnetic waves. Fortunately, we can.

Finally, let's match up the details of Maxwell's verbal description, quoted above, with our pictorial version. You see, in the picture, that the electric and magnetic fields are perpendicular (or, in other words, at right angles) to each other, and that the direction of motion is perpendicular to both. That's exactly the geometry Maxwell describes in words. And the rapid, alternate (i.e., up-and-down) oscillations he mentions are just what you will observe, at any fixed point, as the wave progresses.

## *Purified Light, Revisited*

We have pure electromagnetic wave solutions of the Maxwell equations for any wavelength, traveling in any direction.

Pure electromagnetic waves with wavelengths in a specific, narrow range—from about 370 to 740 nanometers, quantitatively—are the raw material for human vision. They correspond to the pure light revealed in Newton's prismatic spectrum. In musical terms, the human visual range spans one octave (one doubling of wavelength). Each spectral color corresponds to a definite wavelength, as shown in plate P.

The vast bulk of the electromagnetic spectrum evades our vision altogether. We do not, for example, see radio waves, and without radio receivers we're left unaware of their existence. On the other hand, almost all of the Sun's electromagnetic radiation that penetrates Earth's atmosphere is concentrated near the visible part of the spectrum, so that's the most useful part for earthbound creatures to be attuned to. It's where the signal is turned on, so to speak.

For the moment, let's concentrate on the resource that our Sun's illumination gives us in abundance, and consider just the visible part of the spectrum.

Does our perception take full advantage of that resource? No. Not by a long shot.

What would a full analysis of the signal entering our eyes consist of? The answer to that question has two aspects, which are quite distinct. One is the *spatial* aspect. The signal contains information about the directions of light rays as they arrive from different objects. We use that information to form images. The other is the *color* aspect. It captures a different sort of information. We can have images in black and white, and we can have color patterns—in the extreme case, simply eye-filling, uniform colors—that don't make images.

## COLOR, TIME, AND HIDDEN DIMENSIONS

Through our discussion of electromagnetic waves and their spectrum of possibilities, we've laid the groundwork for a profound, and very pretty, perspective on what color is. While imagery reveals information about what's happening in space, color is telling us about what's happening in *time.* Specifically, color is giving us information about rapid variations in the electromagnetic fields that enter our eyes.

To avoid a possible confusion, let me emphasize that the time information carried by color is very different from, and complementary to, the time information we use in our everyday reconstruction of the order of events in time. Roughly speaking, our eyes give us snapshots twenty-five times a second, and our brains create the illusion of a nonstop movie from those snapshots. That construction underlies our everyday sense of the flow of time. In the process of gathering light for those snapshots—as photographers say, during the exposure time—the light is simply added up, or integrated. Because the light arriving at different times within each time frame is lumped together, information about arrival time, within each frame, is lost.

Color, as we sense it, is a way of preserving some very useful information about the temporal microstructure of the signal that survives the averaging process. Colors give us information about the variation of electromagnetic fields over *much* smaller time intervals, in the range $10^{-14}$–$10^{-15}$—a few millionths of a billionth—of a second! Because no everyday objects can move very far, or do anything noteworthy, during such tiny intervals of time, the two kinds of time information, the kind encoded in the changes from one snapshot to another, and the kind encoded in colors, are quite independent.

For instance, when we perceive pure spectral yellow, our eyes are telling us that the incoming electromagnetic waves are pure waves that repeat themselves about 520,000,000,000,000 times per second. When we perceive spectral red, the message is that repeats occur about 450,000,000,000,000 times per second.

Or rather, our eyes *would* be telling us those things if they didn't conflate the possible message "spectral yellow" with a manifold of other possible mixtures of colors that likewise *look* yellow, and the possible message "spectral red" with a (different) manifold of mixtures that look red. The actual message they convey is ambiguous because many possible inputs give the same output!

A true analysis of the incoming signal as to color must reveal the same information as Newton's prismatic analysis. A true analysis, that is to say, would resolve the incoming signal into its pure spectral components, each having its own independent strength. To report the result of such an analysis, we would need to specify a *continuous infinity* of numbers, one for the strength of each pure spectral component. Thus space of potential color information is not merely infinite, but *infinite-dimensional.* Instead, our eyes' projection of this information captures, as Maxwell discovered, just three numbers.

In short: The space of color information is infinite-dimensional, but we perceive, as color, only a three-dimensional surface, onto which those infinite dimensions project.

To complete the story, I should also mention another kind of electromagnetic information that's on offer, but also disregarded, within the signal entering our eyes. Referring back to plate W, you will notice that the electric fields (red) are oscillating in a vertical direction, while the magnetic fields (blue) are oscillating in a horizontal direction. There is also another solution, where you rotate the whole pattern by 90 degrees, so the electric fields are horizontal and the magnetic fields vertical. This rotated solution oscillates at the same rate as the original solution, so it represents the same spectral color. But it is physically distinct. The new property that corresponds to this distinction is called the polarization of the wave. Thus the electromagnetic information entering our eyes at each image point is infinite-dimensional twice over, because for each spectral color there are two possible polarizations, each of which can occur with an independent strength. Human vision overlooks that doubling because human eyes cannot distinguish between different polarizations of light.

## *Color Receptors*

The central result of Maxwell's color matching experiments, that mixing three basic colors can generate any perceptual color, reveals a deep fact about the "what" of human perception, but it also raises the question "How?" A beautiful and instructive answer to the "how" question emerged in the mid-twentieth century, as biologists explored the molecular basis of color vision. (It's fun to notice that physicists figured out the biology, and then biologists figured out the physics.)

The central result in the molecular story of vision is that three different kinds of protein molecules (rhodopsins) extract our information about color. When light impinges on one of these molecules, there is a certain probability that the molecule will absorb a unit of light—a photon—and change shape. The shape changes unleash little pulses of electricity, which are the data our brains use to construct our sense of vision.

Now, the probability that a given unit of light will be absorbed depends both on its spectral color and on the properties of the receptor molecule. One kind of receptor is most likely to absorb light in the red part of the spectrum, another peaks in green, and the third in blue, though their tuning isn't sharp (see plate Y). At common levels of illumination there are lots of photons, so there are many absorption events, and these probabilities get translated into three accurate measures of the amount of power the incoming light contains, averaged over three different spectral ranges.

In this way, we become sensitive not only to the overall amount of light arriving, but also to its composition. If it is spectral red light, it will stimulate firing from the red-peaked receptors more than the others, and result in a completely different signal than spectral blue light (which, of course, stimulates the blue-peaked receptors most).

On the other hand, any sort of incoming light that has the same ability to stimulate each of the three kinds of receptors—in other words, that gives the same three weighted averages—will be "seen" as the same by each of our color receptors, and will therefore result in the same visual perception. It takes three numbers to get a match: on that, the molecules agree with the color tops!

## *Varieties of Color Vision*

Now that we know what to look for—what's on offer, in the incoming signal—we can survey the biological world, counting receptors and measuring their absorption properties, to get new perspectives on color perception.

Mammals, in general, have poor color vision. The red of the bullfighter's cape is for the benefit of the human spectators, not the bull, because bulls perceive only shades of gray. Dogs do better; they see a two-dimensional space of colors. We can reconstruct a dog's-eye view of the world, as in plate X, based on dogs' *two* color receptors.

Color-blind people see only a two-dimensional space of perceived colors. They are missing one kind of receptor protein or have mutated proteins that give poor discrimination. Color blindness affects few women, but among men it is fairly common. Something like one in twelve northern European men has it. A color-blind person will be able to match any color in the outer circle of a color top with just two base colors—red and green, say—in the inner circle (see plate U). There are also women who see a four-dimensional space of colors: *tetrachromats*. They have an extra receptor protein, which is a mutation of the usual ones. They can discriminate among mixtures of spectral colors that most people perceive as indistinguishable. This ability seems to be rare, and has not been much studied.

In low light, we all become color-blind. Color enters our perceived world as the Sun rises, and leaves it as the Sun sets. This is, of course, a common observation. We make it, literally, every day. But I find that it becomes a fascinating experience, when I'm alert to it, on long summer evenings.

On the other hand, many types of insects and birds have four or even five color receptors, including sensitivity to ultraviolet, and also sensitivity to polarization. Many flowers are patterned and colored vividly in the ultraviolet, to seduce their pollinators. Within the sensory universe of color, they explore dimensions that evade our awareness.

And then there's the mantis shrimp. "Mantis shrimps" are not a single species, but a group of several hundred distinct species that share many traits and follow a similar lifestyle. They are remarkable creatures in many ways. Growing to about a foot long, they are solitary sea predators. They come in two broad varieties, spearers and smashers. Both types strike with astonishing quickness and force. They are difficult to keep in aquaria, due to their ability to break the glass walls.

But the most remarkable characteristic of a mantis shrimp is its visual system. Depending on the species, they can see between twelve and sixteen color dimensions. Their range of sensitivity extends into the infrared and the ultraviolet (see plate Y), and also encodes some information about polarization.

Why mantis shrimps have developed their peculiar genius for color is a fascinating question. One plausible answer is that they use it to pass secret messages to other mantis shrimps. What to me seems likely is that they use their own bodies to make impressive color displays—mostly lost on us!—to advertise their fitness to potential mates. It's like the peacock's tail on steroids. In support of these ideas, we can note that some mantis shrimp species look very colorful indeed—even to us, as we see in plate Z!—and that the species with the most advanced color vision are generally the most colorful.

How do such small crustacean brains cope with such a gush of sensory input? That question is a subject of current research. It seems probable to me that they use the technique known to information engineers as "vector quantization"—a piece of jargon that I'll now unpack. Humans fill in their three-dimensional color space very finely. We are able to distinguish nearby points in that space, and thus to experience millions of distinct color perceptions. Mantis shrimp probably use a much coarser representation, with input from large regions of their sixteen-dimension space all giving the same output. Where we can resolve points within a relatively small space, they locate blobs dividing up a much bigger space. We make a very crude (three-dimensional) projection of the infinite-dimensional electromagnetic input, but survey that projection accurately, while the mantis shrimp makes a more sophisticated projection, but surveys it crudely.

## *Space-Sense and Time-Sense*

Having examined the "what" and "how" of color vision, we're ready to address the question "Why?" Two "why" questions naturally arise:

> *Why* do humans, and many other creatures, care so much about ultra-rapid oscillations in electromagnetic fields?

Now if I'd posed that question in the form "*Why* do humans, and many other creatures, care about color?" so many answers would spring to mind that the question would seem ridiculous.

But if we ask it in the first form, which should amount to the same thing, it raises a profound point. Information about rapid oscillations in electromagnetic fields is important to us, as biological creatures, because it is important to the electrons in materials. Those electrons often respond in very different ways, depending on their material environment, to electromagnetic oscillations with different frequencies. Because of that, light that originates in the Sun, after interacting with matter and then getting transmitted to us, carries information about the intervening matter, imprinted by its electrons.

In plain English: The colors of objects encode what they are made of. You knew that, of course, from experience. But now you also know, in terms of fundamentals, just what it is you've experienced!

*Why* is vision so different from hearing? After all, both senses are concerned with the information carried to us in vibrations, arriving as waves. Vision is concerned with vibrations of electromagnetic fields, hearing with vibrations of air. But the ways we perceive chords of light and chords of sound are profoundly, qualitatively different.

Let me be precise about that. When we receive several pure sound tones sounding together, we hear chords in which the tones retain their individual identity. Within a C major chord, you can hear the C, the E, and the G separately, and you'll certainly notice a qualitative difference if any one of them is absent, or if it is notably louder than the others. And you can have more complicated chords, with more separate tones, each sounding different, practically without limit (eventually they start to sound like sludge, but it's always sludge with distinct components).

On the other hand, as we've been discussing, when we receive several pure light tones—i.e., spectral colors—together, we perceive a new color in which the identity of the originals is submerged. For instance, mixing green and red gives a perceptual yellow, which is indistinguishable from (the perception of) spectral yellow. It's as if you played C and E together, and as a result heard D!

Clearly, hearing does a better job handling its time-based material.

The physics of hearing is the physics of sympathetic vibration, as we discussed earlier. There's a good physical reason why light must be treated differently. The oscillations of electromagnetic fields in visible light are far too rapid for any practical mechanical system to follow. So the strategy we use in hearing, where vibrations of air get channeled to set up sympathetic vibrations in our heads, won't work. To get in tune with the vibrations of light, we need to use much smaller, nimbler responders.

For light, the useful receptors are individual electrons. But in the subatomic world of electrons, quantum mechanics comes in, and the rules of the game change. The transfer of information from light to electrons can come about only through the transfer of some of the light's energy. According to the quantum rules, however, such energy transfers occur in discrete "all or none" events—absorption of photons—that occur at unpredictable times. These effects make the information transfer a less faithful transcription, and one that is harder to control.

And that—when it is spelled out more rigorously—explains why our perception of light's time-structure, as encoded in color, is cruder than our perception of sound's time-structure, as encoded in musical harmony. Quantum mechanics is to blame. By having several different kinds of receptors, attuned to different features, we rescue some of light's time information. But there is no visual analogue to the vibrating membranes of the inner ear where, for sound, it is all laid out bare, as on the keys of a player piano.

On the other hand, for carrying information about structure in *space,* light has a great advantage over sound. The problem with sound waves, as carriers of spatial information, is just that they're big. Their wavelength, not coincidentally, is comparable to the size of musical instruments, like guitars, pianos, or the pipes of church organs. So they can't resolve structures much smaller than that. For light there is no such problem—the wavelengths of visible light are a bit less than a *millionth* of a meter.

Vision is primarily a space sense, while hearing is primarily a time sense, for good physical reasons.

## OPENING THE DOORS

And now let's make a leap of imagination, ascending from the solid ground of "what," "how," and "why," into the dreamscape of "what if," "how to," and "why not."

Our eyes are wonderful sense organs, but they leave a lot on the table. Based on the spatial information in incoming light—basically, the directions of incoming light rays—they produce a sequence of images of the external world. As we've now discussed in some detail, however, they rescue only a small part of light's incoming time information, and they completely lose track of polarization. Each pixel in our visual field potentially supplies a doubly infinite chord, but we see only color, a three-dimensional projection.

The human mind is our ultimate sense organ. Mind has discovered that there are invisible infinities hidden in light. Our perception of color projects the doubly infinite-dimensional space of physical color onto the three-dimensional wall of our inner Cave. Can we escape that Cave to sample the additional dimensions?

I think we can, and now I'll briefly indicate how. (My philosophy: If mantis shrimps can do it, so can we.)

### *Time and Color Blindness*

For openers, let's consider a simplified version of the problem that's already of practical importance. We know pretty precisely what information color-blind people are missing, namely, the specific average of spectral intensities that their missing receptor protein encodes. How can we put that information back?

Now, in doing that, we want to present the color information where it belongs, within visual images. So we've got to make use of the available receptors to synthesize new ones. And within the images, we want the new information attached to the correct locations. To be concrete, let's call what the missing receptor normally supplies "Green," and our artificial substitute signal *"Green."* Then we want to make sure that the parts of our images that contain lots of Green are supplied with *Green* in proportion.

To meet those requirements—adding information locally, using the existing receptors—we need to put into the signal new structure that the existing receptors can recognize. An elegant way to do that is to manipulate the signal in *time*. For example, we can encode *Green* as added *shimmering, pulsation,* or generally *temporal modulation*—time-varying textures—in the perceptible colors, whose local intensity is proportional to Green in the original image.

Let's reflect on what we're doing here. The missing information, Green, was telling something about the structure of light, as an electromagnetic signal, in time. We're putting it back as *Green,* again a time signal, but slowed down, to match the pace of human information processing. We're using time and brain to open the doors of perception.

For the benefit of people with normal color vision, we commonly encode images in a three-color format, and decode them with three-color projectors. In any context where that's done—in computer displays (including tiny computer displays mounted on eyeglasses or goggles!), or smartphones, or digital projectors, for example—our color-blindness solution can be implemented in software as a modified mapping from input to output.

We can also consider implementing the same general approach in hardware. For example, there are materials called electrochromics whose tendency to absorb light within particular spectral regions can be modulated by applying a voltage. If we fit ordinary glasses with electrochromic layers, and apply a time-varying voltage, we will be opening up new color channels.

## *Ways and Means*

The same general ideas will allow us to open up essentially new dimensions of color vision, as seen in plate AA. Of course, before making this new information accessible, we must first collect it. The fact that digital photography and computer graphics are based on three basic colors, as opposed to more, is not because of any fundamental physical consideration. As we've seen, there's a double infinite of colors out there, awaiting our cleansed perception. The reasons technology has mainly settled on three are:

1. Three basic colors enable us, as Maxwell taught us, to synthesize any perceptual color.
2. Two basic colors won't do that job.
3. It's simplest and cheapest to use the smallest number that's sufficient.

But once we decide to expand into extra dimensions of color space, the technology appears quite feasible (and has been used, in exploratory ways). A straightforward design for four dimensions, suitable for both digital reception and digital transmission, is shown in plate BB.

We can create four (or five . . .) different sorts of color receptors in tight arrays along the same lines we currently create three. On the output side, we can either let three color transmitters do double duty, by superimposing shimmering, or—as in plate BB—set aside a special class of pixels for the new channel. Either way, extra channels become available as we open the output to artificial time variation, whose placement and strength is controlled by the signals from the new receptors.

It will be very entertaining, I think, to experience additional colors in this way.

## "THE GO OF IT"

For me, it has been a joy to commune with Maxwell through his writings and the testimony of his friends. He has become my favorite physicist. Here now, a small impressionistic portrait. According to his friend and biographer Lewis Campbell:

> Throughout his childhood his constant question was, "What's the go o' that? What does it do?" Nor was he content with a vague answer, but would reiterate, "But what's the *particular* go of it?"

Maxwell's correspondence with family and friends is reminiscent of Mozart's. It is full of wordplay, comical drawings, and human warmth. Here is an extract from a letter to his young cousin, Charles Cay. In one line he alludes to his "Dynamical Theory," the masterwork we discussed earlier; in the next, without a pause, he describes observations on his new dog

> I have also a paper afloat, with an electromagnetic theory of light, which, till I am convinced to the contrary, I hold to be great guns.
>
> Spice is becoming first-rate: she is the principal patient under the ophthalmoscope, and turns her eyes at command, so as to show the tapetum, the optic nerve, or any required part.

Throughout his life Maxwell wrote verse. One of his best poems is a song, "Rigid Body," that he sang to his own guitar accompaniment. In each stanza, Maxwell himself complains about the difficulty of calculating how rigid bodies will move; then the rigid body itself replies saying, in essence, "I just do my thing." It goes to the tune of Robert Burns's "Comin' Thro' the Rye," and is sprinkled with Scotticisms:

Gin a body meet a body
Flyin' through the air,
Gin a body hit a body,
Will it fly? and where?
Ilka impact has its measure,
Ne'er a ane hae I,
Yet a' the lads they measure me,
Or, at least, they try.
Gin a body meet a body
Altogether free,
How they travel afterwards
We do not always see.
Ilka problem has its method
By analytics high;
For me, I ken na ane o' them,
But what the waur am I?

(Gin = when; ilka = every; ken = know; na ane = not one; waur = worse.)

## *Death and Life*

In the spring of 1877, when he was not yet forty-six years of age, Maxwell began to feel symptoms of indigestion, pain, and fatigue. Over the next several months his symptoms worsened, and soon it became clear that he was suffering from abdominal cancer, the illness that had taken his mother at a similar age, when Maxwell was a child of nine. Maxwell knew he hadn't much longer to live. According to Campbell:

> During the last few weeks his sufferings were very great, but he seldom mentioned them; and . . . his mind was absolutely calm. The one

thought which weighed upon him, and to which he constantly referred, was for the future welfare and comfort of Mrs. Maxwell.

Maxwell died in 1879, at the age of forty-eight.

As a young man of twenty-three, Maxwell made a thoughtful entry, in his private journal, anticipating the life he was to lead:

> Happy is the man who can recognize in the work of to-day a connected portion of the work of life and an embodiment of the work of Eternity. The foundations of his confidence are unchangeable, for he has been made a partaker of Infinity. He strenuously works out his daily enterprises because the present is given him for a possession.
>
> Thus ought man to be an impersonation of the divine process of nature, and to show forth the union of the infinite with the finite, not slighting his temporal existence, remembering that in it only is individual action possible, nor yet shutting out from his view that which is eternal, knowing that Time is a mystery which man cannot endure to contemplate until eternal Truth enlighten it.

# PRELUDE TO SYMMETRY

*Symmetry, as wide or as narrow as you may define its meaning, is one idea by which man through the ages has tried to comprehend and create order, beauty, and perfection.*

• *Hermann Weyl*

*Nature seems to take advantage of the simple mathematical representations of the symmetry laws. When one pauses to consider the elegance and the beautiful perfection of the mathematical reasoning involved and contrast it with the complex and far-reaching physical consequences, a deep sense of respect for the power of the symmetry laws never fails to develop.*

• *C. N. (Frank) Yang*

*But although the symmetries are hidden from us, we can sense that they are latent in nature, governing everything about us. That's the most exciting idea I know: that nature is much simpler than it looks.*

• *Steven Weinberg*

Over the course of the twentieth century, and continuing today, *symmetry* increasingly has come to dominate our best understanding of the fundamental laws of Nature. So say the masters. The concluding parts of our meditation, which bring us to present-day frontiers and beyond, celebrate great triumphs of symmetry, and prophesy more.

Change Without Change. What a strange, inhuman mantra for the soul of creation! Yet its very unworldliness presents an opportunity: we can expand our imaginative vision by making its wisdom our own.

Our Question asks us to discover beauty at the root of the physical world. To answer its challenge, we must be active on both sides. We must enlarge our sense of beauty, as we enlarge our understanding of reality. For the beauty of Nature's deep design, we shall find, is as strange as its strangeness is beautiful.

Accordingly, we will take a few "symmetry breaks" as we dig toward the physical world's deep roots—lighter interludes exploring the particular form of beauty they ultimately reveal: *symmetry*, expanded and empowered.

## A VOYAGE WITH GALILEO

To begin, we do well to join Galileo on a voyage of imagination.

> Shut yourself up with some friend in the main cabin below decks on some large ship, and have with you there some flies, butterflies, and other small flying animals. Have a large bowl of water with some fish in it; hang up a bottle that empties drop by drop into a wide vessel beneath it. With the ship standing still, observe carefully how the little animals fly with equal speed to all sides of the cabin. The fish swim indifferently in all directions; the drops fall into the vessel beneath; and, in throwing something to your friend, you need throw it no more strongly in one direction than another, the distances being equal; jumping with your feet together, you pass equal spaces in every direction. When you have observed all these things carefully (though doubtless when the ship is standing still everything must

> happen in this way), have the ship proceed with any speed you like, so long as the motion is uniform and not fluctuating this way and that. You will discover not the least change in all the effects named, nor could you tell from any of them whether the ship was moving or standing still. . . . The cause of all these correspondences of effects is the fact that the ship's motion is common to all the things contained in it, and to the air also. That is why I said you should be below decks; for if this took place above in the open air, which would not follow the course of the ship, more or less noticeable differences would be seen in some of the effects noted.

Galileo was here combating what was no doubt the greatest psychological barrier to acceptance of Copernican astronomy. Copernicus put Earth (and everything on it) into rapid motion: daily rotation around its axis, and yearly revolution around the Sun. The speeds involved are enormous, by everyday standards. For the rotation: a little over one thousand miles, or sixteen hundred kilometers, per hour. For the revolution: a little over sixty-seven thousand miles, or one hundred eight thousand kilometers, per hour. But we *feel* as if we're not moving, at all, much less so very fast!

Galileo's answer is that unchanging motion—namely, motion at constant velocity, in a straight line—is undetectable, because it doesn't change any aspect of physical behavior. And in a closed system—like Galileo's ship cabin, or spaceship Earth—experienced from within, motion at a constant velocity, however large that velocity, feels exactly like no motion at all. (The Earth's rotation and revolution involve circles, not straight lines, but the circles are so large that they look much like straight lines over long stretches.)

It's easy to express Galileo's observation as a symmetry. We *change* the world—or a big part of it, like the interior of a large ship—by moving everything at a common velocity, *without change* to the way things behave.

We call that kind of transformation, in Galileo's honor, a *Galilean transformation.* Correspondingly, we call his postulate of symmetry *Galilean symmetry*, or *Galilean invariance.*

According to Galilean symmetry, we can change the state of motion of the Universe, imparting a constant overall velocity—give it a boost, so to speak—without changing the physical laws it obeys. Galilean transformations move the physical world at a constant velocity, and the symmetry states that the content of physical laws is unchanged by such transformations.

# QUANTUM BEAUTY I: MUSIC OF THE SPHERES

The classical science of Newton and Maxwell brought new themes into our meditation that *seem* to conflict with the earlier visions and intuitions of Pythagoras and Plato, from which we began. But in the quantum world of atoms, a strange world that also happens to be our own, a miracle occurs. Old ideas come back to life, wearing splendid new forms. In their resurrected forms those ideas attain new levels of precision, truth, and, surprisingly, *musicality*.

Here is how the new embraces the old:

- *From the heart of matter, music:* There is no logical reason to expect that the mathematics developed to understand music should have anything to do with atomic physics. Yet the same concepts and equations turn out to govern both domains. Atoms are musical instruments, and the light they emit makes their tones visible.
- *From beautiful laws, beautiful objects:* The basic laws do not postulate the existence of atoms. Atoms *emerge* from the laws, and they emerge as beautiful objects (see plate CC). Physical atoms, mathematically described, are three-dimensional objects that will, to the animating spirit of an artist, yield images of exceptional beauty.

- *From dynamics, permanence:* The basic laws are equations that describe how things change in time. But those equations have some important solutions which do *not* change in time. Those solutions, and those alone, describe the atoms which build up our everyday world, and ourselves.
- *From continuity, discreteness:* The wave functions that describe electrons in atoms are fields of probability (probability distributions) that fill space. They are continuous, and cloud-like. But the stable cloud patterns are discretely different, and bear the stamp of Number.

## BACK TO PYTHAGORAS

In the earliest days of modern quantum theory, there was, of course, not yet a textbook. Would-be practitioners, eager to use the new atomic theory, turned instead to a textbook on a different subject: Lord Rayleigh's *Theory of Sound*. For there they found the mathematics necessary to describe how atoms work. It had been developed before, to describe how musical instruments work! Although the symbols stand for different things, basically the same equations appear, and the same techniques can be used to solve them. Pythagoras would be pleased.

### *The Yoga of Musical Instruments*

The physics of musical instruments is the physics of *standing waves.* Standing waves are waves in finite objects, or confined spaces. Thus the vibrations of strings in musical instruments, or of their sounding boards, are standing waves. Standing waves should be contrasted with *traveling waves*. When we speak of sound waves, for example, we usually mean traveling waves that expand, or *propagate*, from a source. The vibrations of a grand piano's sounding board, which are standing waves, push nearby air to and fro. The moving air exerts forces on its neighboring air, which exerts forces on its neighboring air, and so forth, resulting in a disturbance that takes on a life of its own.

Standing waves are the sort of motion you set up in a bathtub's water when you make a splash, or in the vibrations of a gong or of a tuning fork after you strike it. In each of these cases—tub splash, gong crash, or fork bash—after a noisy start, the motion will settle down to become regular in space and periodic in time. That's the point of a tuning fork: It "wants to" vibrate at a characteristic frequency, and so it produces a reliable, pure tone. A typical gong generates a more complicated and interesting sound pattern. We'll come back to that shortly.

We can spotlight the yoga of musical instruments most clearly by considering the musical instrument of ultimate simplicity, which is in fact Pythagoras's instrument: a tense string, nailed down at two ends (figure 24). In the simple one-dimensional geometry of a finite line segment, we can survey the natural standing wave patterns at a glance.

In the figure, the solid lines and dotted lines show the shape of the string at different times, in four natural standing wave patterns. (The size, or amplitude, of the distortion is greatly exaggerated, for the sake of visibility.) At intermediate times, the points on the string move up and down, as the solid pattern evolves smoothly into the dotted, then the dotted into the solid, in a cycle.

A simple geometric requirement brings whole numbers and discreteness into the description of these continuum-based patterns. They have to fit! Going from top to bottom, and comparing patterns, we see that the variation, as we move across the string from left to right, is two, three, then four times as rapid.

You can have natural vibrations that accommodate three cycles, or two, or four, or any whole number, but *not* anything in between. As a re-

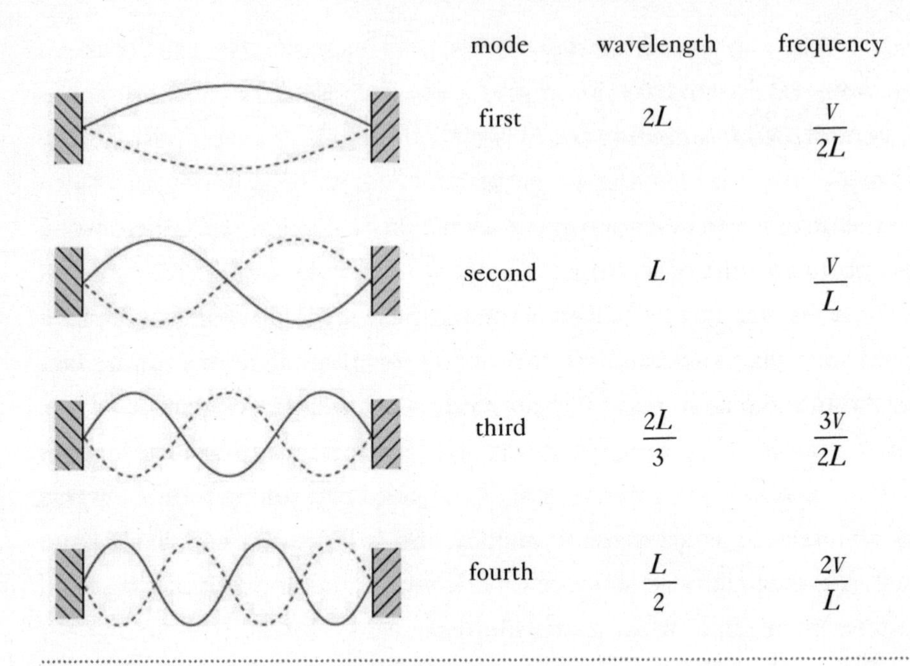


FIGURE 24. IN THE SIMPLE ONE-DIMENSIONAL GEOMETRY OF A FINITE LINE SEGMENT, WE CAN SURVEY THE NATURAL STANDING WAVE PATTERNS AT A GLANCE. THEY HAVE TO FIT! THIS SIMPLE GEOMETRIC REQUIREMENT BRINGS WHOLE NUMBERS AND DISCRETENESS INTO THE DESCRIPTION OF BEHAVIOR IN A CONTINUUM.

sult, the natural frequencies of our instrument are discrete or, as we say, *quantized*.

Unlike the proverbial spherical cow,* our Pythagorean musical instrument is not far removed from reality. More important, the lesson we learn from our simple instrument—that *geometric constraints in finite objects lead to discreteness (quantization) of their natural vibration patterns*, and therefore of their natural frequencies—is perfectly general. In quantum mechanics that lesson becomes the linchpin of atomic physics, as we'll soon discuss.

* The spherical cow joke: Milk production at a dairy farm was low, so the farmer wrote to the local university, asking for help. A team of professors was assembled, headed by a theoretical physicist. Shortly thereafter the physicist returned to the farm, saying to the farmer, "I have the solution, but it only works in the case of spherical cows in a vacuum."

## *Natural Vibrations, Resonant Frequencies*

You also get standing waves on the sounding board of a guitar when you strum a string, or on a plate if you stroke its sides (figure 25). Patterns become visible. The basic idea is the same as what we just discussed for a tethered string. The standing wave is an up-and-down motion that is bigger (or, in the jargon, has larger amplitude) at some places than at others. There are curves along which the vibration strength vanishes, and there is no motion at all. Such points are called nodes, and the curves are called nodal curves. If you sprinkle some sand on the plate, it will accumulate along the nodal curves, and that is what you're seeing in the picture.

The geometry, for these two-dimensional vibrators, is more complicated than that of the string. Reflecting that, the patterns in the natural vibrations are more complicated.

In these examples, to elicit one or another of the simple vibration patterns, as opposed to a mixture of several, we impose forces that repeat regularly or, as we say, periodically, in time. The guitar lets us do this by plucking a string—that's what its strings are for! A different pattern will dominate, depending on how fast the driving forces oscillate (i.e., on their frequency).

Each natural vibration pattern repeats in time. The forces that each moving piece of string, wood, or metal exerts on its neighbors are different in the different patterns, and the pace at which things change is different in each. The patterns that vary rapidly in space will tend to set up bigger forces and therefore faster, higher-frequency motion. Each natural vibration pattern takes place at its own natural frequency.

The natural frequency is also called a resonant frequency, for the following reason. If the frequency of your driving force is close to the natural frequency of some pattern, that pattern will leap out in powerful response. For then, and only then, does the external driving force match up with the internal forces, cycle after cycle, to build up the strength of the

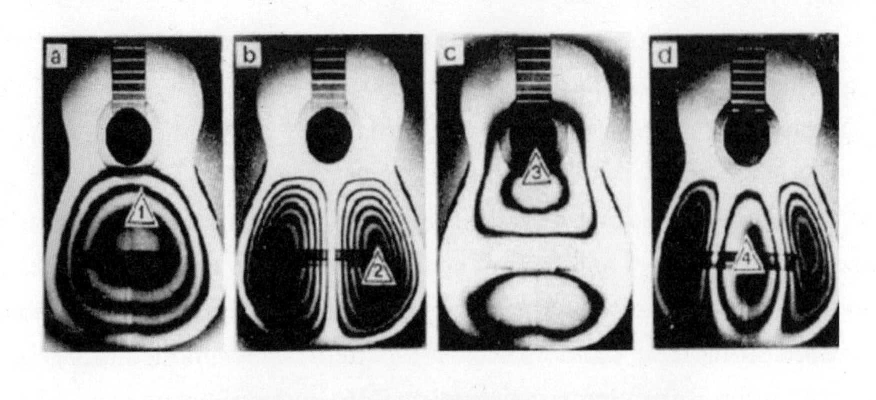


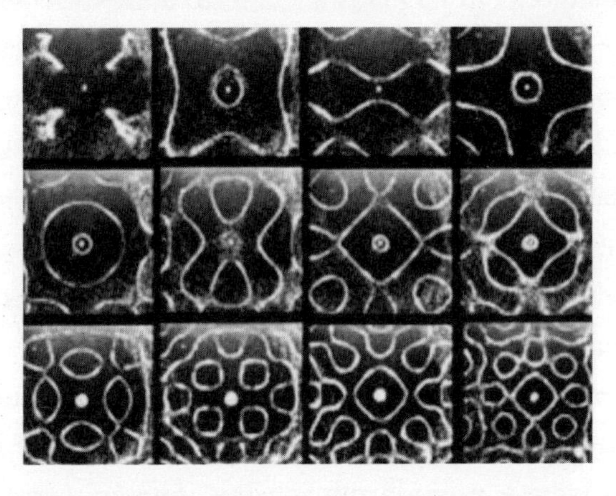

FIGURE 25. THE VIBRATION PATTERNS, OR *STANDING WAVES*, OF A GUITAR'S SOUNDING BOARD MAKE GEOMETRIC PATTERNS THAT REFLECT AN INTERPLAY BETWEEN THE SHAPE AND MOLDING OF THE WOOD AND THE FREQUENCY OF THE DRIVING STRING VIBRATION.

motion. Anyone who's pumped their legs and straightened their body to build up the motion of a swing, or pushed a child on one, knows how important that is.

When you strike a tuning fork or gong, the vibration radiates out from the point of impact, then bounces off the edges and returns, like an echo chamber for vibrations. Complicated motions dissipate their energy rapidly into traveling sound waves and heat, leaving behind one (for a tuning fork) or a few (for a gong) relatively long-lived patterns, each vibrating at its resonant frequency. Those are what you hear as a stable tone, or a slowly evolving chord, after the noisy start. Gongs produce an evolving chord, gradually shedding complexity until it becomes a single tone, because there are several long-lived patterns that decay at different rates.

The vibration patterns, or standing waves, of a guitar's sounding board make geometric patterns that reflect an interplay between the shape and molding of the wood and the frequency of the driving string oscillation, as shown in figure 25. Similar standing wave patterns in square vibrating plates (at bottom) are more symmetric. These patterns bear striking resemblance to the patterns of electron clouds (figure 26, page 187). The resemblance between their governing equations is profound, and more striking still.

### *A Missed Opportunity*

It is a great pity that the Pythagoreans did not follow up on their discoveries with vibrating strings to consider "instruments" one step more complicated, like our two-dimensional plates. A lovely interplay between geometry, motion, and music, going far beyond the simple rules for strings, was there to be enjoyed and explored by ear, eye, and mind. They'd have had a ball.

They'd also have discovered an easier, more accessible path to the basic laws of mechanics than the difficult detour through astronomy that eventually, centuries later, led there. And, as we'll soon see, they would have paved a royal road to quantum theory.

## MUSIC OF THE SPHERES: THIS TIME, FOR REAL

Arthur C. Clarke's third law of prediction reads:

Any sufficiently advanced technology is indistinguishable from magic.

I'd like to add an observation, which our meditation amply justifies:

*Nature's* technology, whereby she builds the material world, is quite advanced.

Fortunately, Nature lets us study her tricks. By paying close attention, we ourselves become magicians.

## *Outrageous Hypotheses*

In the quantum world of atoms and light, Nature treats us to a show of strange and seemingly impossible feats.

Two of these feats seemed, when discovered, particularly impossible. (The early history of quantum theory is complicated. Several other, less direct paradoxes also played major roles in guiding the pioneers' thinking, and people also pursued many blind alleys. Here I've laid out a clear and relatively simple story that is a severe idealization of the actual history. In history, unlike in the deep structure of Nature, Real and Ideal differ greatly. A wise teacher—Father Jim Malley, SJ—gifted me this valuable pearl: "It is more blessed to ask forgiveness than permission.")

One involves light; the other involves atoms.

- Light comes in lumps. This is demonstrated in the photoelectric effect, as we'll discuss momentarily. It came as a shock to physicists. After Maxwell's electromagnetic theory was confirmed in Hertz's experiments (and later many others), physicists had thought they understood what light is. Namely, light is electromagnetic waves. But electromagnetic waves are continuous!
- Atoms have parts, but are perfectly rigid. Electrons were first clearly identified in 1897, by J. J. Thomson. The most basic facts about atoms were elucidated over the following fifteen years or so. In particular: atoms consist of tiny nuclei containing almost all of

> their mass and all of their positive electric charge, surrounded by enough negatively charged electrons to make a neutral whole. Atoms come in different sizes, depending on the chemical element, but they're generally in the ballpark of $10^{-8}$ centimeters, a unit of length called an angstrom. Atomic nuclei, however, are a hundred thousand times smaller. The paradox: How can such a structure be stable? Why don't the electrons simply succumb to the attractive force from the nucleus, and dive in?

These paradoxical facts led Einstein and Bohr, respectively, to propose some outrageous, half-right hypotheses that served as footholds on the steep ascent to modern quantum theory.

The photoelectric effect is the effect that when you shine light (or, even better, ultraviolet radiation) on suitable materials, they emit electrons. Solar panels, which convert light to electricity, exploit this effect.

The idea that light could accelerate electrons, increase their energy, and perhaps occasionally fling them out of atoms is, in itself, not surprising. The electric fields in light *should* do things like that. What was shocking was the way it happened. You might expect that the energy would take a while to build up, so that if you turned down the light, you'd find no electrons coming off at first. Instead, the effect "turns on" immediately. Also, you might expect that the frequency of the light—i.e., its spectral color—is less important than its strength or brightness. Instead, you find that spectral colors toward the red end of the spectrum are ineffectual. If the light is too reddish, you liberate few electrons, however brightly you illuminate the material.

Einstein explained these effects (and others) with his photon hypothesis. According to that hypothesis, light comes in units, *photons*, that cannot be broken down further. The amount of energy in a minimal unit, or quantum, of light is proportional to the frequency of the light. Thus photons from the blue end of the spectrum carry about twice the energy of photons at the red end, and ultraviolet photons carry still more.

The photon hypothesis gives a simple, qualitative explanation of the paradoxical elements of the photoelectric effect. Because each photon delivers either all of its energy or none, there is no need for a gradual buildup, and no onset time. Because the reddish photons deliver less energy, they are less effective; and if they don't have enough energy to liberate an electron, they simply won't.

Einstein's photon hypothesis was not part of a great system, like Maxwell's equations or Newton's celestial mechanics. In fact, it basically *contradicted* what seemed to be the plain implication of Maxwell's equations. It explained some facts at the cost of undermining the existing, hugely successful framework for explaining many others. It was outrageous. In nominating Einstein for membership in the Prussian Academy of Sciences, in 1913, Planck wrote:

> That he may sometimes have missed the target in his speculations, as, for example, in his hypothesis of light-quanta, cannot really be held too much against him, for it is not possible to introduce really new ideas even in the exact sciences without sometimes taking a risk.

Einstein had introduced the light-quanta, or what we now call photons, in 1905—eight years previously! Eight years later, in 1921, when he received the Nobel Prize, it was Einstein's work on light-quanta that was specifically cited. By then, it had proved its worth.

To address our second paradox, the paradox of rigid, stable atoms, Niels Bohr introduced the idea that atoms can exist only in *stationary states*. In classical mechanics the possible orbits range over a continuum, as we saw in Newton's Mountain. Bohr proposed that in an atom the electrons orbit around the nucleus, bound by electric forces, but that only a discrete subset of the orbits is possible. For the simplest atom, hydrogen, he proposed a simple, definite rule to pick out which orbits those were. (For experts: The requirement is that the integral of momentum, weighted by length—the so-called action integral—taken over the orbit should be a *whole number* times Planck's constant.) When an electron follows one of the "allowed" orbits, we say the atom is in a stationary state. The electron stays in that particular orbit, as long as you don't kick it *too* hard, because the other possible orbits are all quite different—and a small nudge can't get you there! Finally: atoms don't collapse because all of the allowed orbits keep the electrons safely away from the nucleus.

Bohr's stationary state hypothesis was not part of a great system, either. In fact, it too *contradicted* what seemed to be the plain implication of a very successful theory, namely Newton's mechanics. Who was Bohr to tell electrons where they could and couldn't be, or what velocities they might or might not have? It was outrageous—it explained some facts, but at the cost of undermining the existing, hugely successful framework for explaining many others.

Bohr's rules for hydrogen could be, and of course were, tested experimentally. Their success made his outrageous hypothesis credible.

Both Einstein and Bohr were very well aware of what they were doing—and of what they were *not* doing—in putting forward their outrageous hypotheses. They were not proposing consistent "theories of everything," or even grand syntheses along the lines of Newton's celestial mechanics or Maxwell's electromagnetism. Rather, in the exploratory spirit of Pythagoras, or of Newton's work on light and of Maxwell's on perception, they were simply identifying striking patterns of facts that might find deeper explanation in the long run.

An important part of good scientific strategy is to distinguish between problem areas that might be ripe for a grand synthesis, and problem areas where a more opportunistic approach will be more fruitful. A successful theory of *something* can be more valuable than an attempted Theory of *Everything*.

## *"The Highest Form of Musicality"*

Atoms of a given kind—for instance, hydrogen atoms—will absorb some colors of spectral light much more efficiently than others. (More generally, they will absorb electromagnetic waves of some frequencies much more efficiently than others.) The same atoms, when heated up, will emit most of their radiation in those same spectral colors. The pattern of preferred colors is different for different kinds of atoms, and forms a sort of fingerprint through which we can identify them. An atom's pattern of preferred colors is called its *spectrum*.

In his atomic model, Bohr postulated that the electrons in an atom can exist only in a discrete set of stationary states. The possible values of the electrons' energy, therefore, also form a discrete set. And here is how Bohr connected that idea to reality, through another outrageous hypothesis. He assumed that in addition to its "allowed" regular motions, in stationary states, an electron would occasionally make a *quantum jump* between one stationary state and another. Why? How? Don't ask. But the quantum jump process is accompanied by emission or absorption of a photon. Quantum jumps make atomic spectra.

In his otherwise iconoclastic model, Bohr held one principle sacred: the conservation of energy. He insisted that energy should be conserved, even in the quantum jump process.

Now, the energy of a photon, according to Einstein, is proportional to its frequency, and its frequency is encoded in its color. And so Bohr's ideas come together in a predictive package: the colors of an atom's spectrum reflect its possibilities for transitions between stationary states, and its colors reveal the differences between energies of stationary states. Bohr's model, by predicting those energies, predicted the colors in hydrogen's spectrum. And it worked!

Einstein, reflecting on Bohr's work, wrote:

> That this insecure and contradictory foundation was sufficient to enable a man of Bohr's unique instinct and tact to discover the major laws of the spectral lines and of the electron-shells of the atoms . . . appeared to me as a miracle—and appears to me as a miracle even today. This is the highest form of musicality in the sphere of thought.

Einstein was wrong about that, though. The best music was still to come.

## THE NEW QUANTUM THEORY: ATOMS AS MUSICAL INSTRUMENTS

Bohr's success left theorists with a problem of reverse engineering. His model provided a "black box" description of atoms, which described "what" they did, but not "how." By sketching the answer to an unknown question, Bohr had set up a great game of Jeopardy. Physicists had to find the equations for which Bohr's model was the solution.

After epic struggles, played out over more than a decade of effort and debate, an answer emerged. It has held up to this day, and its roots have grown so deep that it seems unlikely ever to topple.

### *What Is Quantum Theory?*

Describing the behavior of matter on atomic and subatomic scales turned out to require not merely adding to what was known before, but also the construction of a radically different framework, in which many ideas thought to be secure had to be abandoned. The framework known as quantum theory, or quantum mechanics, was mostly in place by the late 1930s. Since then our techniques for dealing with the mathematical challenges quantum theory poses have vastly improved, and we have achieved much more detailed and penetrating understanding of the main forces of Nature, as we'll see in subsequent chapters. But those developments have taken place *within* the framework of quantum theory.

Many physical theories can be described as reasonably specific statements about the physical world. Special relativity, for example, is basically the dual assertion of Galilean symmetry together with invariance of the speed of light.

Quantum theory, as presently understood, is not like that. Quantum theory is not a specific hypothesis, but a web of closely intertwined ideas. I do not mean to suggest quantum theory is vague—it is not. With rare and usually temporary exceptions, when faced with any concrete physical problem, all competent practitioners of quantum mechanics will agree about what it means to address that problem using quantum theory. But few, if any, would be able to say precisely what assumptions they have made to get there. Coming to terms with quantum theory is a process, through which *the work will teach you how to do it.*

Let's get started.

## *Wave Functions, Probability Clouds, and Complementarity*

In quantum theory's description of the world, the fundamental objects are not particles occupying positions in space, nor the fluids of Faraday and Maxwell, but *wave functions.* Any valid physical question about a physical system can be answered by consulting its wave function. But the relation between question and answer is not straightforward. Both the way that wave functions answer questions and the answers they give have surprising—not to say weird—features.

Here I will focus on the specific sorts of wave functions we need to describe the hydrogen atom and discover its musicality. (For more see "Terms of Art," especially the entries *Quantum theory* and *Wave function.*)

We are interested, then, in the wave function that describes a single electron bound by electric forces to a tiny, much heavier proton.

Before discussing the electron's wave function, we'll do well to describe its *probability cloud*. The probability cloud is closely related to the wave function. The probability cloud is easier to understand than the wave function, and its physical meaning is more obvious, but it is less fundamental. (Those oracular statements will be fleshed out momentarily.)

In classical mechanics, particles occupy, at any given time, some definite position in space. In quantum mechanics, the description of a particle's position is quite different. The particle does not occupy a definite position at each time; instead, it is assigned a probability cloud that extends over all space. The shape of a probability cloud may change over time, though in some *very* important cases it does not, as we'll soon see.

As its name suggests, we can visualize the probability cloud as an extended object, which has some non-negative—that is, positive or zero—density at each point. The density of the probability cloud at a point represents the relative probability of finding the particle at that point. Thus the particle is more likely to be found where the density of its probability cloud is high, and it is less likely to be found where the cloud's density is low.

Quantum mechanics does not give simple equations for probability clouds. Rather, probability clouds are calculated from wave functions.

The wave function of a single particle, like its probability cloud, assigns an amplitude to all possible positions of the particle. In other words, it assigns a number to every point in space. The wave function's amplitude is a complex number, so the wave function is an assignment of a complex number to every point in space. (If that's unfamiliar territory, you might want to consult "Terms of Art," or just interpret it as poetry. We won't be leaning on the detailed properties of complex numbers, beautiful as they are.)

To pose questions, we must perform specific experiments that probe the wave function in different ways. We can perform, for example, experiments that measure the particle's position, or experiments that measure the particle's momentum. Those experiments address the following questions: Where is the particle? How fast is it moving?

How does the wave function answer those questions? First it does some processing, and then it gives you odds.

For the position question, the processing is fairly simple. We take the value, or amplitude, of the wave function—a complex number, recall—and square its magnitude. That gives us, for each possible position, a positive number, or zero. That number is the probability for finding the particle at that position, as we've already discussed.

For the momentum question, the processing is considerably more complicated, and I won't try to describe it in detail. To find out the probability to observe some momentum, you must first perform a weighted average of the wave function—the exact way to do the weighting depends on what momentum you're interested in—and then square that average.

Answering these questions requires different ways of processing the wave function, which turn out to be mutually incompatible. It is impossible, according to quantum theory, to answer both questions at the same time. You can't do it, even though each question on its own is perfectly legitimate and has an informative answer. If someone were to figure out how to do it, experimentally, he'd have disproved quantum theory, because quantum theory says it can't be done. Einstein tried repeatedly to devise experiments of that kind, but he never succeeded, and eventually he conceded defeat.

Three major points are:

- You get probabilities, not definite answers.
- You don't get access to the wave function itself, but only a peek at processed versions of it.
- Answering different questions may require processing the wave function in different ways.

Each of those three points raises big issues.

The first raises the issue of *determinism*. Is calculating probabilities really the best we can do?

The second raises the issue of *many worlds*. What does the full wave-function describe, when we're not peeking? Does it represent a gigantic expansion of reality, or is it just a mind tool, no more real than a dream?

The third raises the issue of *complementarity*. To address different questions, we must process information in different ways. In important examples, those methods of processing prove to be mutually incompatible. Thus no one approach, however clever, can provide answers to all possible questions. To do full justice to reality, we must engage it from different perspectives. That is the philosophical principle of *complementarity*. It is a lesson in humility that quantum theory forces to our attention. We have, for example, Heisenberg's uncertainty principle: You can't measure both the position and the momentum of particles at the same time. Theoretically, it follows from the mathematics of wave functions. Experimentally, it arises because measurement requires active involvement with the object being measured. To probe is to interact, and to interact is potentially to disturb.

Each of these issues is fascinating, and the first two have gotten a lot of attention. To me, however, the third seems especially well-grounded and meaningful. Complementarity is both a feature of physical reality and a lesson in wisdom, to which we shall return.

## *Stationary States as Natural Vibrations*

The equation that describes how the wave function of an electron evolves in time is called the Schrödinger equation. The Schrödinger equation, regarded as a piece of mathematics, is closely related to the equations we use to describe musical instruments.

The hydrogen atom, viewed as a musical instrument, is like a three-dimensional gong that is stiff on the outside—far from the proton—but easier to set in motion near the middle. This means that our instrument's "vibrations," whose strength the wave function's magnitude encodes, will tend to focus toward the middle. So the wave function tends to concentrate toward the middle, as of course does its associated probability cloud. That is the rigorous quantum-mechanical version of saying that the proton attracts the electron!

Now we're ready to understand how modern quantum mechanics, based on wave functions and the Schrödinger equation, both captures and transcends Bohr's "highest musicality."

The most important step in understanding how any musical instrument operates, from a physical perspective, is to understand its natural vibrations. They correspond to its "notes," the patterns of vibration that the instrument can sustain over substantial lengths of time, and that are easy to excite (that is, to play).

In that spirit, because the Schrödinger equation for an electron in an atom looks very much like the equation for vibrations of a musical instrument, we should consider the solutions that look like natural vibrations. A natural vibration of the *wave function* turns out to mean something extremely simple and appealing for its *probability cloud*—namely, that it doesn't change at all!

(In more detail, using complex numbers: When we speak of a vibrating string, as in figure 24 on page 172, what "vibrates"—that is, what changes over time—is the positions of the elements of the string. For a wave function, what changes is the pattern of complex numbers it assigns to different points in space. In the natural vibrations, the change is simple: the magnitude of the complex numbers stays fixed, but their phases change, all by the same amount. Hence the squares of their magnitudes, which are what appear in the probability cloud, don't change at all.)

These natural vibrations of the wave function, which correspond to unchanging probability clouds, have the properties that Bohr anticipated in his "stationary states." The electron will persist in any of those patterns indefinitely, and no other patterns have that property. Furthermore, one can calculate the energy attached to these natural vibrations, and they turn out to match the energies of Bohr's "allowed orbits."

Let's take a look at some of these stationary states. Figure 26 shows their probability clouds. In all cases, the proton is at the center, and what you're looking at is the two-dimensional profile of a three-dimensional cloud. The brightness of the cloud indicates its size as a mathematical function, so the electron is more likely to be found, in each of these states, where the cloud is brighter. The more compact clouds correspond to stationary states with lower energy.

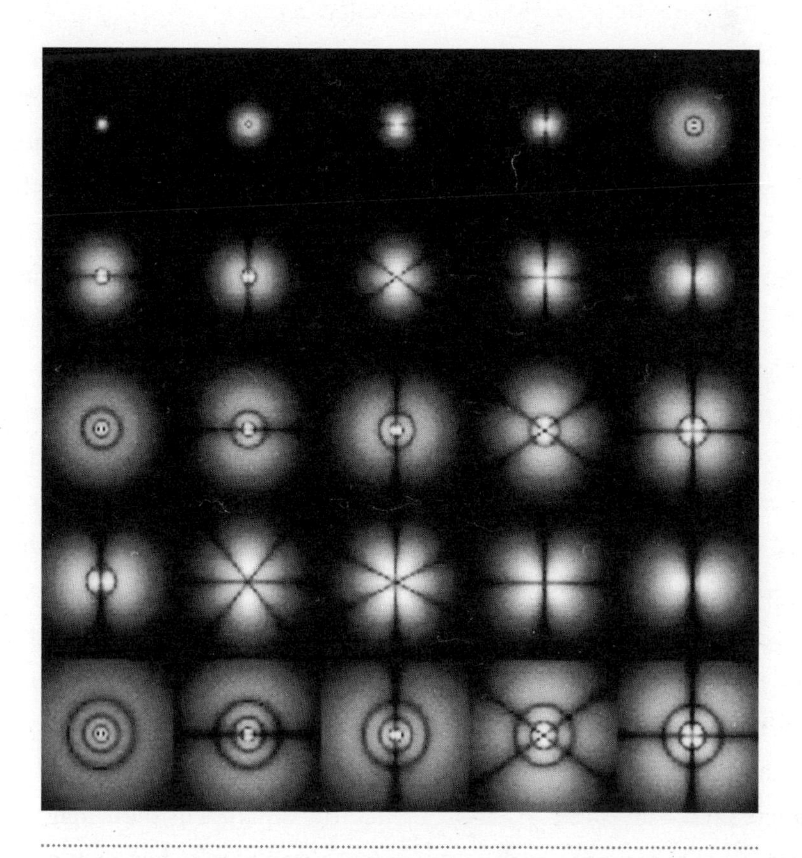

FIGURE 26. EACH IMAGE SHOWS A SNAPSHOT OF THE PROBABILITY CLOUD OF AN ELECTRON, TAKEN IN ONE STATIONARY STATE OF HYDROGEN OR ANOTHER. THE ELECTRON IS MORE LIKELY TO BE FOUND WHERE THE CLOUD IS BRIGHTER. IN EACH CASE, A SINGLE PROTON IS AT THE CENTER OF THE CLOUD. (THE SAME ORBITAL SHAPES ALSO APPLY TO ELECTRONS IN OTHER ATOMS, SUCH AS CARBON.)

To do justice to the wave functions themselves, as opposed to their projected probability clouds, requires more work, but it gives richer rewards. In plate CC we see just one of the stationary states. The surfaces are surfaces where the magnitude of the wave function has a constant value. They are cut away, so you can see inside. The colors indicate the phase of the wave function, as a complex number. You must imagine this picture as a snapshot. As time progresses, the colors cycle. Atoms are psychedelic!

The modern quantum theory, despite its greater complexity, has some overwhelming advantages over Bohr's pioneering model:

- In modern quantum theory, transitions between stationary states occur as a logical consequence of the equations. Physically, they arise due to the interaction between electrons and the *electromagnetic fluid*. Because that interaction is fairly feeble, compared to the basic electric forces that bind electrons, we often do well to include it as a correction, while retaining the stationary states as a starting point. In this treatment, we find that the transitions are not true discontinuities, though they do occur rapidly.
- Bohr's rule for determining which orbits are "allowed" was clearly formulated only for single electrons. In the years of Jeopardy, roughly 1913–25, there were many attempts to guess the rules for more complicated situations. But when Schrödinger (and, earlier, Heisenberg) came up with his equation, it was so obviously superior, and even so "obviously right," that it more or less immediately emerged as the consensus, and quickly evolved into modern quantum theory. And judging from its overwhelming, continuing success, Nature seems to like quantum theory too.
- It's more musical!

The process that replaces quantum jumps is especially interesting. In it, the electron gives birth to electromagnetic energy, in the form of a photon, where initially there was none. This occurs when the electron encounters spontaneous activity in the electromagnetic fluid and, by imparting some of its own energy, amplifies that activity. In that way, the electron transitions to a state of lower energy, a virtual photon becomes a real photon, and there is Light.

## *Cold*, *Austere*, and *Gorgeous*

Before proceeding further, I'd like to pause a moment for a brief quarrel with a hero of my youth, Bertrand Russell, who wrote:

> Mathematics, rightly viewed, possesses not only truth, but supreme beauty—a beauty cold and austere, like that of sculpture, without appeal to any part of our weaker nature, without the gorgeous trappings of painting or music, yet sublimely pure, and capable of a stern perfection such as only the greatest art can show.

I don't disagree with that statement, exactly, but I think its puritanical tone is misguided (and, coming from Russell, very odd). Cold and austere beauty can be wonderful, but gorgeous trappings can be wonderful too. They are complementary. The Schrödinger equation is as cold and austere as they come. Yet it is also the source of plate CC!

## *Artisanal Atoms*

In recent years, the frontiers of atomic physics have moved from observation to control and creation. The students of the Artisan have graduated, to become artisans in their own right.

At one frontier, atomic engineers have found ways to trap isolated atoms. This allows clean looks at basic quantum processes. For instance, one can monitor the sudden changes in state that occur when a single atom emits or absorbs light, and watch Bohr's "quantum jumps" in real time. Atomic engineers can also manipulate such atoms, exposing them to electric or magnetic fields, or to light. This allows exquisite control. Individual atoms are wonderful materials for engineering because they are essentially frictionless, and their properties can be tuned (using the fields) and reliably anticipated (using the theory). Atoms make the world's best clocks, for example. At present, the best atomic clocks are accurate to about one second per billion years.

Another frontier is making new kinds of atoms. Quantum dots are artificial structures which are based on the same principles as natural atoms, but tailored to human specifications. In effect, they are new kinds of musical instruments, designed to work on light rather than sound. Basically, a quantum dot consists of a small number of electrons confined in a small space, where they've been trapped by artfully sculpted electric fields. Quantum dots empower enormous flexibility in the design of detectors and generators of light. That could be very useful for expanding color perception, as we discussed earlier, and for many other applications.

The pioneers of atomic physics never dreamed of manipulating single atoms, let alone making artificial ones. In their early writings you can even find statements that deny the possibility of quantum engineering. Bohr, in particular, emphasized the division between a fully accessible "classical world" and a distinct "quantum world" that could be observed (in limited ways!) but not engineered. But their research, originally driven by the search for beauty and by simple curiosity, has engendered marvelous, and infinitely promising, new technologies.

There is a lesson here.

Many kinds of rewards are given to people for tangible services rendered. These rewards take the form of salaries, profits, social status, and so forth. But the accumulated wealth of basic science and art often derives from efforts whose ultimate value isn't immediately obvious. Even in cases where some breakthrough is clearly important, it may be years before the work yields any economic benefit; or the benefit might be entirely cultural and never become economic in the usual sense. People who work toward increasing this special kind of wealth are devoting their careers to long-term investment in the improvement of life for humanity as a whole.

And what hardheaded businessperson or consumer will pay for that? Yet history teaches us that such devotion to the long term, and to the common good, pays off handsomely. A wise society will cherish opportunities to foster such devotion.

## BACK TO PLATO

The details of Plato's atomic theory, based on atoms in the form of Platonic solids, are entirely wrong. Yet Plato's atoms serve as a beautiful and appropriate metaphor for the real thing, because they capture central truths.

Matter really *is* made from a few kinds of atoms. The atoms really *do* exist in vast numbers of perfectly identical copies. The properties of materials really *are* determined by the properties of the atoms that make them. And, what for Plato was most important: *atoms embody Ideals.*

In Plato's original theory, atoms embodied the beautiful geometry of symmetry. In the modern theory, atoms are the solutions of beautiful equations. (One more layer down, as you'll see, we get back to symmetry!) If you have a sufficiently powerful computer, and give the right equations, that computer will be able to predict any property of the atom that can be measured. Nothing else is required. In that precise sense, the atoms *embody* the equations.

## THE BEAUTY OF CONSTRAINT

The fundamental laws of today's physics are dynamical laws. They are laws, in other words, that govern how things change in time. They convert inputs (the conditions at one time) into outputs (the conditions at another time). But they are happy to work with any inputs, and so do not impose structure.

On the face of it, then, atoms as we know them are unlikely products for dynamical equations to produce. Atoms of a given kind—atoms of hydrogen, say—are structures that exist in vast numbers of identical copies. And they neither evolve nor erode, nor, in a stable environment, display any properties that vary in time. Looking back in history, thanks to the finite speed of light, we may see, through their spectra, that the atoms in galaxies a long time ago, and far far away, behave in the same way as the atoms we see on Earth today. We can also, with exquisite precision, compare the spectra in neighboring labs, or the same lab two weeks later.

In human manufacturing, use of interchangeable parts was a revolutionary innovation, and hard work to achieve. How did Nature achieve it? How could uniformity, if achieved by careful adjustment, stand up to the ravages of time? And if the building blocks are supremely stable, and resistant to change, how did they arise in the first place?

Maxwell was alert to, and intrigued by, this issue, seeing in it evidence of benevolent Creation. As he put it:

> Natural causes, as we know, are at work, which tend to modify, if they do not at length destroy, all the dimensions of the earth and the whole solar system. But though in the course of ages catastrophes have occurred and may yet occur in the heavens, though ancient systems may be dissolved and new systems evolved out of their ruins, the molecules out of which these systems are built—the foundation stones of the material universe—remain unbroken and unworn.
>
> They continue this day as they were created—perfect in number and measure and weight, and from the ineffaceable characters impressed on them we may learn that those aspirations after accuracy in measurement, truth in statement, and justice in action, which we reckon among our noblest attributes as men, are ours because they are essential constituents of the image of Him who in the beginning created, not only the heaven and the earth, but the materials of which heaven and earth consist.

Newton wondered at the Solar System's stability (and thought that it required a Creator's occasional repair work). Here, for similar but even more compelling reasons, Maxwell wonders about the stability of material structures, as evidenced by their accurate similarity, and the possibility of accurate chemistry.

## *Atoms Versus Solar Systems*

If not through divine supervision, how *do* the atoms of modern chemistry, with their exactly reproducible, stable properties, emerge from equations that, fundamentally, are equations for change?

To appreciate the force of that question, let us contrast a similar-looking question that has a very different answer. It is the question, we saw, that inspired Kepler: What determines the size and shape of our Solar System?

To Kepler's question, the modern answer is, basically, "It is an accident. No fundamental principles fix the size and shape of our Solar System." There are many possible ways that matter can condense into a star surrounded by planets and moons, just as there are many possible poker hands. Which one you get is the luck of the draw. Indeed, astronomers are now exploring systems of planets around other stars than our Sun, and finding that they are arranged in many different ways. All these systems evolve according to the laws of physics. But those laws are dynamical. They do not fix the starting point. Newton's dynamical worldview wins out over Kepler's aspiration for the geometrically ideal.

Does this mean that anything goes? No indeed. We *can* relate many features of the Solar System's size and shape to fundamentals. Some can be traced to its origin in the gravitational collapse of a large cloud of dust and gas. (We see this process occurring in other parts of the Galaxy, notably the Orion Nebula.) That the bulk of the mass ends up in a central star, like our Sun, is a logical consequence. For universal gravity encourages the accumulation of matter, and very large accumulations of mass produce enough central pressure to ignite nuclear burning, making a star. The facts that so impressed Newton—that the planets all orbit in roughly the same plane (the ecliptic) and in the same direction—reflect their role as repositories of angular momentum, spun off as the original gas cloud condensed. Other features reflect the long influence of history, wearing down the rough edges, so to speak. The fact that the same side of our Moon always faces Earth is one such feature: rotation of the Moon would raise powerful tides, which act as friction. Presumably, in the distant past, there was such rotation, but it has been damped out. (For similar reasons, the length of Earth days is increasing. Geological records, which show daily fluctuations in tidal deposits, indicate that during the Cambrian era, 650 million years ago, days were roughly twenty-one hours long.)

We can also roughly "predict" the size and shape of Earth's orbit around the Sun, from quite a different kind of consideration. That is: if the size and shape of that orbit were very different, intelligent life would not be around to observe it! For under those circumstances, life in anything close to the form we know it would be impossible (or at least very difficult). Among other problems: if the orbit is much smaller, surface water boils away; if the orbit is much bigger, surface water freezes; if the orbit is not nearly circular, one has wrenching changes in temperature.

Arguments like that, which promote the conditions for our existence into principles, are called anthropic arguments. Taken in their most general form, anthropic arguments beg many questions. First of all, who is the "our" in "our existence"? If we require everything that is necessary for the existence of, say, Frank Wilczek—or alternatively, *you*, the reader—we will be making principles out of many special circumstances that we really don't want to regard as fundamental features of the Universe, the Solar System, or even Earth. A more reasonable approach, possibly, is to base anthropic "predictions" on the looser requirement that some kind of intelligence capable of making observations and predictions should emerge. Even that formulation, however, raises difficult questions on the ragged edge of biology (What conditions allow the emergence of intelligence?) and philosophy (What is intelligence? What is observation? What is prediction?).

Our constraints on the size and shape of the Earth's orbit offer a modest, fairly straightforward, and benign example of anthropic reasoning. We'll encounter more adventurous, and controversial, examples later.

As Maxwell recognized, if atoms and molecules operated on the same principles as the Solar System, the world would be very different. Every atom would be different from every other, and every atom would change over time. Such a world wouldn't have chemistry as we know it, with definite substances and fixed rules.

It is not immediately obvious what makes atomic systems behave so differently. In both cases we have a massive central body attracting several smaller ones. The forces in play, gravitational or electrical, are broadly similar—both decrease as the square of the distance. But there are three factors which make the physical outcome very different, giving us stereotyped atoms but individualized solar systems:

1. Whereas planets differ from one another (as do stars), all electrons have exactly the same properties (as do all nuclei of a given element, or more precisely a given isotope).
2. Atoms obey the rules of quantum mechanics.
3. Atoms are starved for energy.

The first item in this explanation begs the question, of course. We're trying to explain why atoms can be the same as each other, and we start off by asserting that some other things, electrons, are all the same as each other! We'll come back to that later.

But having the same parts doesn't guarantee the same outcome, by any means. Even if all planets were the same as one another, and all stars were the same as one another, there would still be many possible designs for solar systems, and they'd all be subject to change.

We've seen how quantum mechanics brings discreteness, and fixed patterns, into the description of continuous objects that obey dynamical equations. It's the story, you'll recall, that unfolds in figures 24 (page 172), 25 (page 174), and 26 (page 187), and plate CC.

To close the loop, we need to understand why the electrons in atoms are usually found in just one among their infinite variety of patterns. That's where our third item comes in. The pattern with lowest energy—the so-called ground state—is the one we generally find, because atoms are starved for energy.

*Why* are atoms starved for energy? Ultimately, it is because the Universe is big, cold, and expanding. Atoms can pass from one pattern to another by emitting light, and losing energy, or absorbing light, and gaining energy. If emission and absorption were balanced, many patterns would be in play. That's what would happen in a hot, closed system. Light emitted at one time would be absorbed later, and a balanced equilibrium would set in. But in a big, cold, expanding Universe, emitted light leaks into vast interstellar spaces, carrying away energy that is not returned.

In this way we find that dynamical equations, which by themselves cannot impose structure, do so through jujitsu (gentle skill), focusing the power of other principles. They guide the constraining powers of quantum mechanics and cosmology. Cosmology explains their poverty of energy, and quantum mechanics shows how poverty of energy imposes structure.

## IMAGE AND INSPIRATION

I think that plate CC is an extraordinary work of art. It uses some clever tricks of shading and perspective to give the impression of three-dimensionality in what is, after all, a two-dimensional image. It also uses a cutaway design, and a tasteful choice of the surfaces displayed (i.e., surfaces of equal probability), to reveal the intricate structure.

Hydrogen atoms have only a single electron. Moving up one step in complexity, to helium atoms, we have two electrons. A quantum atom with two electrons is a much more complicated object to visualize, and I'm not aware that it's ever been done very well. The challenge is that for each possible position of one electron, the wave function of the other is a *different* three-dimensional object. So really, the natural home of the total wave function, for the two-electron system, is a space of $3 + 3 = 6$ dimensions. It is quite a challenge to present such an object in a way that human brains find meaningful. The ideas I mentioned in connection with expanding the space of color perception could be useful here, too.

Ambitious artist-scientists, in the spirit of Brunelleschi and Leonardo, will take this challenge as an *opportunity* for creativity. They will reveal deep aspects of reality that are both beautiful and mind-expanding. Plate CC is, I hope, a sign of things to come.

Worthy images of atoms will share, in their mixture of regularity and variation, the qualities of mandalas. They will offer, too, an awe-inspiring perspective on the assertion at the heart of mystical spirituality: That Art Thou. Because, you know, it is.

# SYMMETRY I: EINSTEIN'S TWO-STEP

With his two theories of relativity—special and general—Albert Einstein (1879–1955) brought a new style into thinking about Nature's fundamental principles. For Einstein beauty, in the specific form of *symmetry*, takes on a life of its own. Beauty becomes a creative principle.

## *Mythic Background*

Describing his approach to science, Einstein said something that sounds distinctly prescientific, and hearkens back to those ancient Greeks he admired:

> What really interests me is whether God had any choice in the creation of the world.

Einstein's suggestion that God—or a world-making Artisan—might not have choices would have scandalized Newton or Maxwell. It fits very well, however, with the Pythagorean search for universal harmony, or with Plato's concept of a changeless Ideal.

If the Artisan had no choice: Why not? What might constrain a world-making Artisan?

One possibility arises if the Artisan is at heart an artist. Then the constraint is desire for *beauty*. I'd like to (and do) infer that Einstein thought along the line of our Question—Does the world embody beautiful ideas?—and put his faith in the answer "yes!"

Beauty is a vague concept. But so, to begin with, were concepts like "force" and "energy." Through dialogue with Nature, scientists learned to refine the meaning of "force" and "energy," to bring their use into line with important aspects of reality.

So too, by studying the Artisan's handiwork, we evolve refined concepts of "symmetry," and ultimately of "beauty"—concepts that reflect important aspects of reality, while remaining true to the spirit of their use in common language.

## SPECIAL RELATIVITY: GALILEO *AND* MAXWELL

If Einstein was Pythagoras reborn, he had learned much in the meantime (thanks to many cycles of reincarnation). Einstein did not, of course, abandon the discoveries of Newton, Maxwell, and the other heroes of the Scientific Revolution, nor their respect for observed reality and concrete facts. Richard Feynman called Einstein "a giant: his head was in the clouds, but his feet were on the ground."

In his theory of *special relativity*, Einstein reconciled two ideas of his predecessors that appear to be contradictory.

- The observation of Galileo that overall motion at constant velocity leaves the laws of Nature unchanged. This idea is fundamental to

Copernican astronomy, and is deeply embedded in Newton's mechanics.

- The implication of Maxwell's equations that the speed of light is a direct result of basic laws of Nature, and cannot change. This is an unambiguous consequence of Maxwell's electrodynamic theory of light, a theory confirmed in the experimental work of Hertz and many others.

There is tension between those two ideas. Experience suggests that any object's apparent speed will change, if you yourself are in motion. Achilles catches up to the tortoise, and even outruns it. Why should light beams be different?

Einstein resolved that tension. By critically analyzing the operations involved in synchronizing clocks at different places, and how that synchronizing process is changed by overall motion at constant velocity, Einstein soon realized that the "time" assigned to an event by a moving observer is different from the "time" it is assigned by a fixed observer, in a way that depends on its position. In referring to events they observe in common, one observer's time is a mixture of the other's space and time, and vice versa. This "relativity" of space and time was what was new in Einstein's special theory. Both of the theory's assumptions were already out there, and widely accepted, before his work—but no one had taken both seriously enough to demand and force their reconciliation.

Because the Maxwell equations contain the speed of light, special relativity's second assumption—that the speed of light is invariant under Galilean transformations—follows directly from Einstein's motivating idea, to preserve both the Maxwell equations and Galilean symmetry. But it is a much weaker assumption.

In fact, Einstein proceeded to turn the argument around, by showing that one could derive the complete system of four Maxwell equations from one of them, by making Galilean transformations to recover the general case. (By putting charge in motion, you get currents, and by putting electric fields in motion, you get magnetic fields. Thus the law governing how unmoving electric charges generate electric fields, after Galilean transformations, gives the general case.) That profound trick was a taste of the future. *Symmetry, rather than a deduction from given laws, became a primary principle, with a life of its own.* One can constrain the laws by requiring them to have symmetry.

## TWO POEMS IN LIGHT

### *Reweaving the Rainbow*

There is one physical consequence of special relativity that I find beautiful above all others. It brings together many of our profoundest themes, yet appeals directly to sensory experience. The early chapters of our meditation, recounting the physics and the history of light and color, have prepared us to enjoy it together here, now.

Let us consider how a pure beam of light, with a definite spectral color, changes if we view it from a platform moving at constant velocity; in other words, if we make a Galilean transformation. Naturally, we still see a light beam. And that beam still progresses through space at the previous rate: The speed of light is invariant. If we began with a pure beam, with a definite spectral color, it will still appear to us as a pure beam, with a definite spectral color. But . . .

It's a *different* color! If we move in the same direction as the beam (thus, running away from its source), or if its source is receding from us, its color shifts toward the red end of the spectrum (or, if it started red, it will shift into the infrared). If we move in the opposite sense, the color shifts toward the blue end of the spectrum (or into the ultraviolet). The faster we move, the more pronounced the effect.

The former effect is commonly encountered in cosmology because distant galaxies are moving away from us—or, as we say, the Universe is expanding. In that context, it is known as the redshift. It was through the observed redshift of known spectral lines that the expansion of the Universe was discovered.

For us, the great conclusion is this: *all the colors can be obtained from any one of them, by motion,* or, as we say, by making Galilean transformations. Because Galilean transformations are symmetries of the laws of Nature, any color is fully equivalent to any other. They all emerge as different views of the same thing, seen from different but equally valid perspectives.

Here we must have an image! In plate DD, you see the wave pattern associated with a pure light beam—a spectral color—emitted from a source moving to the right at seven-tenths the speed of light. If you are to the right, so the beam is approaching you, you will perceive its color as blue; if you are to the left, so the beam is receding, it appears red. In this snapshot, the source is near the center.

Newton thought he'd proved that each spectral color was inherently different from any other, and no alchemy could transmute one into another. Newton's experimental work established that light of each spectral color will persist in that color despite reflection, refraction, or many other potentially transformative processes.

But he goofed! If only Newton had tried racing past his prisms at tens of thousands of meters per second, he would have seen the error of his ways. I'm joking, of course. It is common, but horrifying, to find popularizers and observers of science saying things of that sort in earnest—as if anything besides the latest True Theory of Everything is so much trash. It's a lightweight version of the sort of thinking behind intolerant, totalitarian ideologies. The real point I want to emphasize is exactly the opposite: how nearly right Newton's conclusions were, and how useful they remain.

Yet it is beautiful to discover that there's another chapter to the story, where we discover deep unity beneath, and supporting, the diversity of appearance. All colors are one thing, seen in different states of motion. That is science's brilliantly poetic answer to Keats's complaint that science "unweaves a rainbow."

### *Reanimating Color*

The physical essence of color, like the physical essence of tone, is a signal that varies in time.

The time variation of light is too fast for our substance to follow. Its frequency is too high. And so, to make the best of a difficult situation, our sensory system processes the information and encodes a small part of it in perceived color.

That code, by the end of the day, bears little trace of its origin! When we perceive a color, we see a symbol of change, not anything that changes.

But we can put back more of the underlying information, concretely, by bringing back time variation, rescaled to suit human abilities. Through this act of transformative restoration, we shall expand the doors of perception.

## GENERAL RELATIVITY: LOCALITY, ANAMORPHY, AND ENABLING FLUIDS

With special relativity, as we have discussed, Einstein elevated Galilean symmetry, or invariance, into a primary principle—a requirement that all the laws of physics must obey. Maxwell's equations satisfy that requirement, as they stand. Newton's laws of motion do not, but Einstein supplied a modified mechanics that does. For bodies that move much slower than the speed of light, Einstein's version reproduces Newton's successful answers.

Newton's theory of gravity, however, is much harder to accommodate.

Newton's theory is built around the concept of mass, but in special relativity, mass loses its secure place. In particular, mass is not conserved. (If those concepts are unfamiliar, please consult the entries *Mass* and *Energy* in "Terms of Art.")

So if you're asking gravity to respond to mass, as Newton's theory does, you're giving ambiguous instructions. A relativistic theory of gravity requires new foundations.

In the end Einstein solved the problem, in his general theory of relativity, through a heightened concept of symmetry. He made symmetry—specifically, Galilean symmetry—*local.*

We can best understand local symmetry (which we have in general relativity) by contrasting it with rigid symmetry (which we have in special relativity).

According to *rigid* Galilean symmetry, or invariance, we can change the state of motion of the Universe, adding a constant overall velocity, without changing the laws of physics. On the other hand, if we change the *relative* motion between different parts of the Universe, by imposing velocities that change in space or in time, we must expect changes in the laws. If you wave a magnet near a compass needle, the needle moves!

*Local* Galilean symmetry, or invariance, postulates that a much wider class of transformations leave the laws unchanged. It says, precisely, that we can choose the added velocity to be different at different times and places. That should sound outrageous, because it's exactly what we just said *doesn't* work!

But there's a way to make it work, by expanding the theory. After many attempts, on different occasions over several years, to describe the essential idea in an accessible way, I've discovered one that I'm happy with. Not coincidentally, it builds on our earlier thoughts, and brings in ideas from art.

We've been using artistic perspective as our prototype of symmetry. You can view the same scene from different places, thereby getting different perspectives. The images the scene projects can and will differ in many ways, but they all convey the same scene. Change in perspective, without change in scene, is a shining example of symmetry.

In a like manner, we can view the world from a different "perspective" by giving it a constant velocity or, what amounts to the same thing, viewing it from a moving platform. When we do that, many things will come to look different, but—according to special relativity—the same laws of physics will hold good. In that sense, it will still be (a picture of) the same world.

Now let us consider more general ways to view the scene, besides changing our perspective. This brings us into the domain of anamorphic art, beautifully exemplified in plate EE. Anamorphic art uses lenses, curved mirrors, and other devices to produce images that are rearranged in interesting, patterned ways. The range of images that can represent a given scene is vastly larger and includes some very distorted appearances.

We could, more "physically," consider viewing the world through a translucent, but light-bending, material—water, say. We could even imagine that the water might be denser in some places than in others, so the amount of bending varies. (It's hard to do that with real water, but no matter.) In this situation, the images we sample at different places could be distorted, and look very different indeed. We might find them hard to interpret.

If we didn't understand the effect of the water, we'd be tempted to think that the images must convey different scenes. But if we know about water, and allow for its effect, we can count many more possible images as valid representations of our scene. We might distribute the water in different ways, for example, to mimic the effect of a fun-house mirror. We could even put our water in motion so that the images might also change with time. In short: *by imagining a space-filling fluid, and allowing for its possible effects, we are able to consider a wide variety of transformed images as representations of the same scene, viewed through different states of the fluid.*

In a similar way, by introducing just the right kind of material into space-time, Einstein was able to allow the distortions of physical law, which are introduced by Galilean transformations that vary in space and time, to be accomplished as modifications of a new material. That material is called the *metric field* or, as I prefer to say, *metric fluid.* The expanded system, containing the original world plus a hypothetical new material, obeys laws that remain the same even when we make variable changes in velocity, though the state of the metric fluid changes. In other words, *the equations for the expanded system can support our huge, "outrageous" local symmetry.*

We might expect that systems of equations that support such an enormous amount of symmetry are very special, and hard to come by. The new material must have just the right properties. Equations with such enormous symmetry are the analogue of the Platonic solids—or, better, the spheres—among equations!

When Einstein worked out those equations, having enriched the world with a new material, he found that he'd gotten his long-sought theory of gravity. The equations reveal that the metric fluid, which he'd introduced in order to enable local Galilean symmetry, gets bent by the presence of matter, and in turn influences how matter moves. So the metric fluid serves, essentially, the same role for gravity that Maxwell's electromagnetic fluid serves for electromagnetism. We call its minimal excitations, or quanta, *gravitons*, analogous to the photons of electromagnetism.

In this construction the role of symmetry, as a world-governing principle, has been taken to a new level. Symmetry has become creative. The assumption of local symmetry has dictated the detailed structure of a rich and complex theory of gravity that describes Nature successfully. To make a go of local symmetry you must introduce the metric fluid, and its consequent gravitons.

I should add that this approach to general relativity, putting local symmetry front and center, is somewhat unorthodox. Usually other concepts, motivating the introduction of the metric fluid in other ways, are invoked. But local symmetry is what's essential, and the minimalist approach serves us well, when we come to make theories of the other forces.

Einstein, in speaking of his theory, used a different terminology. Some of that terminology contains residues of his "searching in the dark" period, and seems ambiguous or confused, at least to me. But basically his *general covariance* corresponds to our *local Galilean symmetry.* In the one-sentence summary of our discussion, it seems appropriate to honor his choice:

Gravitons are the avatars of general covariance.

# QUANTUM BEAUTY II: EXUBERANCE

The analysis of matter reduces it to electrons and atomic nuclei (and ultimately, as we'll see, one step further, down to electrons, quarks, and gluons). It's also fair to add photons to this list of ingredients, as they are the material of the electromagnetic fluid. From that paltry set of ingredients, following a few strange but strict and highly structured rules, the infinitely various material worlds of chemistry, biology, and everyday life emerge.

What's the go of it?

This chapter is brief, but important in advancing our Question. For here we seal the link

$$\text{Ideal} \rightarrow \text{Real}$$

between the strange music of quantum theory and the real world of materials. In later chapters we'll refine our understanding of the Ideal foundations, in the style

$$\ldots \text{Ideal} \rightarrow \text{Ideal} \rightarrow \text{Ideal} \rightarrow \text{Real}$$

But that last link, forged here, will remain firm, and essentially unchanged.

The world of chemistry is vast and fascinating. But our goal is not to make an encyclopedia. To address our Question, it is enough that we should forge that last link. To make that task both manageable and fun, I've decided to focus on what might seem to be a ridiculously limited version of chemistry, using just one element: carbon. As you'll see, this one-element corner of chemistry already creates a wonderland.

## WHAT DO ELECTRONS WANT?

What do electrons want?

That question makes sense, because all electrons, unlike all people, have the same properties. And their "desires" are easy to enumerate. There are basically three, the first two of which we already encountered in our previous chapter.

- Electrons are subject to electrical forces, which attract them to positively charged nuclei but repel them from one another.
- Electrons are described by space-filling fields—their wave functions—which prefer to vary smoothly and gently. They settle into specific standing wave patterns, or "orbitals," that find an optimal compromise between the attraction of nuclei and their natural wanderlust. I like to imagine electrons explaining themselves to nuclei this way:

  "I find you attractive, but I need my space."

- The third important property of electrons has to do with their relations with one another. We did not encounter it when we discussed hydrogen atoms, because each hydrogen atom has only one electron. This third property is a little more complicated than the other two. It is called the Pauli exclusion principle, after Wolfgang Pauli, the Austrian physicist who first formulated it in 1925. The Pauli exclusion principle is a purely quantum-mechanical effect. Without referring to the quantum description of physical reality based on wave functions, it can't even be stated!

When Pauli proposed it, the exclusion principle had no theoretical basis. You could call it an inspiration, or you might say it was guesswork. Both of those views are correct. Like Bohr's vision of stationary states and quantum jumps—and, for that matter, Pythagoras's rules of harmony!—Pauli's exclusion principle came from listening to music (for Bohr and Pauli, the music of atomic spectra) and identifying, in its patterns, governing rules. Today we recognize Pauli's exclusion principle as an aspect of the quantum theory of identical particles, with deep roots in relativity and quantum fluid theory, but it started as inspired guesswork.

For our present purpose, which is appreciating *how* the exuberant, spontaneous creativity of electrons arises from simple rules, Pauli's original formulation is all we need. Rule number three, in the crude form we'll use it, is this: No more than two electrons can be in the same stationary state.

(Why no more than *two*? That seems odd! It is a consequence of the electron's intrinsic spin. Two electrons can both be in the stationary state, only if they are spinning in opposite directions. A more satisfying formulation of Pauli's principle says that no more than *one* electron can be in the same stationary state, but includes spin as part of the description of the state.)

# CARBON!

Those three rules unfold into the worlds of materials science, chemistry, and the physical substrates of biology, including most of heredity and metabolism. To bring that overwhelming exuberance within manageable proportions, I've decided to focus on a tiny corner: the material world of pure carbon. As you'll see, even this tiny corner is strange, rich, and varied. It also brings us to several major research frontiers.

Chemistry based on carbon is often called organic chemistry, because carbon is a major ingredient of all proteins, fats, and sugars, which together with nucleic acids form the starring cast of biology. But those biological molecules also contain other elements besides carbon, and those elements are crucial to their function. Pure carbon compounds play no role in natural biology. So here we are looking into a special chapter in the great book of organic chemistry—the chapter devoted to inorganic organic chemistry.

## *Carbon Atoms, One at a Time*

Carbon compounds are produced by combining carbon atoms, so let's start with those. The carbon nucleus contains six protons, so it has six units of positive electric charge, which attracts six electrons before it is neutralized. When those electrons try to *minimize* their energy, our three rules come into play. The electrons would like to have wave functions in the stationary states, or as the chemists say, orbitals, with the *lowest* possible energy. Those are the nice, round, compact orbitals in the image in the upper left corner of figure 26 (page 187). Mr. Pauli's principle, however, tells us we can only accommodate two electrons that way.

The other four must make use of other kinds of spatial orbitals. Going one step to the right, we find another round orbital. It is less compact, so it gets less benefit from the attractive charge of the central nucleus. Electrons in that orbital are less stably bound to the nucleus than the two "inner" electrons—a key fact for what follows. That second round orbital can accommodate another two electrons, so we've now got room for 2 + 2 = 4 of the six. To accommodate the other two, we need to look a little further.

Going one step farther to the right, there is another kind of orbital that isn't round, but which has more of a dumbbell shape. The dumbbell shape can be oriented in any direction, so actually there are three independent orbitals of this type. So once we put those orbitals in play, there's plenty of room for our two extra remaining electrons.

It turns out that these two new kinds of orbitals have pretty nearly the same energy, so without prohibitive cost in energy the electrons can mix them. The important distinction is between two inner electrons, which are very tightly bound to the nucleus, and four outer electrons that are held much more loosely. When there are other atoms nearby, those four are targets for sharing. By adjusting their orbitals just a bit, those four electrons can reach out to, and benefit from the attraction of, more than one nucleus.

## *Carbon Atoms, in Bunches*

When we have carbon in bunches, there are two especially nice, symmetric ways to share the electrons. They are shown in figure 27.

On the left we have the unit of diamond structure, which displays perfect three-dimensional symmetry. The four orbitals extend to the vertices of a tetrahedron—the simplest Platonic solid, you'll recall.

On the right we have the unit of graphene structure, which displays perfect two-dimensional symmetry. The three planar orbitals extend to the vertices of an equilateral triangle—the simplest regular polygon. The white balls, in both cases, will be replaced by other carbon atoms with the same bonding pattern, while the dark balls contribute to a layer of quasi-free electrons. (Strictly speaking, the electron layer has half its density

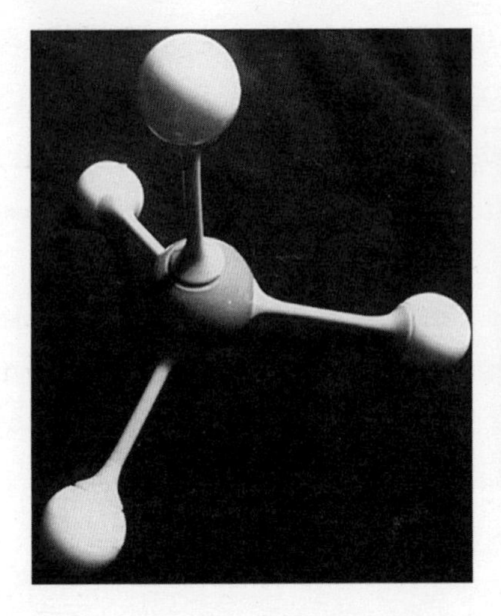

FIGURE 27. PURE CARBON HAS TWO MAIN BONDING PATTERNS, BOTH OPTIMIZED FOR SYMMETRY.

above, and half below, the main carbon plane.) Note that if each orbital is taken up by one electron from each nucleus, we will nicely satisfy Mr. Pauli's principle, with each carbon nucleus sharing four of its electrons altogether. In other figures in this chapter, you will see how these basic elements combine to give a brilliant variety of pure carbon materials.

Not coincidentally, those two especially symmetric bonding patterns turn out to be the ones that lead to favorable (i.e., low) energy. They support the myriad stable ways of combining carbon atoms that we'll now explore.

## *Diamond (3-D)*

The structure of diamond, at the atomic level, is symmetric and harmonious (figure 28, page 216). Each carbon nucleus is the center of four electron orbitals. The orbitals reach out to four neighboring carbon nuclei, whose positions define the vertices of a regular tetrahedron. This arrangement is very efficient, as the electrons manage to avoid each other while visiting two different nuclei. Because the electrons are happy in this arrangement, they resist doing anything different. It's tough to make them break away. That's why diamonds are hard to scratch! Pure diamonds are transparent, basically for the same reason: photons of visible light can't deliver enough energy to make an electron change its state. (Impurities, where some elements other than carbon sneak in, or defects, where the crystal structure is flawed, can lend diamonds color. There is an elaborate system for grading the colors of gem diamonds. Some forms of imperfection are better than others, or than perfection itself. . . .)

## *Graphene (2-D) and Graphite (2+1)*

The most stable form of elemental carbon, at room pressure and normal temperatures, is not diamond, but graphite. Contrary to their advertised reputation, diamonds are not forever: given long enough, they will transform into graphite (but don't hold your breath). Graphite is the black material that makes pencil "lead," and it is also widely used as an industrial lubricant. At an atomic level, graphite is a strongly layered material, consisting of many layers of graphene (figure 29) that are weakly bound to one another. The weakness of the interlayer binding makes it easy for layers to slide by one another, or peel off. That explains the lubricating and smudge-making abilities of graphite. We say graphite is 2+1 dimensional, because the two-dimensional sheets can be stacked indefinitely.

Graphene, the single-layer version of graphite, is the simplest and most glamorous material of this kind.

Graphene was studied theoretically for decades before it was available in any laboratory. Because graphene is so simple and regular, quantum theorists could predict its properties confidently, and in considerable detail. Graphene was expected to be awesome, if it could be produced. But could it be produced?

Graphene was first isolated by Andre Geim and Konstantin Novoselov in 2004. The method of discovery was a bit of nineteenth-century science that somehow escaped into the twenty-first century. They started with

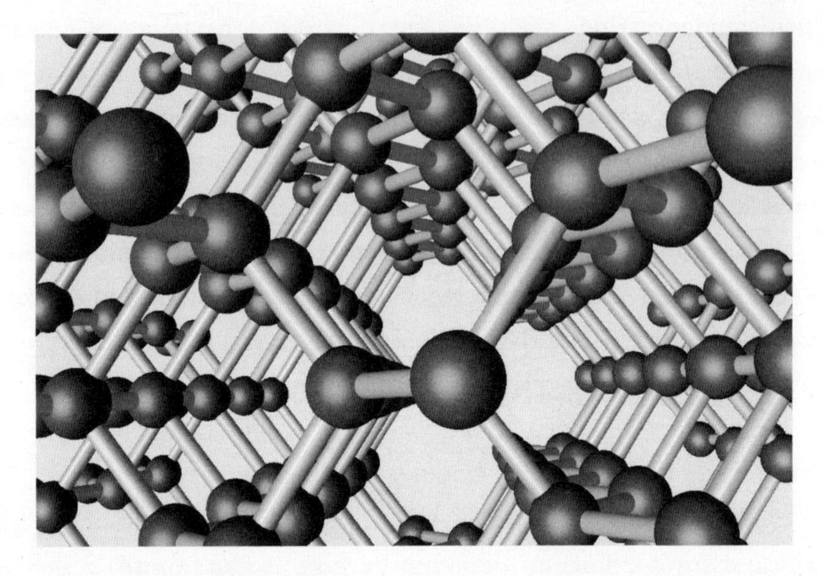

FIGURE 28. THE STRUCTURE OF DIAMOND. EACH CARBON NUCLEUS SHARES ELECTRONS WITH FOUR OTHERS AT THE VERTICES OF A SURROUNDING TETRAHEDRON. THIS IS A SPACE-FILLING, THREE-DIMENSIONAL STRUCTURE.

FIGURE 29. THE STRUCTURE OF GRAPHENE. EACH CARBON NUCLEUS SHARES ELECTRONS WITH THREE OTHERS AT THE VERTICES OF AN EQUILATERAL TRIANGLE. IN THIS HONEYCOMB PATTERN, WHICH CAN EXTEND INDEFINITELY, WE RECOGNIZE ONE OF THE THREE INFINITE PLATONIC SURFACES.

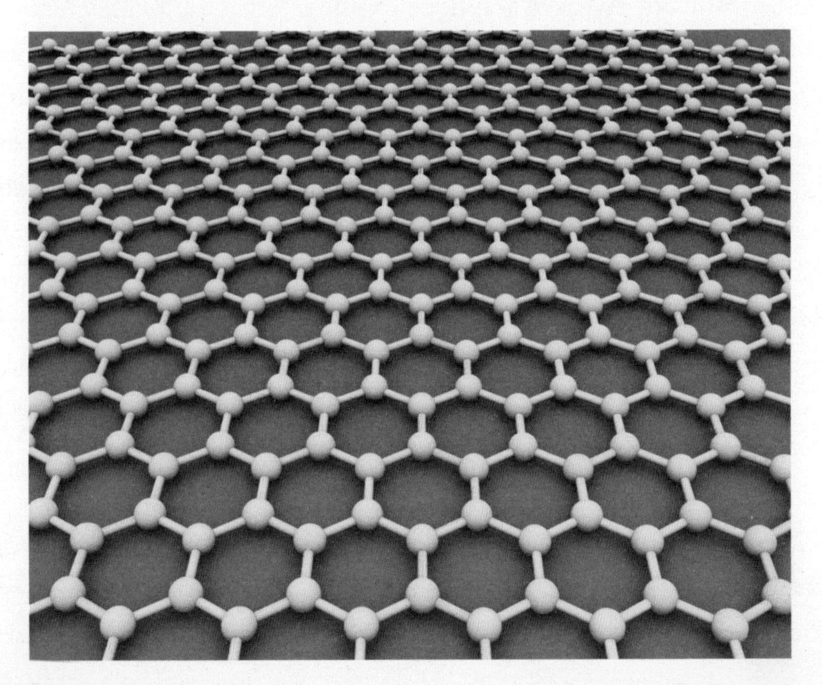

pencil smudges, which typically contain several carbon layers, consisting of graphite. Then they applied adhesive tape to strip some layers off, and transferred the smudge traces to thin microscope slides. The smudge traces made an irregular terrain, with some patches where the carbon was absent, some patches where it had been reduced to one layer—graphene!—others two layers thick, and so on. The different layers show slightly different colors under polarized light, so Geim and Novoselov were able to home in on the graphene patches and to study their properties well enough to demonstrate that they were, in fact, graphene patches. Geim and Novoselov shared the 2010 Nobel Prize in Physics for their work.

Graphene has unique mechanical and electrical properties, which promise many applications. Inspired by graphene's promise, people have figured out some considerably more efficient ways to make it! One optimistic, but maybe not crazy, study forecasts that a $100 billion market in graphene will develop over the next few years.

Here I'll just mention one highlight, which is easy to understand and fits nicely within our tour. Just as for electrons in a diamond crystal, the regular, efficient arrangement of electrons in the plane of graphene is so favorable that it is difficult to break them up. So graphene makes an extremely strong, tough material. At the same time, because it is only one atomic layer thick, a graphene sheet is light and flexible. In explaining their 2010 award, the Nobel committee mentioned that a one-square-meter graphene hammock could support a cat, while weighing about as much as one of the cat's whiskers. As far as I know, that particular experiment hasn't yet been tried.

## *Nanotubes (1-D)*

We can roll up two-dimensional graphene to make one-dimensional tubes, the so-called nanotubes. This can be done in many ways, giving nanotubes with different radii and pitches (see plate FF). Nanotubes that differ only slightly in geometry can have radically different physical properties. It is a triumph of quantum theory that these delicate properties can be predicted unambiguously, purely through calculation, and that they agree with experimental measurements.

## *Buckyballs (0-D)*

Finally, one can imagine closing a graphene sheet upon itself, to make a finite surface. This can be done in many ways. Actually, one can't form a simple closed surface, with each vertex meeting three edges, using hexagons alone. There's no such Platonic solid! A dodecahedron, made from pentagons, comes closest. Also, most important, each vertex in a dodecahedron connects to exactly three others, so the basic structural unit in figure 27 (on page 214, on the right) can be used. The three orbitals need to be bent out of their ideal planar arrangement. Although the dodecahedral molecule $C_{20}$, made from twenty carbon atoms, does exist, larger forms, which incorporate additional hexagons, require less distortion and form more readily. The pretty $C_{60}$ "soccer ball" molecule, shown in figure 30, is especially stable and common.

Balls of pure carbon, with $C_{60}$ the most common but by no means the majority type, form when carbon burns electrically, as in lightning discharges. They also occur, in small amounts, within common candle soot.

Figure 30 shows the structure of the $C_{60}$ molecule, one of a variety known as buckminsterfullerenes or buckyballs. Here graphene has been rolled up twice, in both of its two dimensions, to make a zero-dimensional object (that is to say, no remaining directions extend indefinitely). As in graphene and in nanotubes, the basic unit links a carbon nucleus to three neighbors. The buckyball incorporates a hidden dodecahedron: it has twelve pentagons regularly interspersed among its twenty hexagons, and if you shrink the hexagons to points, you will get a dodecahedron. There are buckyball variants with different numbers of hexagons, but they always have twelve pentagons, for topological reasons. The name honors

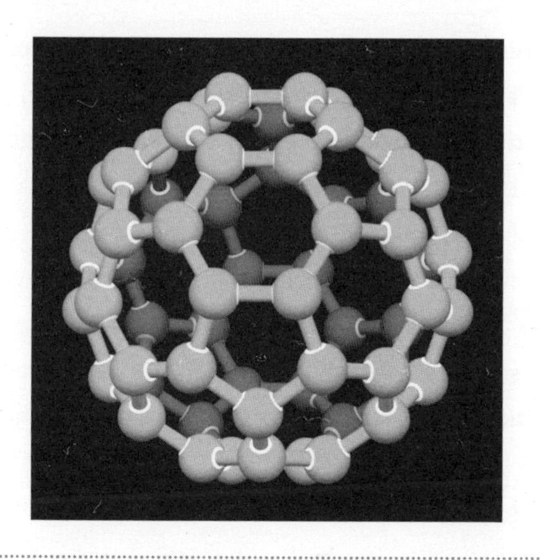

FIGURE 30. STRUCTURE OF THE BUCKYBALL. HERE THE CARBONS HAVE FOLDED UP COMPLETELY, MAKING A FINITE OBJECT. FROM AFAR IT LOOKS LIKE A POINT—THE DIMENSION HAS COLLAPSED TO ZERO.

FIGURE 31. HAROLD KROTO IN HIS ELEMENT, WITH MODEL BUCKMINSTERFULLERENES.

Buckminster Fuller (1895–1983), an inventor and architect whose "geodesic domes" use vaguely similar linked frameworks.

It's appropriate to conclude our brief tour of the exuberant chemistry of pure carbon with figure 31, a photograph of Harold Kroto (1939– ), who shared the 1996 Nobel Prize in Chemistry for his work on buckminsterfullerenes, amid models of his handiwork.

Lively minds will find delight not only in the beauty of diamonds, which dazzle on sight, but also in the hidden, inner beauty of superficially drab smudges, and of soot, and of gossamer sheets that nestle felines.

# SYMMETRY II: LOCAL COLOR

Now many strands of our meditation converge, and we approach an answer to our Question.

In our first symmetry interlude we saw how Einstein, by taking Galilean symmetry local, discovered his theory of gravity: general relativity.

In the next chapter, we shall document how taking symmetry local leads to successful theories of three major forces of Nature: electromagnetism, and the strong and weak nuclear forces. The new symmetries involve making transformations among properties (specifically, "color charges") of particles. In the symmetries' local forms, we allow those transformations to be different, at different places and times.

Here, to motivate the journey, we envision its destination.

## *Anachromy*

Anamorphic art warps the spatial structure of images. It is an excellent representation of the sorts of space-time transformations that general relativity accommodates as symmetries.

The sorts of transformations that the other forces accommodate would be best represented by an art form that is less cultivated, if it exists at all. Anamorphic art leaves the color structure of its images unchanged. In *anachromic* art, on the contrary, we modulate the color structure of images, while leaving their spatial structure unchanged.

At this point a few images are worth many thousands of words.

Plate GG is a pioneering work in anachromic art. In it, we see four versions of a photograph of a display of Barcelona street candy. At the upper left, we have the original, minimally processed photograph. At the upper right, we see the result of a rigid transformation of the colors, where every pixel has been transformed in the same way. (For nerds: $G \rightarrow R$, $B \rightarrow G$, $R \rightarrow B$ in the standard red-green-blue RGB scheme.) In the lower two panels, more complicated color transformations were applied, in which *the nature of the transformation changes from place to place.* At the lower left, we see the effect of relatively mild transformations, while at the lower right we see the effect of more drastic ones.

## AN ANSWER TO OUR QUESTION

Places of worship embody the aspirations of their architects, and the communities they represent, to ideal beauty. Their chosen means of expression feature color, geometry, and symmetry. Consider, in particular, the magnificent plate HH. Here the local geometry of the ambient surfaces and the local patterns of their color change as our gaze surveys them. It is a vibrant embodiment of anamorphy and anachromy—the very themes that our unveiling of Nature's deep design finds embodied at Nature's core.

Does the world embody beautiful ideas? There is our answer, before our eyes: Yes.

Color and geometry, symmetry, anachromy, and anamorphy, as ends in themselves, are only one branch of artistic beauty. Islam's injunction against representational art played an important part in bringing these forms of beauty to the fore, as did the physical constraint of structural stability (we need columns to support the weight of ceilings, and the arches and domes to distribute tension). Depictions of human faces, bodies, emotions, landscapes, historic scenes, and the like, when they are allowed, are far more common subjects for art than those austere beauties.

The world does not, in its deep design, embody *all* forms of beauty, nor the ones that people without special study, or very unusual taste, find most appealing. But the world does, in its deep design, embody *some* forms of beauty that have been highly prized for their own sake, and have been intuitively associated with the divine.

# QUANTUM BEAUTY III: BEAUTY AT THE CORE OF NATURE

Thus far our meditation on quantum reality has revealed that the world of everyday matter, when properly understood, embodies concepts of extraordinary beauty. Indeed, ordinary matter is built up from atoms that are, in a rich and precise sense, tiny musical instruments. In their interplay with light, they realize a mathematical Music of the Spheres that surpasses the visions of Pythagoras, Plato, and Kepler. In molecules and ordered materials, those atomic instruments play together as harmonious ensembles and synchronized orchestras.

Having found such gems of understanding, we are inspired to dig more deeply, confident that we haven't exhausted this vein. Our new insights provide a gratifying, yet still only *partial,* answer to our great Question. They lure us on, as our answers pose new questions, such as these:

- *What* are atomic nuclei?
- *Why* are there electrons?
- *Why* are there photons?

In the two final chapters, we'll take up those questions, and others they lead us to. Our explorations will bring us to the frontiers of current understanding, and then some big steps beyond. We will uncover new concepts and realities that build upon our earlier themes, but also transcend them. As we get closer to the core of matter, we discover new beauties, and capture a vision of how they might all come together, in a mighty synthesis. We'll uncover, in a real sense, the beautiful thing that the physical world *is*—and then, a still more beautiful thing it *might be.*

This chapter is devoted to the cluster of ideas that we presently use to describe the four basic forces of Nature. Two of those forces, gravity and electromagnetism, have already figured prominently in our meditation. Two others, the so-called strong and weak forces, were only discovered in the early twentieth century, as physics came to terms with atomic nuclei.

Atomic nuclei are tiny and hard to study. Understanding them was a long and difficult quest, one that dominated research in fundamental physics for most of the twentieth century, and still continues. For a while things got very complicated and messy. In the end, though, Nature came through! Today we have theories of the strong and weak forces that are worthy to stand beside Newton's (and Einstein's) theories of gravity, and Maxwell's electrodynamics.

As we'll see in this chapter, the concepts and equations we need to describe the strong and weak forces are natural, beautiful enhancements of concepts and equations that arise in describing gravity and electromagnetism. Conversely, our understanding of the strong and weak forces gives us new perspectives on the older theories, pinpointing the essence they all share in common. That shared essence hints at an underlying, deeper unity. In the next chapter, we'll see how that unification seems ripe for fruition.

## *Approaching the Core*

The reigning theories of the strong, weak, electromagnetic, and gravitational forces are often lumped together and called the Standard Model. As I already mentioned in the introduction, that mundane phrase carries modesty too far. For one thing, "*Standard* Model" has an air of conventional wisdom, with strong connotations of blinkered thought and failure of imagination. For another, "Standard *Model*" has the air of "rules of thumb," hinting at some crude, ad hoc construction. None of those odors and insinuations should attach to one of the grandest achievements—I would argue, *the* grandest achievement—of human thought and striving. So I refer to it instead as the Core Theory.

The Core delivers on Newton's program of Analysis and Synthesis. In it, we formulate the basic laws as precise statements concerning the properties and interactions of a few building blocks, and derive the behavior of larger bodies from those basics. We build up matter as we know it, in all its richness, from a few ingredients whose properties and interactions we describe fully and accurately.

The Core provides a secure foundation, in physical law, for all applications of physics to chemistry, biology, materials science, engineering in general, astrophysics, and major aspects of cosmology. Its fundamentals have been tested with a precision more than adequate for these applications, and in more extreme conditions.

The Core does, we shall see, embody beautiful ideas. But those ideas are both strange and deeply hidden. It takes some imaginative growth, and some willing patience, to grasp their beauty.

The challenge of achieving honest understanding, as opposed to crude and/or wishful thinking, is eternal. One of the few stories attaching to Euclid—most likely apocryphal—is his reply to his patron and king Ptolemy I, when asked if there was an easier approach to geometry than the *Elements*. Euclid supposedly replied,

Sire, there is no royal road to geometry.

Nevertheless, I hope I've shown you that there are beautiful things in geometry that you *can* glimpse, through imagery and intuition, without lengthy study.

Similarly here, I will present images and explanations that will allow you to glimpse some beautiful aspects of the Core. Not coincidentally, they're the most central aspects!

The experiments that, historically, served to establish the ideas of the Core pieced together clues from the behavior of a large, bewildering cast of unstable particles, mainly discovered in experiments at high-energy particle accelerators. In conventional accounts of the Core, many complications arise from its surface appearance, embodied in a world of "elementary particles" that turn out not to be so elementary. Amid that complexity, the underlying ideas can get obscured. Fortunately, the Core's core ideas are simpler than the evidence that establishes them. It is important that the evidence exists, of course. But it serves our meditation to focus primarily on the ideas, rather than on their proof.

After those generalities, let me describe the content of this chapter. For ease of digestion, it is served up in four courses.

In the first part we'll explore, through images and metaphors, what I consider the soul of the Core. The central concepts of *property space* and *local symmetry* are very well suited to such treatment. Which is to say, they are beautiful concepts.

With that, at the level of Platonic Ideals, our work is basically done. The remaining parts supply the sort of connections that our Question calls for, i.e.,

$$\text{Ideal} \leftrightarrow \text{Real}$$

In the second part, we'll discuss the strong force in some depth, and in the third part the weak force, more selectively. The full story of the weak force, especially, contains many complications we'll barely touch on. (In the present state of understanding, frankly, they don't seem very beautiful!) In the fourth part I'll very briefly introduce the full cast of characters, and then sum up. At that point, we'll have clear views both of the Core's beauty and of its remaining esthetic deficiencies, setting the scene for our final chapter's adventures.

## PART 1: THE SOUL OF THE CORE

### *Property Spaces*

We humans are, as we've reflected before, intensely visual creatures. A large fraction of our brain is devoted to visual processing, and we're very good at it. We're naturally gifted geometers, adapted to organizing our visual perception in terms of objects moving through space.

So although it is possible to discuss the properties of particles and forces purely in terms of numbers and algebra, without attempting to put the concepts in geometric terms, it is humanly appealing to bring in spatial imagery and geometry. Doing that allows us to repurpose our brains' most powerful modules, and to play with the concepts easily. In other words, it brings out those concepts' beauty.

The central equations of the Core, and the extensions we'll consider in the next chapter, are well adapted to spatial imagery. We must be ready, however, to be flexible, and to make a few adjustments to our everyday notions of space geometry. The key new idea is that of a *property space.*

Molière's Monsieur Jourdain was delighted to learn, from his philosophy teacher, that he had been speaking prose:

**M. Jourdain.** So when I say: "Nicole, bring me my slippers and fetch my nightcap," is that prose?

**Philosophy Master.** Most clearly.

**M. Jourdain.** Well, what do you know about that! These forty years now I've been speaking in prose without knowing it!

In the same way, you have been perceiving extra dimensions, fields, and property spaces* every day, for years, very likely without knowing it. Any time you look at a color photograph, your brain is digesting a three-dimensional (color) property space atop ordinary space. When you watch a color movie or a television program, or interact with a computer screen, you are processing a three-dimensional property space defined over space-time.

Let me explain that audacious, yet manifestly valid, claim.

To be concrete, let's consider the example of a computer screen. How can we represent the information it presents? Or, in practical terms: If we're programming a computer, how do we tell the computer what it must do to bring our viewing screen to life?

We can represent the different picture elements, or pixels, by their horizontal and vertical positions. This requires two numbers $x$, $y$. For each pixel, to support a general color perception, we must specify—as Maxwell taught us!—the intensities of three color sources. These sources are usually chosen to be forms of red, green, and blue, and their intensities are denoted $R$, $G$, $B$. So to tell the computer precisely what it needs to output, at a given time $t$, at each position on its screen, we must specify the six numbers $t$, $x$, $y$, $R$, $G$, $B$. Two of those numbers ($x$, $y$) give the spatial position, as we've said, and the three numbers $t$, $x$, $y$ give the position in space-time. The remaining three numbers describe the color. Simply regarded as numbers, they look very much like the first three! And so it is logical, and proves immensely fruitful, to declare that they define positions in a new space, a *property space*, that sits on top of space-time.

Here are two images—abstract and tangible, respectively—that illus-

* Those three concepts are so closely related as to be basically interchangeable, like "that" and "which." For more, see *Dimension*, *Field*, and *Property Space* in "Terms of Art."

FIGURE 32. THE CONCEPT OF EXTRA DIMENSIONS, PICTURED ABSTRACTLY: OVER EACH POINT IN ORDINARY SPACE, THERE IS AN ADDITIONAL SPACE EMBODYING "EXTRA DIMENSIONS." HERE THE EXTRA DIMENSIONS ARE LITTLE SPHERES.

trate the property space concept (figure 32 and plates II and JJ). In the first figure we picture a simple property space geometrically. Attached to each point in ordinary space, there floats an additional space. Here, the abstract additional space has the shape of a sphere. Our color property space, described above, is most naturally represented as a three-dimensional cube, because the possible intensities each range, as a fraction of their maximum values, from zero to one. That is shown at the top of plates II and JJ. The bottom represents the space you sample when you look at a computer screen (as we've just discussed). As you can see, it is a concrete, colorful embodiment of figure 32!

The color assigned to pixels is described by a position within a three-dimensional *R, G, B* property space, as described earlier. In plate KK, we play on the color space of color and exhibit some of its flexibility and fertility. A normal photographic image is shown on the bottom. We can generate slices of the same raw material by projecting on lower-dimensional subspaces of the property space. On the upper left, we project on green

(*G*) alone, thus reducing the color property space to one dimension. On the upper right, we project on green and red, omitting blue, thus reducing the color property space to two dimensions.

There are uncanny parallels between these different-dimensioned property spaces and the infrastructure of our Core Theories. That fact, which I'll now explain, is what the plate's labels "electromagnetic," "weak," and "strong" allude to.

Electrodynamics, in the language of quantum theory, describes how photons respond to the distribution of electric charge in space and time. In other words, photons sense, and react to, the positions and velocities of electrically charged particles. So photons "see," at each point of space-time, a single number, indicating the amount of electric charge at that point. Photons see, that is, in a one-dimensional property space.

As we'll discuss in detail shortly, the strong force is a kind of electrodynamics on steroids. The equations of our theory of the strong force, quantum chromodynamics (QCD), are similar to the Maxwell equations of electrodynamics, but are based on a three-dimensional strong property space. And in QCD we have not merely one photon, but eight photon-like particles, the gluons, which respond in various ways to what's going on in the strong property space. By an eerie coincidence, the *properties* that the gluons respond to were also christened *colors*, although of course they have nothing directly to do with color in the usual sense. The strong colors, instead, are more akin to electric charge. But we're getting ahead of ourselves. . . .

## YIN AND YANG, FOUR TIMES OVER

John Wheeler had a knack for inventing striking phrases to describe physical ideas. "Black hole" is a memorable Wheelerism, as is "Mass Without Mass," which we'll make use of later. Wheeler had a poetic way of describing the essence of Einstein's theory of gravity, general relativity, which we can build on:

Matter tells space-time how to curve.
Space-time tells matter how to move.

For our later purposes, it will be important to spell out—and then correct!—the thought that space-time tells matter how to move. We'll first do some spelling out of "tells," and then some tailoring of "matter" and "space-time."

*How* does space-time instruct matter to move, exactly? Its instruction, according to general relativity, is very simple: *Keep going as straight as you can!*

On a curved surface, there is a notion of straightest possible paths, or *geodesics*. Geodesics, like straight lines in ordinary Euclidean geometry, have the property that there is no shorter path connecting two of their points. The same mathematical concepts (curvature and geodesics) apply not only to surfaces—which can, after all, be thought of as two-dimensional spaces in their own right—but to space as a whole, and even to space-time. And Einstein's genius, in general relativity, was to cast gravity in the form Wheeler poeticized: gravitational "falling" or "orbiting" is just matter doing its best to move straight (i.e., traveling along geodesics) in a curved space-time.

Wheeler's description is wonderfully suggestive, but it is oversimplified. After all, gravity isn't the only force in town! To make the poetry accurate, and bring out its full potential, we need to introduce some refinements.

## *Mantra of Geometry*

In Wheeler's poem, "matter" is a little *too* poetic. Matter can have several properties (for example, electric charge), but the curvature of space-time responds only to the total density of energy and momentum. So we should say instead:

Energy-momentum tells space-time how to curve.

Also, forces other than gravity influence how matter moves. Those forces will lead to deviations from the straightest possible (geodesic) paths. What we should say is therefore:

Space-time tells energy-momentum what straight is (in space-time).

And so, putting it all together:

Energy-momentum tells space-time how to curve.
Space-time tells energy-momentum what straight is (in space-time).

And now comes the Core Theory of electromagnetism:

Electric charge tells electromagnetic property space how to curve.
Electromagnetic property space tells electric charge what straight is (in electromagnetic property space).

And of the weak force:

Weak charge tells weak property space how to curve.
Weak property space tells weak charge what straight is (in weak property space).

And of the strong force:

Strong charge tells strong property space how to curve.
Strong property space tells strong charge what straight is (in strong property space).

In the full Core Theory, including all four forces, matter has four kinds of properties: energy-momentum, electric charge, weak charge, and strong charge. Particles of matter propagate through a more complex space than Wheeler allowed for, which includes electromagnetic, weak, and strong property spaces atop ordinary space-time. But matter follows, according to the Core, the same yin principle, adapted to this more complex environment:

Keep going as straight as you can!

## *Yin-Yang*

The wonder of the Core is that all four forces sound like recognizable variations of the same theme. It does not seem to me overly fanciful, and it is certainly pretty, to see in the duality

matter || space-time

. . . an instance of the Chinese complementarity

yin || yang

Yin is the yielding principle, associated with earth and water (matter). It "does what comes naturally" (*Oklahoma!*) or "follows the force" (*Star Wars*), following the path of least resistance—the geodesic.

Yang is the animating principle, associated with sky (space-time), light (electromagnetic fluid—see below!), or other driving forces.

The soul of the Core, from this perspective, is yin-yang, four times over.

As a very special feature of this book, we have an original Taiji (yin-yang) figure contributed by a contemporary master of traditional Chinese arts, He Shuifa. It appeared as the frontispiece to this book, and is also represented as plate A.

Taiji has been translated in several ways, of which "Supreme Polarity" may be the most suggestive. Its symbol contains two contrasting elements, yin (dark) and yang (light), and is commonly called a yin-yang figure. Note that these two elements form an inseparable whole, and that each contains, and is contained in, the other.

Our deepest descriptions of physical reality, in quantum theory and in the four Core Theories of forces (gravitation, electromagnetism, strong and weak forces), bring in concepts that call to mind yin and yang. Niels Bohr, an influential founder of quantum theory, saw strong parallels between his concept of complementarity and the unified duality of yin-yang. He designed a coat of arms for himself, in which the yin-yang figures centrally (see figure 42, page 324). Our Core Theories center on the interplay between lightlike space filling fluids (yang) and substances (yin) they both direct and respond to.

## *Mantra of Flow*

A map of the world need not be a globe. We can represent the geometry of a curved surface, such as the surface of Earth, by projecting information about distances onto a flat grid.

More generally, we can represent the geometry of a curved space, or curved space-time, by mapping information about distances onto a flat grid. At each point of the grid, and in each direction emanating from that point, we'll have a number that tells us how far you get by proceeding one step in that direction. In this way, we represent the geometry of our space by assigning a few numbers to each point. This construction defines what is called a *metric field* (or simply *metric*) in mathematics.

In physics, when we consider the geometry of space-time, it is appropriate, in the spirit of Faraday and Maxwell, to speak of a *metric fluid.* This is the concept, in Einstein's general relativity, that replaces the gravitational forces of Newton's theory.

As the term "fluid" suggests, the metric fluid, much like the electromagnetic fluid of Maxwell's theory, takes on a life of its own. For example, it supports self-sufficient disturbances—*gravity waves,* analogous to the electromagnetic waves by which Maxwell accounted for light and Hertz launched radio.

Using these geometry-encoding fluids, we obtain mantras of flow:

Energy-momentum tells the metric fluid how to flow.
The metric fluid tells energy-momentum how to flow.
Electric charge tells the electromagnetic fluid how to flow.
The electromagnetic fluid tells electric charge how to flow.
Weak charge tells the weak fluid how to flow.
The weak fluid tells weak charge how to flow.
Strong charge tells the strong fluid how to flow.
The strong fluid tells strong charge how to flow.

In a sense these mantras of flow are merely rephrasings of the mantras of geometry, but they afford attractive new perspectives:

- In this formulation, yin (matter) and yang (force) appear on an equal footing: each instructs the other. This hints that their apparent duality might resolve into a deeper unity. We shall see, in the next chapter, how that outlandish idea can be realized, through *supersymmetry*.
- This mantra of flow, for electromagnetism, is much closer in spirit to the original ideas of Faraday and Maxwell than our earlier "geometric" mantra. The geometric mantra, by contrast, is closer in spirit to the ideas by which Einstein was led to his theory of gravity, general relativity. This harmony of ideas is a great gift. It is beautiful in itself. And, again foreshadowing our next chapter, it suggests a deeper unity among the forces.
- Most basically: Once geometry—whether of space-time or of property spaces—is encoded as a mathematical fluid, we can easily envisage that the fluid flows and takes on a life of its own.

## AVATARS OF LOCAL SYMMETRY

We have now spelled out, and refined, the second line of Wheeler's poem. We have, in other words, discussed how the forces direct matter, or how the yang directs the yin. To close this circle of ideas, we should discuss the principles that govern the opposite direction of influence.

Concretely, our challenge is this: How do we get equations for the curvatures of space-time and the property spaces? The central guiding principle, *local symmetry,* is as beautiful as it is deep. We introduced that idea earlier, in "Symmetry I," and now we'll briefly review and then build further upon it.

Recall that soon after he developed the special theory of relativity, in 1905, Einstein realized that it could not be reconciled with Newton's theory of gravity. He struggled with this challenge for a full ten years, calling it "years of anxious searching in the dark."

Einstein achieved enlightenment by discovering suitable equations for the curvature of space-time, which completed his new theory of gravity, general relativity. He discovered them by demanding that they embody what he called *general covariance,* which is the space-time version of local symmetry.

To understand the Core's local symmetry more deeply, let us begin by recalling the basic idea of symmetry of equations, which we introduced earlier, in our discussions around Maxwell's equations. We say an equation (or system of equations) has symmetry if there are changes one can make among the quantities that appear in the equation, without changing its content. Demanding symmetry gives us a way of finding equations that are special, because most equations, chosen at random, are *not* symmetric. This is also, speaking subjectively, a way to find particularly *beautiful* equations.

(Some people find this use of the word "symmetry," to describe a property of equations, jarring, because it seems rather distant from the everyday meaning of that word. If you have that difficulty, you might want to keep "invariance" in mind as a supplement or substitute. After some deliberation I decided to stick with "symmetry" because it is deeply embedded in the literature, and not without resonance. Whatever you call it, the big idea remains Change Without Change.)

Conventional—that is, nonlocal, or (the word I'll use) *rigid*—symmetry of physical laws typically involves changing the Universe as a whole, rigidly. Thus, for example, we postulate that the content of the laws of physics will not change if we change the position of everything that appears in them by a common amount—say, moving things a meter, in the same direction, everywhere (and for all times). If you think carefully about that, you'll realize that is a precise (though perhaps peculiar) way of saying that the laws do not recognize a preferred location in space or, more simply, that the laws take the same form everywhere. But if we move the positions of some things more than others, we will change their relative positions. That definitely changes the content of force laws—e.g., Newton's law for gravity, and Coulomb's similar law for electric forces—which depend on relative distances.

Local symmetry brings in transformations that vary in space and time. It is because we can choose the transformations *locally,* without worrying about the Universe as a whole, that we use the word "local" in describing this possibility. Consider, again, the sort of transformation we just discussed in the preceding paragraph: simply moving things over. On the face of it, as we saw, that can only be a symmetry of the laws of physics, if we imagine moving everything by the same amount, in the same direction. If we change relative distances, we change the force laws! However—and this is the yoga of local symmetry—if we have a metric fluid, *and we make appropriate adjustments in the metric fluid at the same time we make the motions,* then we can keep the relative distances, and the force laws, intact!

Anamorphic art, as shown in plate EE, provides a wonderful metaphor—or, I should say, model—for local symmetry. As we've discussed earlier, perspective/projective geometry is the art/science of the Change Without Change one encounters when viewing the same object (no change) from different vantage points (change). In this way, we recognize that many different images can represent the same object. But we can get more complex images, still using the same underlying object, if we allow for the presence of distorting media—curved mirrors, say, or lenses and prisms . . . or, in general, structure that varies from place to place, and bends light. *By allowing for the presence of media, we come to regard a much wider range of images as representing the same object.* Local symmetry is the same idea, but applied to equations rather than objects.

In asking for local symmetry, we make heavy demands on our equations. We are asking that very distorted-looking versions of those equations have the same consequences as the originals. To make that possible, we need to assume that space-time (including any property spaces it supports) is filled with appropriate fluids. Depending on how you want to read the situation, you might say that the fluids are responsible for, or alternatively that they compensate for, the apparent distortions. (They're *responsible for* the apparent distortions, if you read from object to perception; they *compensate for* the apparent distortions, if you read from perception to object!) In any case, we need those space-time–filling fluids if we're going to have local symmetry. And if they're going to be successful, versatile compensators, the fluids have to have very particular properties. In other words, they will have to obey very special equations.

In fact, it was by demanding a local version of special relativity that Einstein got the equations for the metric field that are the core of general relativity! And it is by demanding local versions of rotations in property spaces that C. N. Yang and Robert Mills found the equations that bear their name and govern the weak and strong fluids. Yang and Mills built on the work of Hermann Weyl, who showed that *Maxwell's* equations for the electromagnetic fluid can be derived in that way.

When we pass from the fluids to their associated subatomic particles, or quanta, we realize that the existence of gravitons, photons, weakons, and color gluons—the quanta of the metric, electromagnetic, weak, and strong fluids, respectively—and their properties, are *unavoidable and unique consequences of various local symmetries.* The usual jargon for those local symmetries, in the physics literature, is:

- General covariance, for the local version of special relativity
- $U(1)$ gauge symmetry, for the local version of rotation in electric charge property space

- *SU*(2) gauge symmetry, for the local version of rotation in weak charge property space
- *SU*(3) gauge symmetry, for the local version of rotation in strong charge property space

The historical origin of the term "gauge symmetry" is quite interesting. It is discussed in the endnotes.

We can summarize our discussion memorably, and with justice, as follows:

Gravitons are the avatars of general covariance.
Photons are the avatars of gauge symmetry 1.0.
Weakons are the avatars of gauge symmetry 2.0.
Color gluons are the avatars of gauge symmetry 3.0.

Let us celebrate this extraordinary fruition of

$$\text{Ideal} \leftrightarrow \text{Real}$$

with a fitting image (plate LL). When objects containing symmetric details are photographed with a fish-eye lens, the symmetry of the different details is represented in different ways, depending on position. Such images can convey the spirit of local symmetry, in an appropriate, strangely beautiful visual form.

Finally let us, with figure 33, shift our focus from the results of theories with local symmetry to the process of their creation. It is a three-step process. We must choose the objects we want to depict (the substances), the ways we will allow them to look (the transformations), and the media that will support those transformations (the fluids). This drawing, showing the making of anamorphic art, is an update of plates K and L. Our modern Artisan is a rigorous craftsman, but now we know that his thoughts are more imaginative, his tools are more varied—and his attitude is more playful—than the Artisan that Blake envisaged.

FIGURE 33. THE PROCESS OF CREATING ANAMORPHIC ART.

## WHAT FROM WHERE

When a particle moves in a property space, we would say, in ordinary language, that it has changed into a different kind of particle. A "red" quark, say—i.e., a quark with one unit of red color charge—can change into a "blue" quark. But now we have a different way to view the situation, which goes deeper. From this new perspective, we see that those two particles—the red quark and the blue quark—are really the same entity, occupying different positions! *What* encodes *where.*

Because color charge is what color gluons specifically respond to, color gluons decide what they should do by "looking at" where particles are—or, more generally, what the distribution of wave functions, or fields, looks like—in color property space. For those gluons, what's important is location and location—location in that property space, as well as location in space-time. Conversely, when we monitor the behavior of color gluons, we're receiving messages from color charge space. Property spaces, first introduced as aids to imagination, thereby evolve into tangible elements of reality.

## PART 2: THE STRONG FORCE, CONCRETELY

### *Unveiling Atomic Nuclei*

The key discovery leading to modern, successful atomic models was made by Hans Geiger and Ernest Marsden in 1911. Working in Rutherford's laboratory and following his suggestion, Geiger and Marsden studied the deflection of alpha particles, emitted in the radioactive decay of radium, by a thin layer of gold foil. They saw some big ones. Rutherford said of this episode:

> It was quite the most incredible event that has ever happened to me in my life. It was almost as incredible as if you fired a fifteen-inch [artillery] shell at a piece of tissue paper and it came back and hit you. On consideration, I realized that this scattering backward must be the result of a single collision, and when I made the calculations I saw that it was impossible to get anything of that order of magnitude unless you took a system in which the greater part of the mass of the atom was concentrated in a minute nucleus. It was then that I had the idea of an atom with a minute massive center, carrying a charge. . . .

Rutherford proposed a definite, remarkably simple model that explains the observations. He proposed that within each atom there is a tiny nucleus containing all of its positive charge and almost all of its mass. That could account for the rare but powerful backscatter—the nucleus doesn't want to move (because it's heavy) and it's capable of pushing back (because it has concentrated charge). Rutherford put this model to work, and validated it, by accounting *quantitatively* for the large-angle deflections. The remainder of the atom, according to Rutherford, consists of much lighter, negatively charged electrons, somehow dispersed over a much larger volume.

This was an epochal result. It showed that understanding the atomic structure of matter could be divided conveniently into two tasks. One task—what we now call atomic physics—is to consider a heavy, positively charged nucleus as given, and then to determine how electrons are bound to it. We discussed that domain of quantum beauty earlier.

The second task—what we now call nuclear physics—is to understand what those inner cores of atoms are made of, and the laws that govern them.

It quickly became clear that electric forces alone cannot account for nuclear physics. Indeed, a purely electric model could not deal with the concentrated positive charge within atomic nuclei. If not overbalanced by another, more powerful force, electric repulsion blows nuclei apart. What about gravity? Acting on such meager masses, it is completely negligible. New forces, unknown to classical physics, had to be at work.

Nuclear physics posed two challenges, one existential, the other dynamical. The existential challenge is to identify the ingredients of nuclei, and the dynamical challenge is to understand the forces that those ingredients exert on one another. The census of ingredients was settled in a few years, and rather simply. One ingredient was more or less obvious. The hydrogen nucleus is stable, (apparently) indivisible, and carries one (positive) unit of electric charge. It is the lightest of all nuclei, and other light nuclei have masses that are close to whole-number multiples of its mass. So that *proton*—so named by Rutherford—is one ingredient.

A second ingredient was discovered by James Chadwick in 1932. The neutron is an electrically neutral particle only slightly heavier than a proton. Its discovery enabled a simple yet useful picture of what atomic nuclei are: namely, that nuclei are collections of protons and neutrons, bound together. With that picture, many observed facts fell into place. For example, the nuclei of different chemical elements differ only in the number of protons they contain, because that number determines the electric charge of the nucleus, which controls its interaction with the atom's surrounding electrons, which in turn govern its chemistry. Different numbers of nuclear protons give us atoms of different chemical elements. With neutrons as a second player, we solve the riddle of isotopes. Atoms containing isotopic nuclei have the same chemical properties, but differ in weight. Their nuclei contain the same number of protons, but different numbers of neutrons. Thus the simple proton + neutron model of atomic nuclei explains both the census of chemical elements and the existence of isotopes.

The next step, people thought, would be to figure out what forces act between protons and neutrons, holding them together. As we mentioned, new forces are required, because electromagnetism wants to blow nuclei apart, and gravity is so feeble as to be negligible.

Experiments probing nuclear forces soon led in unanticipated directions. Almost all the experiments followed the strategy of the original Geiger-Marsden experiment. To investigate, say, the force between protons, one shoots beams of protons at other protons (i.e., a hydrogen target) and keeps track of what comes out. By observing deflections through different angles, one can try to infer the responsible force. Using beams of protons with different energy, and with the protons spinning in different directions, enriches the analysis. Experiments of this kind soon revealed that the forces between protons and neutrons do not obey a simple equation. They depend not only on distance, but also on velocity and spin, in complicated ways.

More profoundly, experiments soon undermined the hope—the hope of our Question—that protons and neutrons are simple particles, or that any beautiful "force," in the traditional sense of that word, could do justice to the reality of their interaction. For when high-energy protons bash into other protons, the typical result is not merely deflection of the two impacting particles. Instead, a torrent of particles emerges!

In fact, the experiments aimed at revealing a simple force instead revealed a new and unexpected world of particles. $\pi$, $\rho$, $K$, $\eta$, $\rho$, $\omega$, $K^*$, $\varphi$ mesons and $\Lambda$, $\Sigma$, $\Xi$, $\Delta$, $\Omega$, $\Sigma^*$, $\Xi^*$, $\Omega$ baryons are among the lightest and most accessible. (There are dozens more.) These particles are without exception highly unstable, living no more than a microsecond (and in most cases, much less). Their existence and properties must be inferred through the study of their decay products, in detectors at high-energy particle accelerators like those at Brookhaven National Laboratory, Fermilab, and CERN. These new particles are called, collectively, *hadrons*.

Like the classification of butterflies or equine paleontology, the census of the hadronic zoo, and the characteristics of its specimens—masses, spins, lifetimes, decay patterns—are fascinating to connoisseurs. To advance our quest for beauty in the fundamentals, however, we must pass to broader concerns. For future use, let me briefly summarize the two most important lessons the zoo has to offer.

*Hadrons comprise two kingdoms, baryons and mesons.* Protons and neutrons are the prototype of baryons. All baryons share several properties. They all feel strong short-range forces in one another's presence, or in the presence of mesons, and (for experts) they are all fermions. Mesons also share common properties. They all feel strong short-range forces in one another's presence, or in the presence of baryons, and (for experts) they are all bosons.

*Protons and neutrons are neither simple nor fundamental.* It's a useful step to analyze atomic nuclei into protons and neutrons, but protons and neutrons are not simple or basic—their interactions are complicated, and they are just two members among a much larger family of resembling particles. To put them in proper perspective, and complete the analysis of matter, a new, larger vision is required.

## *The Quark Model*

The quark model was invented by Murray Gell-Mann and George Zweig in a brilliant display of imagination and pattern recognition.

According to the quark model, baryons are bound states of three more fundamental entities: three kinds, or "flavors," of quarks, up $u$, down $d$, and strange $s$. (For present purposes, I will defer consideration of the much heavier, highly unstable quarks $c$, $b$, $t$.)

How do just three flavors of quarks—$u$, $d$, $s$—generate hundreds of different baryons? The point is that a given trio of quarks, say $u$, $u$, $d$, can exist in many different states of motion, analogous to Bohr's quantized orbits for electrons in atoms, or the stationary states of figure 26 (page 187) in "Quantum Beauty I." These discretely different states have different energies, and therefore—using $m = E/c^2$—different masses. Thus they appear, operationally, as different particles! In this way, we find that many different particles reflect the same underlying material structure, captured in different states of internal motion.

Similarly, the quark model postulates that mesons are bound states of a quark and an antiquark. A given quark-antiquark pair, say $u\bar{d}$, in various states of motion, generates many different mesons.

The quark model also gives a plausible explanation for the complexity of hadronic forces. Even if individual quarks have simple interactions, when bound states containing three quarks, or a quark and an antiquark, come together, there is ample opportunity for cross-talk and cancellations. Indeed, it is for such reasons that ordinary chemistry, based on interactions of atoms, emerges complex and exuberant from underlying forces that, between individual electrons, are extremely simple.

The quark model was a major step in organizing the hadron zoo. It provides a picture of hadrons analogous, in its explanatory power, to Bohr's model of atoms. But the quark model also shares the limitations of Bohr's model. While correct in spirit, and historically important, the quark model is logically incomplete, and only semimathematical. It also faced a big problem, as we'll now discuss.

The quark model gave a successful *descriptive* account of many features of protons, neutrons, and their kindred hadrons. But it postulated some very odd properties for the quarks. Perhaps the weirdest of those properties is confinement, as humorously depicted in the cartoon shown as plate MM, taken from the poster that commemorated my Nobel Prize. Quarks are supposed to be the building blocks of protons, but despite very great efforts, no individual particles with the properties of quarks (such as carrying a fractional ⅔ or –⅓ of the proton's electric charge) were ever detected. So quarks in groups of three could build up protons, wherein the forces between them appeared modest. But for some reason they can never escape—they are confined.

In order to account for this behavior, we appear to need some spring-like or rubber-band-like force between quarks that pulls more powerfully as the mediating spring or rubber band gets stretched over a growing distance. Springs and rubber bands, of course, are complicated physical objects themselves, so it's unseemly to postulate them in a fundamental theory. Doing so begs the question: What's the spring made of?

One generally expects fundamental forces to grow weaker with distance, as gravitational and electromagnetic forces do, so confinement posed a major puzzle. Many physicists couldn't bring themselves to take quarks seriously because of it.

## BREAKING THROUGH: QUANTUM CHROMODYNAMICS

Maxwell's equations for electrodynamics, Newton's (and then Einstein's) equations for gravity, and Schrödinger's (and then Dirac's) equations for atomic physics set high standards for beauty, precision, and accuracy. Neither the complicated equations (tables, really) summarizing nuclear forces, nor the rough ideas of the quark model, came close to living up to those standards.

Yet beautiful, precise, accurate equations for the strong force were out there. They went unused for many years before we were able to exploit them. They are equations that build on Maxwell's equations, and fulfill the visions we sketched in part one of this chapter.

Almost twenty years elapsed between Yang and Mills's formulation of the equations and the emergence of quantum chromodynamics, as their embodiment in reality. This history is a stunning example of

Ideal → Real

In the domain of the strong interaction, there can be no doubt that the answer to our Question,

*Does the world embody beautiful ideas?*

. . . is simply:

*Yes, it does.*

## *Maxwell on Steroids*

Quantum chromodynamics (QCD) uses ideas and equations that are a grand generalization of Maxwell's equations for electromagnetism, expanded to bring in even more symmetry. I like to say that QCD is like QED (quantum electrodynamics) on steroids.

QED features one kind of charge—electric charge. Electric charge can come in positive units, as in protons, or negative units, as in electrons, but in either case we quantify it by using a single number (positive or negative). QCD, by contrast, contains three kinds of charge. They are, for no good reason, called colors: let's say red, green, and blue, to be definite.

QED features one force-mediating particle, the photon, that responds to electric charge. QCD, by contrast, contains *eight* force-mediating particles, called color gluons. Two of them, like the photon, respond to color charge. (Why not three? See the next paragraph.) The other six mediate *transformations* of one color into another. Thus there is a gluon that changes a unit of red charge into a unit of green charge, another that changes a unit of green charge into a unit of blue charge, and so forth.

The *bleaching rule* is a beautiful feature of QCD that is physically important, fairly easy to state, and very easy to demonstrate mathematically, but difficult to motivate intuitively. (At least, I haven't found a good way.) According to the bleaching rule, the net effect of having a unit of red charge, a unit of green charge, and a unit of blue charge in the same place is nothing: They cancel. (For experts: Here I'm assuming they're in an antisymmetric configuration.) This is vaguely reminiscent of how the three spectral colors red, green, blue can add up to neutral white—hence "bleaching"—though of course the physics is completely different. It is due to the bleaching rule, which renders one combination of charges impotent, that we get only two, and not three, kinds of color-responsive gluons.

Each quark carries one unit of color charge. The color of a quark is an independent property, which we must specify in addition to properties like electric charge or mass, and it is no less important. Unlike electric charge or mass, however, *the color of a quark is not a single number, but a triple. More precisely, it encodes position in a three-dimensional property space.* The existence of these new kinds of charge is the heart of QCD. It is a fact so central, so beautiful, and so important for later developments that we owe ourselves a review of its grounding in reality.

## *The Strange Reality of Quarks and Gluons*

Quarks were first actually "seen" in experiments performed by Jerome Friedman, Henry Kendall, and Richard Taylor at the Stanford Linear Accelerator in the late 1960s. In essence, they took snapshots of the interiors of protons. By using (virtual) photons of very high energy, they were able to resolve very small distances and times.

Those snapshots were very revealing! Three revelations, in particular, stand out in retrospect:

*Protons contain quarks:* Because the snapshots were made using photons, they tracked the distribution of electric charge inside the proton. They showed that electric charge is concentrated in very small, pointlike structures, not diffuse. The startling discovery of Rutherford and Geiger-Marsden returns!—but now within proton, rather than atomic, interiors. The quantity of charge in those pointlike structures, and other properties, matched expectations derived from the quark model.

*Inside protons, quarks are nearly free:* Most of the snapshots show three quarks, and nothing else, and the position of each quark is found to be nearly independent of the positions of the others. That indicates that within the proton, the force between quarks is feeble. On the other hand, many other experiments had indicated that quarks never escape from protons as individual particles. So we need a force that is relatively feeble at short distances, but gets powerful at long distances. That central paradox of strong interaction dynamics, which we mentioned earlier, here becomes sharply etched.

*There's a lot more to protons than three quarks:* A few snapshots captured traces of additional quark-antiquark pairs. That's not too surprising: because there's plenty of spare energy inside protons, and quarks have very little mass, to make them is as easy as $m = E/c^2$—with a very small $m$! More profound is what was *not* seen in the snapshots. If you add up all the energy of quark motion, as derived from the observations, you get only about *half* of what you need to account for the proton's total mass. Because photons are blind to electrically neutral particles, the obvious interpretation is that there is an important electrically neutral component to protons, in addition to the electrically charged quarks. This microcosmic "dark matter" problem was the first indication that there's a lot more to protons than three quarks. As we'll see shortly, color gluons provide the missing ingredient.

Later experiments, at higher energies, revealed a different, vividly tangible aspect to the reality of quarks and gluons. To see it, please now examine plate NN.

To describe what emerges from ultra-high-energy collisions, whether of electrons with positrons (as in plate NN) or of protons with protons (as at CERN's Large Hadron Collider), it is easiest to begin by pretending we've produced quarks, antiquarks, and gluons—even though those particles don't "exist" (they're confined)—and work from there to what we actually observe. (This will be crystal clear in a moment.)

The point is that a fast-moving quark, antiquark, or gluon materializes, in the laboratory, as a *jet* of hadrons, all moving in nearly the same direction. The total energy and momentum of the particles in a jet add up to the original energy of the quark, antiquark, or gluon that triggered it, because energy and momentum are conserved. So if we are willing to squint and to "go with the flow"—keeping track of its energy and momentum, while forgetting that it is divided up among many hadrons—we get to see the underlying fundamental particles. This is very helpful for interpreting the results, because we can do a *much* better job predicting the production of quarks, antiquarks, and gluons, which obey simple equations, than we can for hadrons, which are complicated.

If you go to a conference on high-energy physics these days, you will hear experimentalists speak routinely of producing nonexistent particles (quarks, antiquarks, or gluons) and measuring their properties. It's become the standard language of the field. What they mean, of course, is that they've observed the corresponding jets. In this way, the mathematically Ideal becomes tangibly Real.

## *Self-Sticky Glue*

Light passes freely through light. Were that not true, the visual messages we receive from the world would be scrambled by scattering, and much more complicated to interpret. In QED, that basic fact makes good sense: photons respond to electric charge, but photons themselves are electrically neutral.

The most dramatic *qualitative* difference between QCD and QED is that unlike photons, *color gluons interact with one another*. Consider, for instance, the color gluon that changes a unit of red charge into a unit of blue charge. Let's call it $\bar{R}B$. When such a gluon is absorbed, the total red charge of the absorber goes down by one unit, and the total blue charge of the absorber goes up by one unit. Because those charges are conserved, we conclude that $\bar{R}B$, regarded as a particle, carries minus one unit of red charge, and plus one unit of blue charge. It is *not* neutral. Other gluons that change or respond to red or blue charge will interact with $\bar{R}B$. And so it is for all the others: the eight color gluons form a complex of mutually interacting particles.

When we pass from those quanta to the fields they build up, the interactions have a dramatic effect. The gluon lines of force attract one another! The fields, instead of spreading influence evenly through space, concentrate into tubes (see plate OO)—and compare with figure 20 (page 122).

The self-stickiness of color glue is the key to confining quarks. Gluon flux tubes are emergent "rubber bands," ready for the job of enforcing confinement! As you increase the separation between a color charge and its opposite, they must be connected by a longer flux tube. It costs a finite amount of energy, per unit of extra separation, to supply the new fields. As a result there's a resisting force. And that force doesn't get any smaller as you pull farther away. Because it would take an infinite amount of energy to liberate the color charge completely, that can't happen, and it's confined.

Gluon self-stickiness is also an appealing way to introduce, and visualize, the concept of asymptotic freedom. Because self-stickiness focuses the color fields far away from a quark, they exert stronger forces than they otherwise would, like an army that concentrates its strength. Conversely, we can start with *weaker* forces than we imagined, at the source, to explain a given force far away. That is the essence of asymptotic freedom: *a feeble force at short distance can bring forth a strong force at large distance.* It is exactly the sort of behavior we need, as you may recall, to explain the Friedman-Kendall-Taylor proton snapshots.

We can also interpret asymptotic freedom in terms of how we *probe* the force. High-energy probes are sensitive to the behavior of the force at short distances. Near freedom at *short distances* maps onto feebleness of interactions, and simplicity of behavior, at *high energy.*

The emergent simplicity of QCD at high energy is a splendid gift from Nature to physicists seeking fundamental understanding. It brings, in fact, a shower of gifts.

## GIFTS OF UNDERSTANDING

*The early Universe is comprehensible.* Very early in its history, close to the Big Bang, the Universe was a high-energy place indeed. Thanks to asymptotic freedom, we can model its content with confidence.

*We can read the message of high-energy collisions.* Because the dominant force gets simpler at high energies, we can calculate its implications precisely. That allows us to interpret the output of ultraviolent collisions between protons cleanly, and to scrutinize them for new effects. For example, the Large Hadron Collider became a tool for discovery of the Higgs boson, as described later in this chapter. In the near future, we'll find out whether promising, ambitious theories for the unification of forces describe reality, as we'll discuss in the next chapter.

*Different forces come to seem less different.* The striking *mathematical* resemblance between the equations of QCD and QED becomes a close *physical* resemblance between their consequences when we consider behavior at very high energy (or very short distances). The strong force of QCD becomes both simpler and less powerful, until quarks behave very much like electrons, and gluons behave very much like photons. The effect of the steroids wears off, you might say. With both mathematical and physical resemblances exposed, the possibility of a unified theory looms large. The symmetry-based mathematics of QCD opens the door to unification, and asymptotic freedom pushes us through. As we follow this idea out, bringing in the weak force and gravity as well, we'll find that it explains several otherwise mysterious "coincidences." We'll explore unification of all the forces, as the current frontier of our Question, in our next chapter.

Thank you, Mother Nature, for these gifts!

## *Leverage, and the Trembling of the Veil*

It is very difficult to construct theories of particles and interactions that are consistent both with the principles of quantum mechanics and with the principles of special relativity. That's a good thing! It means that if we believe in quantum mechanics and special relativity, we get a lot of leverage. The available theories are rigid—they can't be changed much without becoming inconsistent. That makes them powerful. And there aren't many of them. That makes it possible to survey them.

Given that leverage, the right kind of fact can generate enormous consequences.

Asymptotic freedom, it turns out, is just such a fact. The experimental discovery that the strong force, acting between close-by quarks, is not so strong after all was very difficult to reconcile with other things we know. In most theories that are consistent with quantum mechanics and special relativity, like repels like, so focusing of forces does not happen. The op-

FIGURE 34. "GIVE ME A PLACE TO STAND AND I WILL MOVE THE EARTH." —ARCHIMEDES

posite behavior—forces getting stronger at short distances—is much more common. So when David Gross and I, and independently David Politzer, discovered that it *is* possible, it was the sort of moment the Kabbala describes as the "trembling of the Veil of the Temple," when the shroud that keeps the divine world hidden from our vision stirs.

Gross and I went on, based on a few other facts—most important, the fact that we can bind three quarks to make a baryon, where the color charges cancel (the bleaching rule!)—to single out the theory we now call quantum chromodynamics (based on local symmetry and a three-dimensional property space) as the only possible theory of the strong force. Even now, when I reread our declaration:

> Finally let us recall that the proposed theories appear to be uniquely singled out by nature, if one takes both the SLAC results and the renormalization-group approach to quantum field theory at face value.

I relive the mixture of exhilaration and anxiety that I felt at the time.

QCD itself was, historically, the first gift of asymptotic freedom.

## *A New Kind of Physics*

For several decades now, it has been convenient to divide physics into two branches: theory and experiment. Both, in principle, seek a better understanding of the physical world, but they use different tools.

In recent years, marked by explosive growth in computer power, a third branch has budded and thrives. We might call it "numerical experimentation," or "simulation," or simply "solving hard equations." It has elements of both theory and experiment, but is distinctly different from both. This new kind of physics has been especially important and successful in QCD.

QCD gives us perfectly definite equations that we can teach to computers. Once we've done that, we've got access to extremely fast, tireless, honest, and relentlessly accurate assistants who like nothing better than to calculate. Let's take a quick look at two highlights of what's been achieved by following this approach. They provide a brilliant conclusion to our consideration of the strong force.

First, let us return to the question from which we began: What are atomic nuclei? That question's essence, we've seen, lies in its simplest case: What is a proton? Knowing, as we do, the governing equations, we can compute a detailed portrait. In this way, we discover that our innermost substance is a thing of beauty (plate PP) and subtlety (plate QQ).

Finally, as a fitting climax to our discussion of QCD, let us document the origin of (most) mass. The modest-looking figure 35 summarizes a monumental scientific achievement and, for our Question, marks a milestone.

On the horizontal axis, you see the names of mesons and baryons. Again, although there's a lot to be said about those particles, and the details are fascinating to specialists, for present purposes it's enough to note

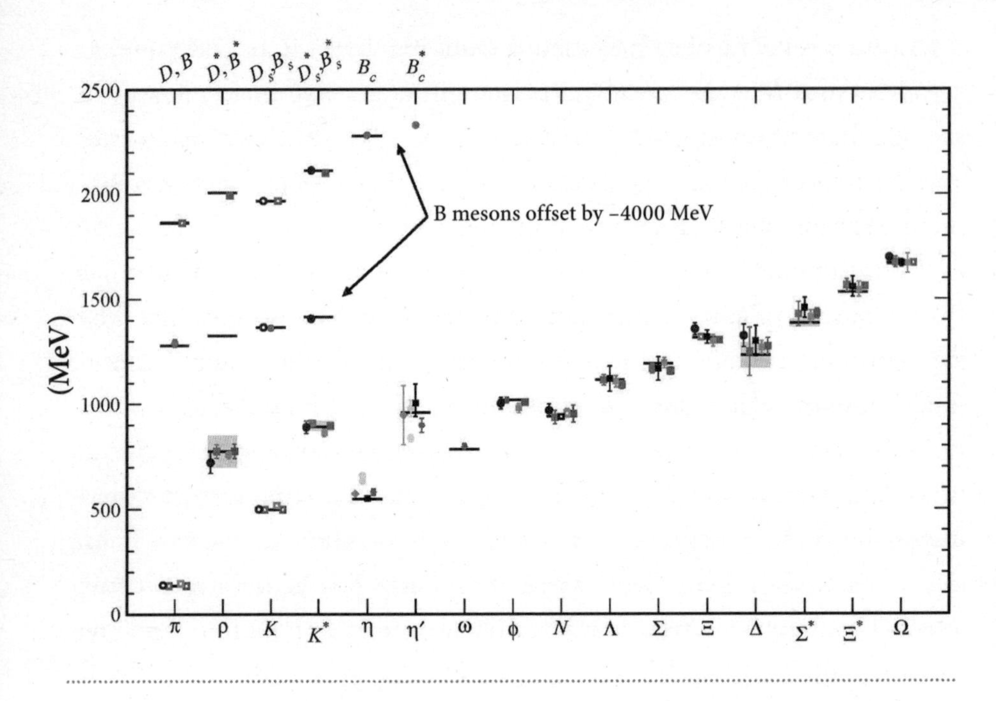


FIGURE 35. SUCCESSFUL CALCULATION OF HADRON MASSES, BASED ON QCD: THE ORIGIN OF (MOST) MASS.

that there are lots of hadrons, and they have different names (consisting of different Greek and Latin letters, and an occasional asterisk or prime) and different masses. Above each name you find a horizontal line segment indicating the *experimentally measured* value of that particle's mass. (Some of the particles are very short-lived, and this smears their mass over a significant range. In such cases, for example ρ, you will see a gray rectangle surrounding the central segment.) In the neighborhood of each segment, an array of several shaded dots, with vertical lines through them, indicates the *calculated* values of the particle's mass, extracted directly from the equations of QCD in work by different research teams. The vertical lines reflect a range of uncertainty in the calculations, introduced by limitations on computer time and other factors. The calculations, I should mention, are extremely demanding. They employ very clever algorithms, and they run on some of the most powerful computer systems in the world, for long periods of time.

All the results for the "main sequences" of mesons $\pi$, $\rho$, $K$, $K^*$, $\eta$, $\eta'$, $\omega$, $\varphi$ and baryons $N$, $\Lambda$, $\Sigma$, $\Xi$, $\Delta$, $\Sigma^*$, $\Xi^*$, $\Omega$ are outputs, given three inputs: the average mass of the up and down quarks, the mass of the strange quark, and the unit of color charge. As you can see, there is outstanding agreement between measurements and calculations.

I want to emphasize that much more comes out of these calculations than is put into them. The equations of QCD are tightly constrained by symmetry, and they contain few opportunities for adjustment. To pinpoint these calculations we need to specify only three inputs: the average mass of the up and down quarks, the mass of the strange quark, and the unit of color charge (an overall measure of interaction strength). So if anything fails to fit, there's nowhere to hide! We'd better find, emerging from our calculations, all the hadrons that are observed, with the masses they are observed to have. And, most important, we'd better not find anything that isn't observed—in particular, no isolated quarks or gluons!

From this trial by ordeal, the theory emerges triumphant.

Among the masses computed is the mass of a particle called *N*. That's not just another mass, because "*N*" stands for *nucleon*, meaning proton or neutron. (The difference between proton and neutron masses is too small to be visible on this scale.) That mass, we find, depends very little on the quark masses which, in the relevant cases, are relatively tiny.

Therefore: *almost all of the nucleon's mass, and therefore almost all of the mass of ordinary matter in the Universe, arises from pure energy*, according to:

$$m = E/c^2$$

The nucleon's mass results from the kinetic energy of confined quarks, and from the field energy of the gluon fields that confine them. We get Mass Without Mass, emerging directly from the purely conceptual, symmetry-rooted equations of QCD.

Does the world embody beautiful ideas? You bet it does. And so do you.

## PART 3: THE WEAK FORCE

Quantum chromodynamics (QCD) governs the basic dynamics that build protons, neutrons, and the other hadrons out of quarks and gluons, and the forces that bind together nuclei—the so-called strong force. Quantum electrodynamics (QED) runs the worlds of light, atoms, and chemistry, as we've discussed.

Neither of those two great theories, however, incorporates processes whereby protons transform into neutrons, and vice versa. Yet such transformations occur. How can we account for them? To explain these events, physicists had to define one more force in addition to those of gravity, electromagnetism, and the strong force.

This new addition, this fourth force, is called the weak force. The weak force completes our current picture of physics: the Core.

Life on Earth is powered by a tiny fraction of the energy released from the Sun, captured as sunlight. The Sun derives its power by burning protons into neutrons, releasing energy. The weak force, in this very specific sense, makes life possible.

### *Weak Force Basics*

A full description of the weak force would require introducing two large casts of characters—a bewildering host of particles, and a long honor roll of discoverers—and would bring in many details that are tangential to our main themes. Here I will confine myself to a brief, simplified description of two highlights, selected for their fundamental interest and for later use. We're aiming toward the summarizing images in plates RR and SS, and TT and UU, which provide the platform for our reach toward ultimate unification. You may want to consult those images, as we go along, for orientation.

*Conversion of Quarks:* Because protons and neutrons are, as we've discussed, complex composites of more basic quarks and gluons, we should track proton ↔ neutron conversions to their more basic source. The deep structure underlying those conversions is the quark process:

$$d \rightarrow u + e + \bar{\nu}$$

Because neutrons are based on *udd* quark triads, while protons are based on *uud*, the quark transformation $d \rightarrow u$ enables a neutron to transform into a proton. That transformation is accompanied by emission of an electron $e$ and an antineutrino $\bar{\nu}$. So our basic, quark-level interaction is realized, at the level of hadrons, as

$$n \rightarrow p + e + \bar{\nu}$$

This slow decay (lifetime fifteen minutes) is the fate of isolated neutrons. (They are stabilized only when bound inside nuclei.)

Basic rules of quantum mechanics tell us we will also get valid processes if we change any particle into its antiparticle and take it to the other side of a reaction, or if we reverse the direction of the reaction's arrow. By applying those rules to our $d \rightarrow u + e + \bar{\nu}$, we find possibilities like

$$d + \bar{u} \rightarrow e + \bar{\nu}$$
$$d + \bar{e} + \nu \leftarrow u$$

. . . plus a host of others. These give rise to many different forms of nuclear decay (radioactivities), destabilize other hadrons, and drive many transformations in cosmology and astrophysics (including the synthesis of all the chemical elements, starting from a primordial mix of protons and neutrons). As an example of the possibilities: The first of these processes, $d + \bar{u} \rightarrow e + \bar{\nu}$, leads directly to decay of a $\pi^-$ meson (which is based on a quark-antiquark pair $d\bar{u}$) into an electron and an antineutrino.

*Handedness and Parity Violation:* A very profound aspect of the weak force, called parity violation, was discovered theoretically by T. D. Lee and

C. N. Yang in 1956. To describe it, we must introduce the concept of particle *handedness*. It applies to particles that are moving and also spinning.

If an object rotates around an axis, we can assign a direction to the axis, as follows: Imagine our spinning object as an ice skater. If her rotation brings her right hand toward her abdomen, we choose the direction from her toes to her head; if it brings her right hand toward her back, we choose the direction from head to toes.

The particles we're interested in have a small intrinsic spin. They're forever twirling, like tireless ice skaters. So we can apply our logic to them, and derive a direction associated with their spin. If our particle is moving in that direction, we say that particle is right-handed. If it is moving in the opposite direction, we say it is left-handed. The handedness of a particle, in other words, orients its spin with respect to its velocity.

What Lee and Yang proposed is that left-handed quarks, electrons, and neutrinos (and also muons and $\tau$ leptons) participate in the weak interaction, as do right-handed antiquarks, antielectrons (i.e., positrons), and antineutrinos (and also antimuons and anti-$\tau$ leptons), but that particles with the opposite handedness do not. Experiments bore out their proposal.

## *Another Color Anamorph: From ?? to !*

The transformative aspect of the weak force, and several other of its more specific aspects, suggested to Sheldon Glashow, and to Abdus Salam and John Ward, that it might be possible to describe that force, too, as an embodiment of local symmetry.

We can see how this might work, using the ideas and imagery we've developed. We want our basic weak process—let's take it in the form $u + e \rightarrow d + \nu$, to be definite—to occur through motions in a property space. The property space should have (at least) two dimensions so that we can have *u* and *d* be the same entity at different positions, and likewise *e* and *ν*. Then we'll be able to view our whole process, which on the face of it involves changes in the identity of particles—their *what*—as a change in location—their *where*. This is "what from where" in action!

The theory built on local symmetry goes further, by providing the fluid that drives motion in property space. The most elementary action of that fluid is what happens when its smallest units, or quanta, are created and destroyed. So our process, at the most basic quantum level, can happen this way:

*u* emits weakon $W^+$, turns into *d; e* absorbs weakon $W^+$, turns into *ν*

Or, alternatively, this way:

*e* emits weakon $W^-$, turns into *ν; u* absorbs weakon $W^-$, turns into *d*

The weakon $W^+$ is usually called a "W plus boson," with the superscript indicating its electric charge. The weakon $W^-$, the "W minus boson," is its antiparticle. When you spell out the local symmetry, you find there is a third, electrically neutral weakon *Z*, the "Z boson."

In proposing this local theory, Glashow, Salam, and Ward were following our Jesuit credo "It is more blessed to ask forgiveness than permission," for they ignored, strategically, another aspect of Yang-Mills theory. The local symmetry of Yang-Mills theory requires that $W^+$, $W^-$, and *Z* have *zero mass*. The corresponding predictions of zero mass for gravitons, photons, and color gluons all correspond to reality, and represent a great triumph for local symmetry. But in the weak force theory, that prediction fails. If the weakons had zero mass, they would have been easily visible in collisions at accelerators, or even in chemical reactions, like photons. Basically, the weak force wouldn't be weak!

In short, for the weak interaction, local symmetry seems to be just a little too good for this world.

. . .

To reconcile Ideal with Real, we need to introduce another idea—and it's a beauty! The new idea is *spontaneous* symmetry breaking, introduced in this context by Robert Brout and François Englert, and independently by Peter Higgs (and also Gerald Guralnik, Carl Hagen, and Tom Kibble). It allows us to have our cake and eat it too. More specifically: we can keep the equations of local symmetry, with their lovely "what from where" account of the weak force, while allowing them to have non-zero masses, as observed. We'll proceed to contemplate their audacious, powerful idea at length, after an obligatory paragraph of cartoon history wrapping up our account of the weak force proper.

It was Steven Weinberg who synthesized the two lines of thought—symmetry, and symmetry breaking—to produce the fully satisfactory theory of the weak force that appears in the modern Core. But it was not at all obvious, at first, that this theory would give correct, or even finite, answers, when quantum fluctuations were taken into account. Gerard 't Hooft and Martinus Veltman demonstrated that it does, and in so doing introduced methods of calculation that made the theory more precise and useful. Freeman Dyson had earlier performed a similar service for QED, where it is much easier (but still difficult).

## HIGGS FLUID, HIGGS FIELD, HIGGS PARTICLE

On a water-covered planet in a galaxy far, far away, fish have evolved to become intelligent—so intelligent that some of them become physicists, and begin to study the ways things move. At first the fish-physicists would derive very complicated laws of motion, because (as *we* know) the motion of bodies through water is complicated. But one day a fish genius—Fish Newton—proposes that the basic laws of motion are much simpler and more beautiful: in fact, they are Newton's Laws of Motion.

She proposes that the observed motions look complicated due to the influence of a material—call it "water"—that fills the world. After a lot of work, the fish manage to confirm Fish Newton's theory by isolating molecules of water.

According to the Higgs mechanism, we are like those fish. We are immersed in a cosmic ocean, which complicates the observed laws of physics.

The equations for particles with zero mass, including the Maxwell equations, the Yang-Mills equations, and Einstein's equations in general relativity, are especially beautiful. As we've discussed, they can support an enormous amount of symmetry—local symmetry. Photons have zero mass, as do the color gluons of quantum chromodynamics and the gravitons of gravity. In order to have beautiful equations, and in order to have uniformity in our description of Nature, we'd like to build the world from zero-mass building blocks.

Unfortunately, several kinds of elementary particles refuse to cooperate with our desires. Specifically, the *W* and *Z* bosons, which mediate the weak interactions, have substantial masses. (That is why the weak interactions are short-ranged, and why they act feebly at low energies.) Their masses are vexing, because in other respects, as we've just reviewed, the *W* and *Z* bosons appear remarkably photon-like.

Is there a solution to this difficulty? Consider that the behavior of photons can be affected by the properties of material they move through. A familiar example is that light slows down when traveling through glass or water. That phenomenon, whereby light becomes more sluggish than usual, is very roughly analogous to light acquiring inertia. Less familiar, but for present purposes more profound, is the behavior of photons inside superconductors. The equations that describe photons in superconductors are *mathematically identical* to the equations for a massive particle. Within a superconductor, photons effectively become particles with non-zero mass.

The essence of the Higgs mechanism is the idea that "empty space"—that is, space devoid of particles and radiation—is in fact filled with a material medium that renders the *W* and *Z* bosons massive. This idea lets us keep the beautiful equations for massless particles, while observing a decent respect for the opinion of reality. We need a material that does, for *W* and *Z* bosons, what superconductors do for photons. Indeed, the hypothetical cosmic medium must produce masses on a much larger scale: the masses of *W* and *Z* in (not) empty space are roughly $10^{16}$ times those of photons in superconductors.

Physicists have been invoking the Higgs mechanism for many years, and by using it have gone from success to success. Many aspects of the interactions of *W* and *Z* bosons, besides their masses, were predicted accurately by using the beautiful equations of massless particles and gauge symmetry with their consequences modified by a space-filling material. In this way, we built up a convincing case for the existence of our own "cosmic ocean." But ultimately that case rested on circumstantial evidence. There was no clear answer to an obvious question: What's it made of?

No known substance could be the cosmic ocean. No combination of the known quarks, leptons, gluons, or other particles has the right properties to make it. Something new is required.

In principle, the cosmic Higgs ocean could have been a composite of several substances, and the substances themselves could be complicated. The literature of theoretical particle physics contains hundreds, if not thousands, of proposals of that kind. But among all the logical possibilities, there is a so-called minimal model—the simplest and most economical. In that minimal model, the cosmic material is made from just one ingredient. Though the terminology in this subject is both confusing and evolving, here when I refer to the "Higgs particle" I will mean the unique new particle that is introduced to complete the minimal model.

We can infer a lot about how the Higgs particle interacts with other forms of matter. After all, because we're embedded in the cosmic ocean, we've been observing the properties of Higgs particles en masse since time immemorial. In fact, all the properties of that particle are predicted uniquely, once its mass is known. For instance, both its spin and its electric charge must be zero, because it's got to look like a quantum of "nothing." Since we knew what we were looking for, it was possible to design an intelligent strategy to search for the Higgs particle. The key process, through which the Higgs particle was discovered, is depicted in figure 36.

The first step is to produce it. The dominant production mechanism is quite remarkable. Ordinary matter couples very feebly to the Higgs particle *H*. (That is why electrons and protons can be much lighter than

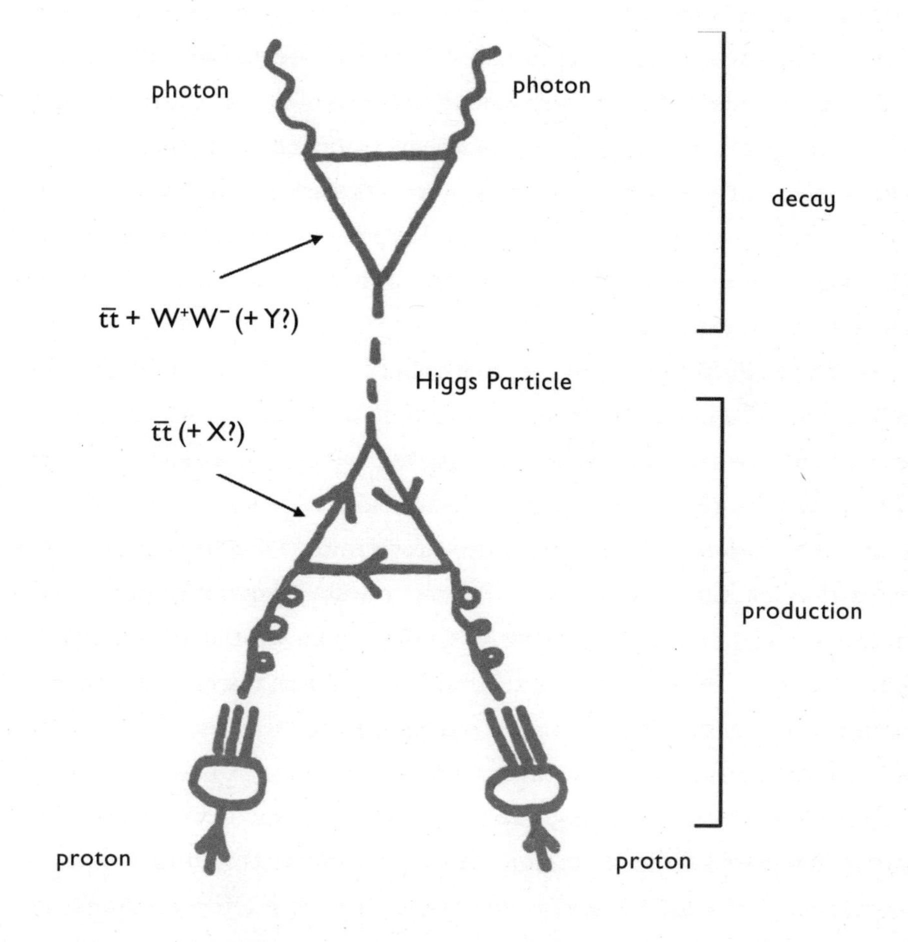


FIGURE 36. THIS SKETCH DEPICTS THE PROCESS THROUGH WHICH THE HIGGS PARTICLE WAS ACCESSED THROUGH GLUONS AND FIRST OBSERVED EXPERIMENTALLY. IT IS A TOUR DE FORCE THAT PUTS MANY ASPECTS OF THE CORE AND DEEP PRINCIPLES OF QUANTUM THEORY TO WORK SIMULTANEOUSLY.

*W* and *Z*—they don't feel its drag.) In fact the dominant coupling arises indirectly, by an indirect process, "gluon fusion," a process that I discovered in 1976, during a memorable walk I'll describe below. It appears at the bottom of figure 36.

Gluons don't couple to the Higgs particle directly. The coupling is a purely quantum effect. It is characteristic of quantum mechanics that spontaneous fluctuations, or "virtual particles," occur. Usually those fluctuations come to be and pass away without discernible effect, aside from their influence on the behavior of nearby real particles. In the most important gluon fusion process, gluons inject energy into a virtual pair consisting of a top quark $t$ and an antitop antiquark $\bar{t}$. The $t$ and $\bar{t}$ quark and antiquark couple powerfully to the Higgs particle—that's a big reason why they are heavy—so there's a fair chance that they will bring forth that particle before expiring.

The most efficient way to get from colliding protons to the Higgs particle is for two gluons, one from each proton, to collide. The rest of the protons materialize as a messy background, typically containing dozens of particles.

*H* decay into two photons, $H \rightarrow \gamma\gamma$, shown toward the top of figure 36, arises through similar dynamics. Photons do not couple directly to the Higgs particle, but communicate with it through virtual $\bar{t}t$ and $W^+W^-$ pairs. Although this is quite a rare decay mode, it was the primary discovery mode for *H*, because it has two big advantages from an experimental point of view.

The first is that the energy and momentum of high-energy photons can be measured quite accurately. We can combine these, according to the kinematics of special relativity, to determine the "effective mass" of a photon pair. If the photons of a pair result from decay of a particle with mass *M*, then their effective mass will be *M*.

The second is that energetic photon pairs are rather difficult to produce by ordinary (non-Higgs) processes, so the background is manageable.

Exploiting both these advantages, experimentalists designed their search strategy: measure the effective masses of many photon pairs, and look for an enhancement at one particular value, relative to nearby ones.

And—to make a long story short—it worked!

There's a bonus: Because the background can be calculated reliably, the size of the enhancement, relative to background, gives a measurement of the production rate of $H$, times its branching ratio into $\gamma\gamma$. One can then check whether the measured enhancement agrees with the predictions for the minimal $H$. This is especially interesting, because those rates open a new window on the unknown. Specifically: there might be other heavy particles, yet unobserved, contributing in their virtual form! So far the observations are consistent with the unembellished minimal model, but greater accuracy is both attainable and highly desirable.

## *An Enchanted Evening*

Up until 10:00 p.m. or so, the day (in summer 1976) that would turn out to be the most productive in my scientific career seemed anything but promising. My very young daughter Amity had an ear infection, and all day long she was feverish, cranky, and needy. Betsy and I, inexperienced in parenting, newly arrived and on our own in Fermilab's impromptu village, coped as best we could. As the dark midwestern night set in, Amity at last fell into exhausted sleep, and then Betsy too. They looked like angels of peace.

The alertness and energy that coping with a stream of little crises had called forth was still with me, after the crises themselves had passed. Seeking an outlet, I decided, as I often do, to take a walk. The night was brilliantly clear; the sky radiant; the horizon sharp and distant; and even the ground, moonlit, seemed ethereal. With images of earthly angels lingering within me, and celestial spectacle surrounding me, I felt an unlikely elation. It was a time for big thoughts.

Over the preceding few years, theories of the strong, weak, and electromagnetic interactions based on local symmetry had matured from bold adventure to conventional wisdom. As I reviewed this situation, it occurred to me that while the various quarks, leptons, gluons, and weakons—not to mention photons—had received a lot of attention, and were the focus of well-thought-out experimental programs, the symmetry breaking was relatively unexplored. There wasn't even a credible proposal to test the very simplest, "minimal" model featuring a single Higgs particle, as described above.

The basic problem is simple: the Higgs particle, in that model, likes to couple to heavy particles, but the particles of stable matter, which we can study directly or put into our accelerators, are very light. The color gluons have zero mass, as do photons, while the *u* and *d* quarks, and electrons, have negligible mass.

But recently (as of 1976) there had been a lot of interest in heavier quarks. The charmed quark *c* was a fairly recent discovery, and there were excellent reasons to suspect that two additional, still heavier kinds would also exist. (And they do. The bottom quark *b* was discovered not much later, in 1977, while the top quark *t* took until 1995. They had been named, and their properties—with the sole exception of their masses—had been calculated, even before their experimental observation.) So it was natural to consider whether new, heavier quarks could open a portal through which we'd get to the Higgs particle. I realized right away that they might. You can use the same tricks that people had used for charmed quarks to produce mesons based on $\bar{b}b$ or $\bar{t}t$. Those heavier quarks would couple vigorously to Higgs particles. If things broke right—basically, if the heavy quarks had more than half the mass of the Higgs particle—then Higgs particles would be produced in the decays of those mesons. That was my first important realization of the night.

Now it was important to consider competing decays, without Higgs particles, because those might dominate and make the whole thing academic. One of the most important possibilities to consider was decay into color gluons. I couldn't do an accurate calculation in my head, though it seemed OK, from rough estimates. (It is.) More important, this got me thinking: if the heavy quarks can couple to Higgs particles *and* to gluons, then they provide a way to connect gluons to Higgs particles! And at that moment, my brain had hatched the basic process you see in the bottom half of figure 36 (page 267). Again, accurate calculation would be a chore, but I did some crude estimates in my head and found the results encouraging. In particular, I realized that even if the missing quarks were *very* heavy, they'd still contribute—and that if there were even heavier quarks, they'd contribute too. It was clear to me, right away, that this was the dominant way Higgs particles would couple to stable matter. It opened a promising window into the unknown. That was my second important realization of the night.

At that point I'd reached the lab site, and I decided to turn back. I'd had good luck thinking about the minimal Higgs model, so I wanted to consider how the new ideas would apply to more complicated versions. The changes are easy to work out, in any specific case, so I started considering what would be the most interesting complications to consider. An especially interesting possibility is to have some extra symmetry that gets broken spontaneously. This can lead to the existence of new massless particles—a spectacular possibility! That was my third important realization of the night.

Back in Princeton, where I'd been teaching during the year, there'd been enormous excitement about something called instantons—which I won't try to explain here. Instantons break symmetry in particularly interesting ways, and I thought it would be fun to bring those in, so I'd have something to talk about that my colleagues would be interested in hearing. I dimly perceived that the particle that would otherwise have been massless, according to my third realization, would instead get a tiny mass, and would have other interesting properties. That was my fourth important realization of the night, and it brought me home.

Those four insights have had different fates. The first was a victim of bad luck. The *b* quark is not heavy enough, compared to the Higgs particle, while the *t* quark is so heavy and unstable that its mesons are useless.

The second is one of my proudest achievements. More than thirty years later, it was central to the actual discovery of the Higgs particle, as described in and around figure 36.

The third hasn't borne fruit yet, but remains interesting. I eventually called the massless particles "familons," and people continue to look for them.

The fourth turned out to be the most interesting, and possibly the most important. When I got back to the lab the next day and consulted the literature on these things, I discovered a very interesting paper by Roberto Peccei and Helen Quinn. They'd looked at the kind of model I'd been playing with, and had pointed out that it could solve a very important problem, the so-called $\theta$ problem. The essence of it is that there's a number—$\theta$—that the Core says could be anything between $-\pi$ and $\pi$, but which is observed to be very, very small. That's either a coincidence, or an indication that the Core is incomplete. In Peccei and Quinn's model, the "coincidence" got explained as the residue of a new (spontaneously broken) symmetry. Peccei and Quinn didn't notice, however, that their model had a light particle in it! And so I got to name the thing. I had noticed, several years before, that there was a detergent, Axion, whose name sounded like a particle. I resolved that if got the chance, I'd make it so. Now the $\theta$ problem, along the way, involves an *axial* current. That gave me an opening through which to sneak the name past the watchful, conservative editors of *Physical Review Letters,* which I did. (Steven Weinberg also noticed this new particle, independently. He'd been calling it the "higglet." We agreed, Deo gratias, to use "axion.")

The axion has had a long, winding, and still unresolved history. It is a subject I've returned to many times, developing the theory of its production in the early Universe, and suggesting the possible existence of an axion background analogous to the famous microwave background. According to this work, the axion background will be difficult, but not impossible, to observe. A hardy band of brilliant experimentalists are actively searching. Someday soon, the axion may deserve a book of its own, for it has become a leading contender to provide the dark matter of the Universe. Or it may not exist at all. Time will tell.

## PART 4: SUMMING UP

### *The Census of Forces and Entities*

We have four fundamental forces: gravity, electromagnetism, and the strong and weak forces. All are described, theoretically, using local symmetry. The theory of gravity, Einstein's general relativity, is based on local symmetry of space-time, while the theories of the other three forces are based on local symmetry of property spaces.

General relativity is a rich theory, and far from easy to master. But it is based on the interplay between ordinary space-time and energy-momentum, which are universal concepts that don't require a detailed census. We mean no disrespect, therefore, when we acknowledge this interplay in the single word "gravity."

Because the behavior of matter, with respect to the other three forces, is determined by flows in property spaces, we need to describe the geometry of the property spaces it inhabits in order to give an account of matter. I will do that in two stages, represented in plates RR and SS and in plates TT and UU. In the first stage I've passed over some complications that I've added back in the second stage.

In plates RR and SS, you see that there are six distinct blocks. Inside the blocks are names of particles: *u* and *d* quarks in three colors (red, green, and blue *u* quarks, for example), and *e* and *ν* leptons (electron and neutrino). Each block encodes, in a way we'll presently describe, a possible property space for matter. So these six blocks represent six distinct varieties of matter that occupy different kinds of property spaces. Some of the blocks contain several distinct kinds of particles; the largest one (block *A*) has six. From our—and, more to the point, the forces'—perspective, the different particles within a block are really a single entity, seen at different positions in the property space. Our census contains sixteen distinct kinds of particles—a disturbingly large number of fundamental world-ingredients! From our deeper perspective we see that those sixteen particles represent only six distinct *entities*—significantly fewer (but still too many . . . we'll do better in the next chapter).

In the horizontal direction, we represent the three dimensions of strong charge (or "color") space. The blocks that have three columns (*A*, *B*, and *C*) represent entities that can move in a three-dimensional strong charge property space. In the vertical direction, we have the dimensions of weak charge space. The blocks that have two rows (*A*, *D*) represent entities that can move in a two-dimensional weak charge property space. The entity represented by block *A* can move independently in both ways, so it samples $3 \times 2 = 6$ property dimensions.

The numbers attached to each block represent the scale of its one-dimensional electric charge property space.*

Finally, the superscripts *L* and *R* denote, respectively, left-handed and right-handed. Lee and Yang taught us that only the left-handed quarks and leptons participate in the weak force. In our census, this is encoded in the fact that only the blocks with *L* superscript contain two rows. Each particle occurs in both its left-handed and right-handed forms, within different blocks.

Block *F* is especially interesting. It has only one entry: the right-handed neutrino $\nu_R$. It has neither strong nor weak nor electromagnetic charge, so it is invisible to all the nongravitational forces. $\nu_R$ accesses no property space, and must content itself with moving through ordinary space-time.

We now conclude the first stage of our census.

* More accurately, it is the so-called hypercharge property space. For more nuanced discussion of this and several other technicalities, see "Terms of Art" and pages 403–4 in the endnotes.

## *Acknowledging Families*

To complete our census of the Core, we need to add two more ingredients, as indicated in plates TT and UU (where I've also mentioned gravity).

One is the Higgs fluid. In the minimal version of the Core—which, as we've discussed, has proved adequate so far—the Higgs fluid feels weak forces, but not strong forces. It accordingly occupies a two-dimensional property space, as indicated in plates TT and UU.

The other is a mysterious triplication of the entire matter sector. Together with the quarks and leptons we've mentioned so far—the so-called first family—there are second and third families. They fill out blocks of exactly the same kind, but with new entries, as follows:

| FIRST | SECOND | THIRD |
|---|---|---|
| $u$ | $c$ | $t$ |
| $d$ | $s$ | $b$ |
| $e$ | $\mu$ | $\tau$ |
| $\nu_e$ | $\nu_\mu$ | $\nu_\tau$ |

So in addition to the up quark $u$ we have charm and top quarks $c$, $t$; in addition to the down quark $d$ we have strange and bottom quarks $s$, $b$; in addition to the electron $e$ we have muons and tau leptons $\mu$, $\tau$; and in addition to the electron neutrino $\nu_e$ we have muon and tau neutrinos $\nu_\mu$, $\nu_\tau$. (Now we must add subscripts in order to distinguish the different neutrinos.)

The second and third families play a very limited role in our present-day natural world.

But they do exist, and their existence poses theoretical challenges—challenges that to date have not been met. For example, the masses of the particles vary over a wide range, and exhibit no obvious pattern. Their weak decays bring in many additional complications, introducing a dozen or so fudge factors whose values have eluded theoretical calculation. (If you ever feel the need to deflate a physicist who bloviates about his "Theory of Everything," just ask about the Cabibbo angle.)

On page 404 in the endnotes, I've spelled out some more details of these family complications, and pointed to some references where you can learn more. In the remainder of this meditation, we'll focus on aspects of physical reality where beauty is more apparent.

## THE END OF THE BEGINNING

We've now discussed every aspect of the Core: Maxwell's electrodynamics, QCD, and (more schematically) the weak force and gravity, and inventoried the entities they act on.

The Core provides a complete, and now battle-tested, mathematical explanation of how subatomic particles combine to make atoms, atoms combine to make molecules, and molecules combine to make materials, and how all those things interact with light and radiation. Its equations are comprehensive, yet economical; symmetrical, yet spiced with interesting detail; austere, yet strangely beautiful. The Core provides a secure foundation for astrophysics, materials science, chemistry, and physical biology.

With this, in a powerful sense, our Question has been answered. The world, insofar as we speak of the world of chemistry, biology, astrophysics, engineering, and everyday life, *does* embody beautiful ideas. The Core, which governs those domains, is profoundly rooted in concepts of symmetry and geometry, as we have seen. And it works its will, in quantum theory, through music-like rules. Symmetry really does determine structure. A pure and perfect Music of the Spheres really does animate the soul of reality. Plato and Pythagoras: We salute you!

Yet I feel that the answer we've reached thus far brings us not to the end of our quest, but only to the end of the beginning, in two respects.

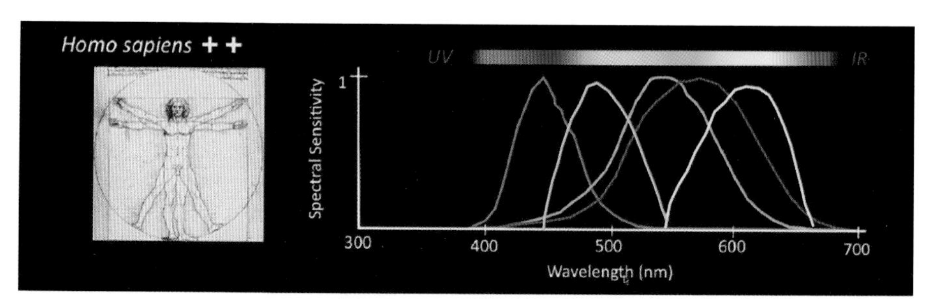


PLATE AA. Using temporal modulation we can add new receptive channels, enhancing human visual perception. We might, for example, add two new artificial channels, rendering color space five-dimensional.

| R1 | R2 | R1 | R2 | R1 | R2 | R1 | R2 | R1 | R2 | R1 | R2 |
|---|---|---|---|---|---|---|---|---|---|---|---|
| R3 | R4 | R3 | R4 | R3 | R4 | R3 | R4 | R3 | R4 | R3 | R4 |
| R1 | R2 | R1 | R2 | R1 | R2 | R1 | R2 | R1 | R2 | R1 | R2 |
| R3 | R4 | R3 | R4 | R3 | R4 | R3 | R4 | R3 | R4 | R3 | R4 |
| R1 | R2 | R1 | R2 | R1 | R2 | R1 | R2 | R1 | R2 | R1 | R2 |
| R3 | R4 | R3 | R4 | R3 | R4 | R3 | R4 | R3 | R4 | R3 | R4 |
| R1 | R2 | R1 | R2 | R1 | R2 | R1 | R2 | R1 | R2 | R1 | R2 |
| R3 | R4 | R3 | R4 | R3 | R4 | R3 | R4 | R3 | R4 | R3 | R4 |
| R1 | R2 | R1 | R2 | R1 | R2 | R1 | R2 | R1 | R2 | R1 | R2 |
| R3 | R4 | R3 | R4 | R3 | R4 | R3 | R4 | R3 | R4 | R3 | R4 |
| R1 | R2 | R1 | R2 | R1 | R2 | R1 | R2 | R1 | R2 | R1 | R2 |
| R3 | R4 | R3 | R4 | R3 | R4 | R3 | R4 | R3 | R4 | R3 | R4 |

PLATE BB. Periodic arrays of four different receptor types, small and closely spaced, can gather imagery whose fine structure supports four color dimensions. Displays based on the same architecture, with at least one of the elements exploiting temporal modulation, can make such information available to view.

PLATE CC. Physical atoms, mathematically described, are three-dimensional objects that will, to the animating spirit of an artist, yield images of exceptional beauty. Here we have a cutaway view of the electron cloud in a particular excited state of hydrogen. (For experts: the $(n, l, m) = (4, 2, 1)$ state.) The surfaces are surfaces of equal probability; the colors depict the relative phases.

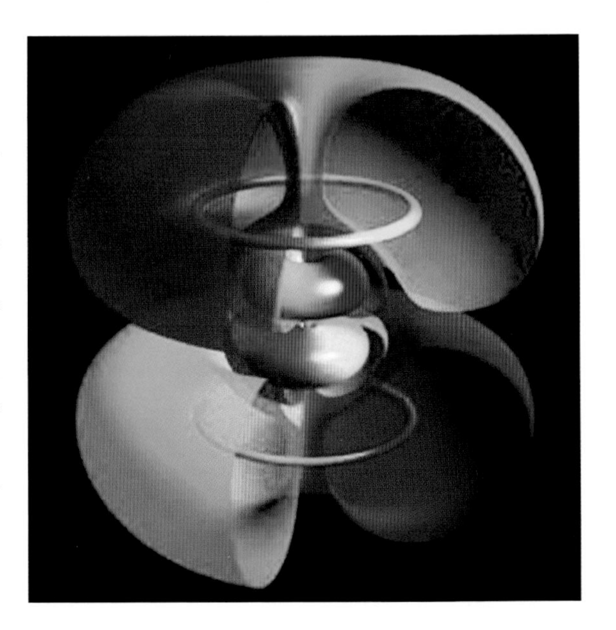

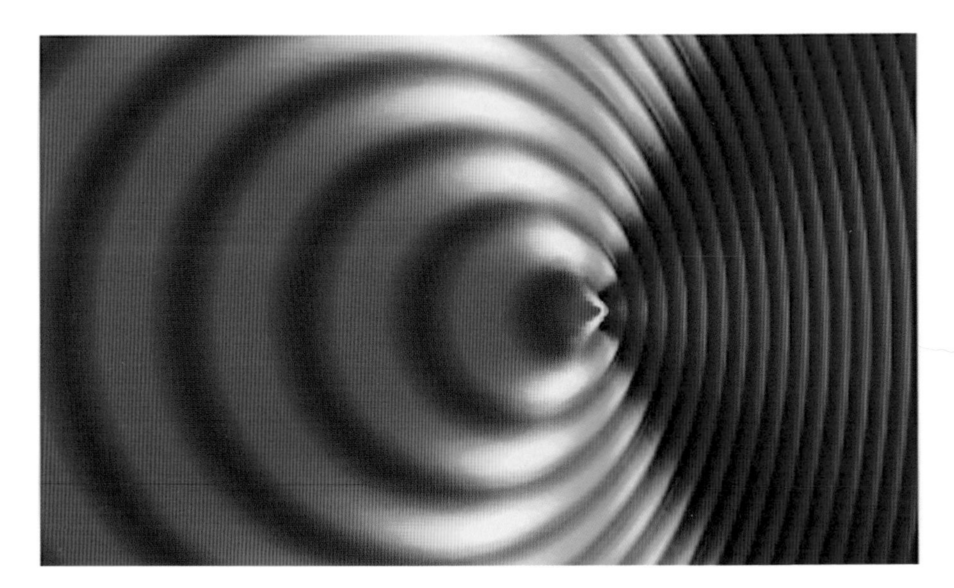

PLATE DD. A light beam will appear to have different colors when viewed by observers in rapid relative motion. Depicted here is a light beam emitted from a source moving to the right at seven-tenths the speed of light. If you are to the right, so that the beam is approaching you, the color is blue; if you are to the left, so that the beam is receding, it appears red. This picture shows a snapshot of the wave pattern, with the source near the center.

PLATE EE. Anamorphic art supports not only changes of perspective but also more general sorts of transformations. The range of images that can represent a given scene is vastly larger and includes some very distorted appearances.

PLATE FF. Rolling up a graphene sheet in different ways yields a variety of line-like, one-dimensional molecules, the nanotubes.

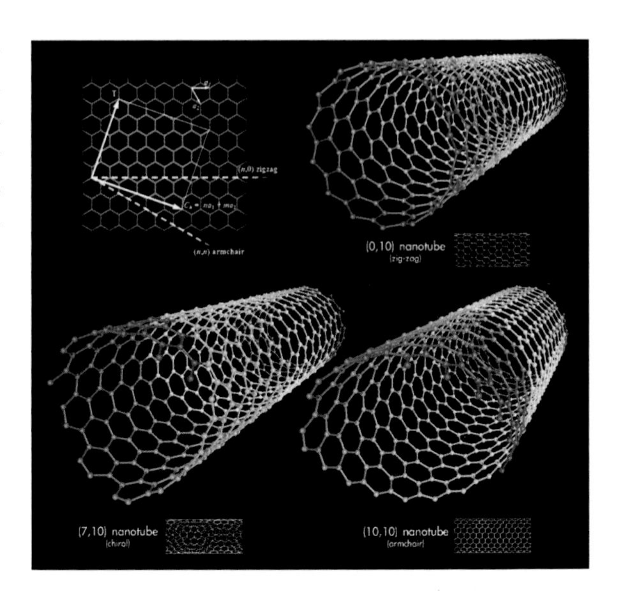

PLATE GG. Illustrating transformations in color space. At the upper left, we have the original image of a candy stall in Barcelona. At the upper right, a simple, rigid transformation in color space has been applied to it. In the two lower panels, two different local transformations have been applied: one mild, one wild.

PLATE HH. Geometry, color, symmetry, anamorphy, and anachromy empower a splendid beauty.

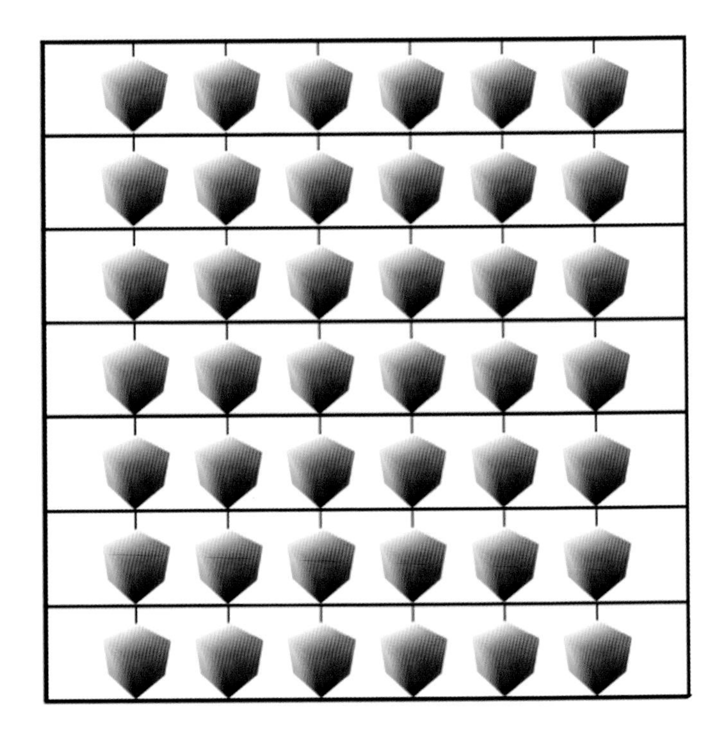

Plates II, JJ. The RGB color cube depicts the choices one has, at each point, for coloring a picture element (pixel). Through color vision we are accessing three extra dimensions.

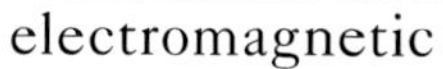

weak

strong

Plate KK. Here the concept of color property space is illustrated by restricting an image to slices of the color corresponding to a line and a face, before accessing the entire cube. The labels allude to the fact that our Core Theories use property spaces of one, two, and three dimensions.

Plate LL. Use of a fish-eye lens adds a second level of anamorphy to this striking modern mosque interior.

Plate MM. Quarks appear to need some spring- or rubber-band-like force that pulls more powerfully as the mediating spring or rubber band gets stretched.

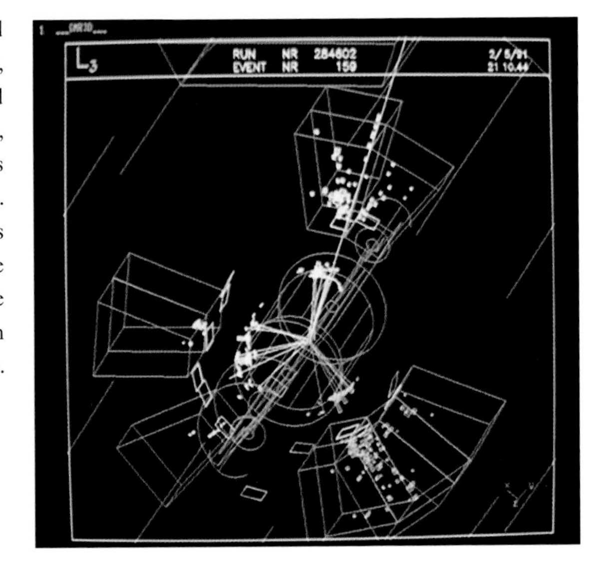


Plate NN. An electron and an antielectron (positron), accelerated to high energy and moving in opposite directions, have annihilated, and this picture captures the aftermath. One sees clusters of particles moving—very rapidly—in three distinct directions. These three jets are avatars of a quark, an antiquark, and a color gluon.

PLATE OO. The "lines of force," in Faraday's sense, that connect a quark and an antiquark form a tight tube. That tube represents a flux of color electric field that flows between the source quark and the sink antiquark. Because the gluons that make up color electric fields are self-sticky, they bundle up. This phenomenon is the key to quark confinement.

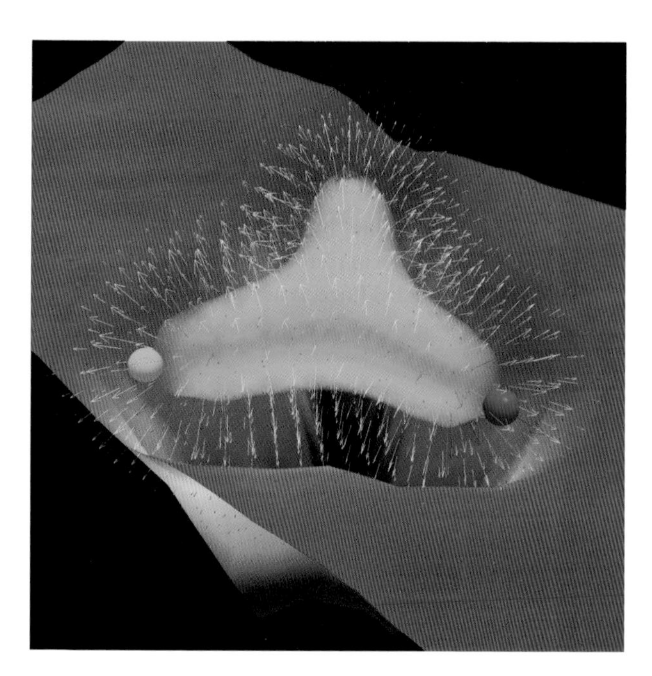

PLATE PP. This figure is an important variant of plate OO. Here the flux distribution connects three quarks. This is the backbone of baryons, such as the protons that form our substance. That Art Thou—*Tat Tvam Asi.*

| | | |
|---|---|---|
| $\left(\begin{matrix} \cdot & \cdot & \cdot \\ \cdot & \cdot & \cdot \end{matrix}\right)^{L}_{1/6}$ | A | $\begin{pmatrix} u \\ d \end{pmatrix}^{L}$ |
| $(\cdot \;\; \cdot \;\; \cdot)^{R}_{2/3}$ | B | $(u)^{R}$ |
| $(\cdot \;\; \cdot \;\; \cdot)^{R}_{-1/6}$ | C | $(d)^{R}$ |
| $\begin{pmatrix} \cdot \\ \cdot \end{pmatrix}^{L}_{-1/2}$ | D | $\begin{pmatrix} \nu \\ e \end{pmatrix}^{L}$ |
| $\circ^{\bar{R}}_{-1}$ | E | $(e)^{R}$ |
| $\circ^{R}_{0}$ | F | $(\nu)^{R}$ |

six
fundamental
"entities"
(plus two repeats)

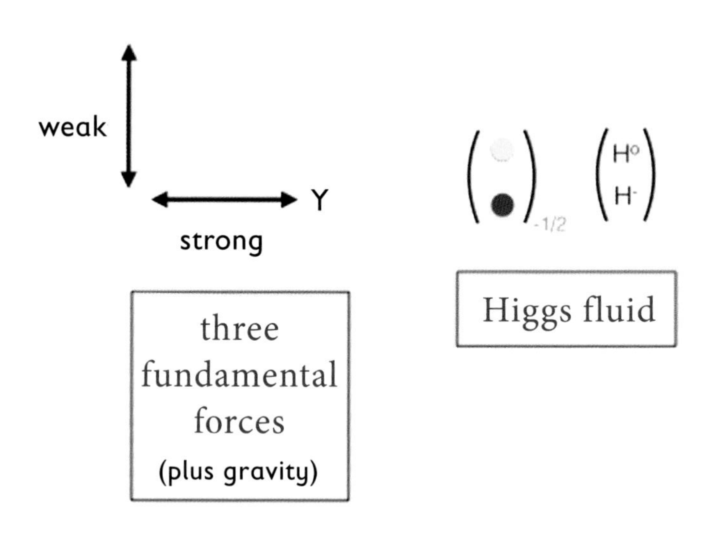


PLATES TT, UU. Summarizing the Core Theory, Step 2.

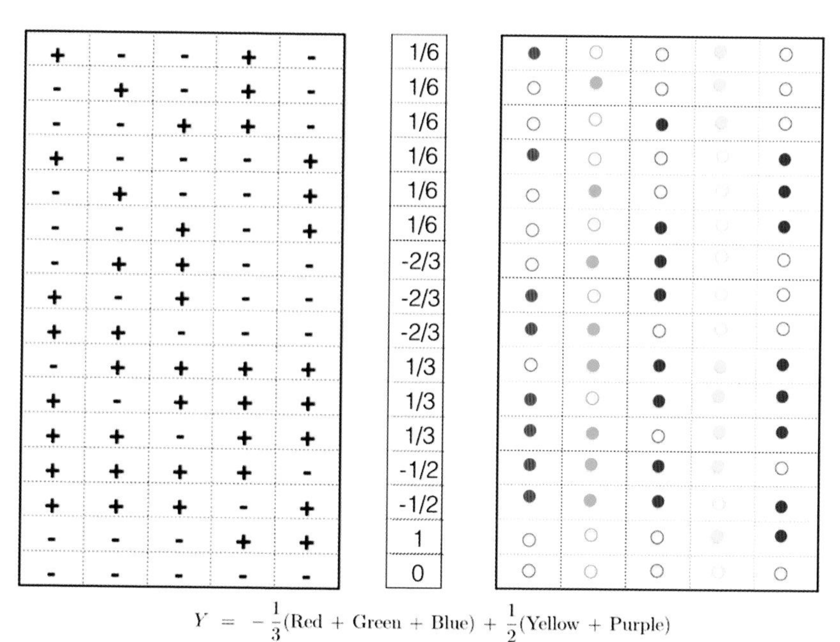

| | | | | | |
|---|---|---|---|---|---|
| + | - | - | + | - | 1/6 |
| - | + | - | + | - | 1/6 |
| - | - | + | + | - | 1/6 |
| + | - | - | - | + | 1/6 |
| - | + | - | - | + | 1/6 |
| - | - | + | - | + | 1/6 |
| - | + | + | - | - | -2/3 |
| + | - | + | - | - | -2/3 |
| + | + | - | - | - | -2/3 |
| - | + | + | + | + | 1/3 |
| + | - | + | + | + | 1/3 |
| + | + | - | + | + | 1/3 |
| + | + | + | + | - | -1/2 |
| + | + | + | - | + | -1/2 |
| - | - | - | + | + | 1 |
| - | - | - | - | - | 0 |

$$Y = -\frac{1}{3}(\text{Red} + \text{Green} + \text{Blue}) + \frac{1}{2}(\text{Yellow} + \text{Purple})$$

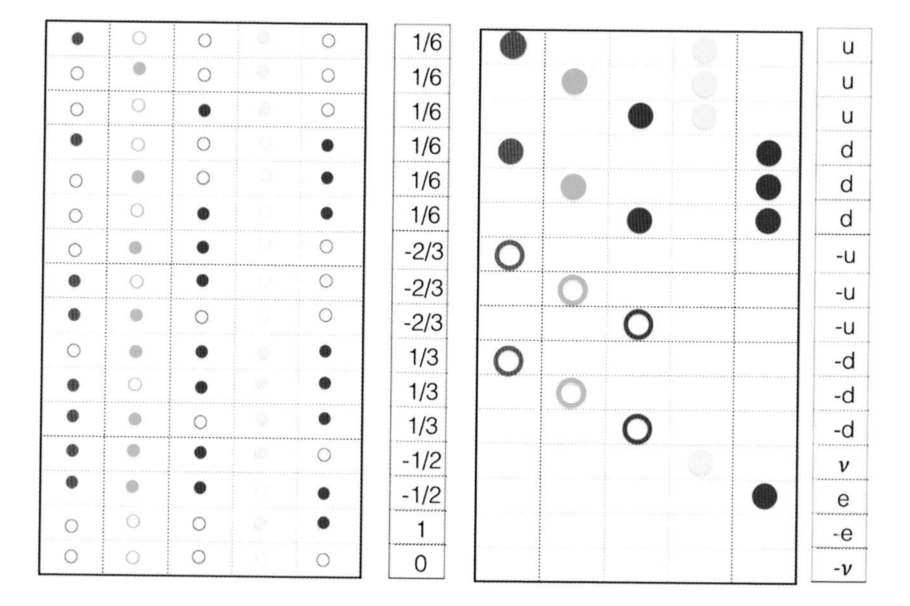


## One Entity, One Force

PLATES VV, WW. By assuming greater symmetry we can clean up the appearance of the Core Theory, as summarized in plates RR and SS, very considerably. We thereby reach this superb icon of Real = Ideal, whose interpretation is spelled out in the text.

PLATE XX. A magnified picture of empty space, viewed with excellent spatial and time resolution.

PLATE YY. Caravaggio's *Incredulity of Saint Thomas*. Thomas is an enthralled inquirer, and his examination is encouraged.

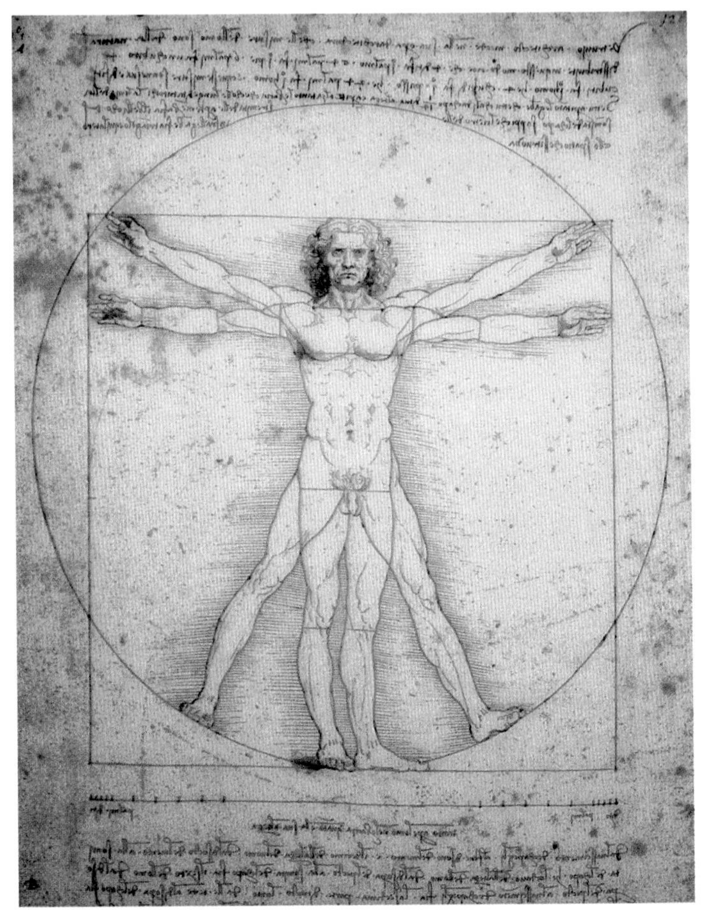

PLATE ZZ. Leonardo da Vinci's iconic *Vitruvian Man*, like Kepler's model of the Solar System, was inspired by beautiful ideas about deep reality that are mistaken. (Or are they?)

PLATE AAA. A highly processed image of the microwave sky, revealing the seeds of structure in the universe.

First, there are loose ends.

As I've already mentioned, we've got family problems. And the astronomers, with their discoveries of dark matter and dark energy, have served up an ample helping of humble pie. Our brilliant theory covers 4 percent of the total mass in the Universe! (Weight isn't everything, but still . . .)

More profoundly, our wonderful answers empower us to imagine and address new, more ambitious questions. Above all this one: Do the disparate parts of the Core arise from a deeper unity? In the remainder of our meditation, we'll consider that question, and (I think) make a very promising start on it.

Second, there are open doors.

As we have come to understand, for all practical purposes, what matter *is*, we are in the position of a child who has just learned the rules of chess, or an aspiring musician who has just learned the sounds of which her instrument is capable. Such basic knowledge is the preparation for mastery, not mastery itself.

Can we use imagination and calculation, rather than trial and error, to design the materials of the future? Can we tune in to the messages the Universe broadcasts in gravity waves, in neutrinos, and in axions? Can we understand the human mind, molecule by molecule, and systematically improve it? Can we engineer quantum computers, and through them fashion truly alien forms of intelligence? To ask these questions is to discover, in the ripeness of one golden age, the seeds of new ones.

# SYMMETRY III: EMMY NOETHER—TIME, ENERGY, AND SANITY

Symmetry, in general, is Change Without Change. But it was the amazing Emmy Noether (1882–1935) who made a tight connection between the *mathematical* symmetry of physical laws and the existence of specific *physical* quantities that do not change. "*X* does not change with time" is quite a mouthful, and also negative, so we commonly say "*X* is conserved" instead. In that language, Noether's theorem says that symmetries of physical law give rise to conserved quantities.

Thus, in Emmy Noether's work, our visionary correspondence

$$\text{Ideal} \leftrightarrow \text{Real}$$

becomes a mathematical theorem.

No doubt a concrete example of a Noetherian

$$\text{symmetry} \Rightarrow \text{conservation law}$$

pair will be welcome at this point. And here we can display a precious gem. It is, I think, the single most profound result in all of physics.

In our example, the symmetry is what is called time translation symmetry. That jargon might seem intimidating, but its meaning is simple: *the same laws of physics that apply today have applied in the past, and will apply in the future.*

The assumption that the same laws always apply might not sound, at first hearing, like an assumption of symmetry, but that's what it is. For it says that you can *change* the meaning of time that appears in the laws of physics by adding or subtracting a constant, *without change* to the laws' content. (In the jargon of mathematics and physics, displacement by a constant amount, whether in space or in time, is called "translation.")

Time translation symmetry is the wisdom of Ecclesiastes:

> What has been is what will be,
> and what has been done is what will be done,
> and there is nothing new under the sun.

What Shakespeare lamented, here,

> If there be nothing new, but that which is
> Hath been before, how are our brains beguil'd,
> Which, labouring for invention, bear amiss
> The second burden of a former child

is not the same thing. Time translation symmetry applies to the laws that relate events, not to the events themselves. To put that formally: time translation symmetry is a property of our dynamical equations, but tells us nothing about initial conditions.

Toward the end of this interlude I will discuss the assumption of time translation symmetry critically, but for now let's take it at face value.

## TIME AND ENERGY

According to Noether's theorem, any symmetry of the laws implies the conservation of some physical quantity. For time translation symmetry, the conserved quantity is—energy!

Energy, as a physical concept, has a bizarre history. I'd like briefly to recount some highlights from that history, first of all because it is interesting, but especially here because it highlights the importance of Emmy Noether's realization.

### *A Brief History of Energy*

Today we recognize that energy makes the world go round. We seek its sources, store it, discuss its price, weigh the trade-offs involved in getting hold of it in different ways, and so forth. Familiarity, however, should not disguise energy's essential weirdness.

The idea that conservation of energy is a fundamental principle emerged only in the mid-nineteenth century. Even then, *why* it should be true was quite mysterious, until Noether's insight. And even today, as I'll explain, I don't think we've got to the bottom of it.

In the conceptual struggles preceding Newtonian clarity, scientists trying to understand motion repeatedly found, in several different kinds of problems, that the *square* of a body's velocity kept popping up as a particularly useful measure of the body's motion. Galileo, for example, found that for bodies moving under the influence of near-Earth gravity, such as thrown stones, cannonballs, or (what he measured carefully) balls rolling down inclined planes and pendulums, a fixed change in altitude always gives a fixed change in the square of the velocity, independent of all other details.

In hindsight, we understand that odd result as an example of the conservation of energy. There are two contributions to the body's total energy, kinetic and potential. The kinetic energy (energy of motion) is proportional to the square of the body's velocity, while the potential energy (energy of position)—for near-Earth gravity—is proportional to its altitude. Conservation of energy says that changes in kinetic energy must be compensated by changes in potential energy, which is another way of stating Galileo's discovery.

It is crucial to our story to mention that Galileo's result was not a straightforward *observation*, but rather an *idealization*. He proved it as a theorem, in a mathematical model that neglected air resistance, friction, and other complications that are always present in reality. By using heavy balls, as opposed to, say, feathers, and taking other precautions, Galileo could do experiments where the effect of those complications is small, and his energy-conserving model is reasonably accurate. But strictly speaking, Galileo's version of (what eventually became) the law of conservation of energy is never exactly true for any real system, as he was fully aware. For Galileo, it was just a curious fact about an idealized model.

In Newton's classical mechanics, conservation of energy became a more general theorem. Yet it remained an idealization, rather than a description of reality. Newton's conservation of energy theorem applies to systems of particles that interact with one another through forces whose strength depends only on their relative distances. In that framework, the theorem tells you what energy *is*—namely, it's the quantity that appears in the theorem, and stays constant in time! Again, one finds that the total energy is composed of kinetic and potential energy pieces. The kinetic energy always takes the same form. You get it by adding up, for each particle, its mass times the square of its velocity, all times one-half. The potential energy is a function of the relative positions, whose precise form depends on the nature of the forces. So far, so good. But frictional forces, in Newtonian mechanics, spoil the conservation of energy. While that fact does not contradict Newton's theorem, because frictional forces fall outside the assumptions of the theorem—friction isn't described by distance-dependent forces between particles—it limits the theorem's application to reality.

When we add Maxwell's electrodynamics, things get more complicated, but the basic conclusion is similar. In the expanded framework we can still derive, given certain assumptions, a mathematical conservation of energy theorem. But, first of all, the meaning of energy gets modified. Specifically, a third form of energy, besides kinetic and potential, must be included. There is also field energy, which—as its name suggests—depends on the strength of the fields. It is only the total energy—kinetic plus potential plus field—that is constant. Worse, even the more complicated version of energy conservation holds only if one neglects both friction and electrical resistance.

I remember, when I was first learning about these things, feeling distinctly underwhelmed and skeptical. It seemed to me that the so-called "law" of conservation of energy was an ugly kludge. Every time some new force or effect got discovered, it would violate the existing "law," and so you'd dream up new kinds of energy to patch things up as best you could. And even so, new leaks might appear. Neither in Newton's mechanics, nor in Maxwell's electrodynamics, is conservation of energy an exact, general principle. It appears, rather, to be a useful but approximate result that applies in limited circumstances. Because it had, as far as I could see, no deep conceptual foundation, and anyway was only approximate, I saw no reason to expect the conservation of energy to be a reliable guide in the exploration of anything essentially new.

The idea that conservation of energy might be a fundamental principle, which holds exactly, emerged gradually in the middle and late nineteenth century. It was a discovery inspired by the needs of technology.

Since the dawn of history, people had tried many techniques to put things in motion, in pursuit of useful tasks such as transporting people and goods, besieging castles, grinding grain, and a host of other applications. During the Industrial Revolution machines became central to economic life, and optimizing their use became big business, so the problem of powering them was studied intensely, both experimentally and theoretically. Thinking about energy and its transformations turned out to be the most fruitful approach. It emerged, specifically, that practical (apparent) violations of energy conservation, due to effects like friction and electrical resistance, always involve *loss* of energy. (With the practical implication that one should focus on the price of energy, in any of its forms, and on minimizing the losses.) That tendency to lose energy, but never to gain it, explained the failure of technologists to produce self-contained, self-powered machines—so-called perpetual motion machines—and, more generally, why machines need power supplies. It was also observed that the *loss* of energy is always accompanied by *production* of heat. Several scientists, with various degrees of clarity, interpreted that situation positively. They suggested that conservation of energy really is a general truth, but that to keep track of it one must recognize that heat is yet another form of energy. Inspired by that vision, James Prescott Joule did a series of exquisite experiments, using falling bodies to drive paddle wheels to heat water, to demonstrate the key idea quantitatively: a given amount of energy (from the falling body) produces a proportional amount of heat.

After that triumph, the scientific world accepted conservation of energy as a working principle. Nature had spoken, and she'd said clearly that there is something very right about the idea.

But absent a deeper justification than "it works," the law remained both mysterious and precarious. "It works" really means "It has worked, so far." One could not be confident that some new discovery might reveal loopholes. It wouldn't be the first time. Conservation of *mass* was a cornerstone of Newtonian mechanics, and served very well indeed as a working principle for over two centuries, both in celestial mechanics and in all sorts of engineering applications. Conservation of mass was tested rigorously and put to work by Antoine Lavoisier in experiments that marked the beginning of modern, quantitative chemistry. Yet in the twentieth century, gross violation of the conservation of mass is a commonplace of extreme physics. At a high-energy electron-positron collider, collisions of two very light particles (an electron and a positron) routinely output dozens of particles whose total mass adds up to many thousands of times the total input mass!

So long as energy appeared as a hodgepodge bundling of many different kinds of things—kinetic (measured in motion), potential (measured in position), field (measured, in principle, by forces on charges and currents), heat (measure by changes in temperature), and others I haven't mentioned—it seemed open to further modification, or maybe even to exceptions.

Emmy Noether brought the concept of energy into sharp focus. By grounding conservation of energy in the uniformity of physical laws in time, she revealed its true nature and unveiled its hidden beauty. Through mathematical magic, Emmy Noether transformed an ungainly frog into a handsome prince.

The emergence, from technology, of the modern concept of conservation of energy, culminating in Noether's explanation of its roots in symmetry, is a marvelous example of

Real → Ideal

Could conservation of energy go the way of conservation of mass? In science only reality is sacred, and reality gets to spring surprises, whether we're ready for them or not. But Noether's theorem raises the stakes: if conservation of energy failed, we'd have to rethink basic concepts we use to formulate the laws of physics, our ideas about the uniformity of time, or both. Most of us read that result as a warning, informing us that thinking about violation of energy conservation, in the absence of prodding from Nature, is unlikely to be fruitful. Why borrow trouble?

## *More Lessons from Noether*

The uniformity of physical laws in space, like their uniformity in time, reflects a kind of symmetry called spatial translation symmetry. According to Noether's theorem, there should be a corresponding conserved quantity. There is: *momentum*. Physical laws should also look the same, when viewed from different orientations. That is another symmetry principle, called rotational symmetry. There should be, according to Noether's theorem, a corresponding conserved quantity, and there is: *angular momentum*. Like conservation of energy, those great conservation laws have a long and distinguished history, having been derived in special cases, under more restrictive assumptions, pre-Noether. Indeed, one of Kepler's laws—the one about planets sweeping out equal areas in equal times—reflects conservation of angular momentum because the rate of sweeping is proportional to angular momentum. But Noether's theorem, by connecting them to simple qualitative aspects of physical reality, offers deep insight into *why* those laws exist.

At the frontiers of modern physics, as we'll soon discuss, Noether's theorem has become an essential tool for discovery. Through it, we relate the theoretical esthetics of possible symmetry, and the question,

Are my equations *beautiful*?

. . . to the hard reality of physical measurement, and the question,

Are my equations *true*?

It's all been very successful and inspiring. Yet I feel that something important is still missing. And I'm not the only one: no less a scientist than Niels Bohr, in the 1920s, when confronted with puzzling experiments in radioactivity, briefly flirted with the idea that energy is not conserved. Lev Landau, another figure revered among physicists, later proposed that stars violate the conservation of energy. (The energy source for stars—nuclear burning—was not clarified until the mid-twentieth century.)

All deductions rely on assumptions, and Noether's theorem is no exception. In fact, the assumptions that go into Noether's theorem are rather abstract, technical, and hard to pin down. (For experts: The theorem is proved for systems whose equations result from varying a Lagrangian.

There are good reasons to admire systems that can be described in that way, but the reasons are complicated, and it's not clear, at least to me, why—or even whether—they're compulsory.) I feel that such an important, easily stated result should have a more direct, intuitive explanation. If I had one, I'd be happy to share it. At this point, all I can say is that I'm still searching!

## EMMY NOETHER, HERSELF

Emmy Noether's great result in mathematical physics, which we've been discussing here, was a youthful show of strength. Her main life's work was in pure mathematics. Her specialty, in fact, was purifying mathematics. She made algebra much more abstract and flexible, so that it could accommodate the elaborate constructions that creative mathematicians had dreamed up for use in algebraic geometry and number theory. By simplifying the foundations, she suggested creative ways to build on them. In pursuit of her passion, she overcame severe challenges and prejudices. David Hilbert wanted to appoint Emmy Noether as a colleague at the world's leading mathematics faculty, in Göttingen. Hilbert wrote, "I do not see that the sex of the candidate is an argument against her. . . . After all, we are a university, not a bathhouse," but his view did not prevail. Emmy Noether hung on, for a time, as an unpaid guest lecturer. But being not only an intellectual woman but also a Jew, with the rise of Nazism she was forced to flee Germany. Hermann Weyl later wrote, in tribute to her spirit in that time of trial:

> Emmy Noether—her courage, her frankness, her unconcern about her own fate, her conciliatory spirit—was in the midst of all the hatred and meanness, despair and sorrow surrounding us, a moral solace.

Others testify to her selflessness, her generosity, and above all her passionate devotion to mathematics. She would often forget herself, accord-

FIGURE 37. EMMY NOETHER, MATHEMATICIAN AND NOBLE BEING.

ing to her student Olga Taussky, "gesticulating wildly" and taking no note when her long hair fell out of its pins into disarray. Considering Emmy Noether's work, and sampling her biography, I am reminded of Novalis's description of Spinoza as "the God-intoxicated man." Emmy Noether was a math-intoxicated woman.

This photograph of Emmy Noether at age twenty seems to capture her spirit.

## SYMMETRY, SANITY, AND WORLD-CONSTRUCTION

Boswell, in his *Life of Samuel Johnson*, recounts the following episode:

> After we came out of the church, we stood talking for some time together of Bishop Berkeley's ingenious sophistry to prove the nonexistence of matter, and that every thing in the universe is merely ideal. I observed, that though we are satisfied his doctrine is not true, it is impossible to refute it. I never shall forget the alacrity with which Johnson answered, striking his foot with mighty force against a large stone, till he rebounded from it, "I refute it *thus*."

David Hume, taking off from Berkeley, produced more sophisticated arguments for radical skepticism. Hume saw no way to justify the assumption that physical behavior is uniform across time. But without that assumption, no prediction is secure—not even, for example, the prediction that the Sun will rise tomorrow. Yet uniformity of behavior is simply, according to Hume, an irrational leap of faith. Bertrand Russell encapsulated Hume's analysis in a memorable joke:

> The man who has fed the chicken every day throughout its life at last wrings its neck instead, showing that more refined views as to the uniformity of Nature would have been useful to the chicken.

And Russell went on to say:

> It is therefore important to discover whether there is any answer to Hume within the framework of a philosophy that is wholly or mainly empirical. If not, there is no intellectual difference between sanity and insanity.

In this section, inspired by Johnson, I'm going to take that on.

To make the world safe for sanity, let's go back to basics, and consider what it means to justify a belief. We'll start with Aristotle's famous syllogism, which launched the study of logic as a subject in its own right. At first sight, that classic:

All humans are mortal.
Socrates is a human.
Therefore, Socrates is mortal.

gives an impression of profundity and logical power. One deduces a new conclusion, with certainty, from old facts.

On reflection, however, it can come to seem empty. After all, we only have the right to assert that "All humans are mortal" if we already know that Socrates—a particular human—is mortal. Thus the reasoning appears to be deeply and utterly circular.

Yet it's hard to deny the feeling that something useful and nontrivial is going on here. The profound point, I think, is that we can feel more certain of the *general* assertion "All humans are mortal" and the identification "Socrates is a human" than we would be of the particular proposition "Socrates is mortal," asserted independently of that information.

The force of the assertion "All humans are mortal" certainly does not come from taking a full census of humanity, followed up by case-by-case verification that each member of that class has died. For one thing, many of us haven't! Rather, it comes from a general understanding of what it means to be human, specifically including the fragility of human bodies, the physiology of human aging, and so forth. An immortal being would have to differ very significantly from "human," in the generally accepted sense of that word, so much so that we would define it as something else. And Socrates, though obviously an unusual person, recognizably was born of human parents, occupied a human body like others, could be wounded in battle, matured and aged on a similar timescale to other humans . . . In short, Socrates fell comfortably within the class "human." So the syllogism applied, even before Socrates died—which, of course, he eventually did.

By the way, wouldn't it have been more evocative, and truly inductive, for Aristotle to have used *this* example?

All humans are mortal.
Aristotle is a human.
Therefore, Aristotle is mortal.

It would have humanized him. Also, when teaching his famous pupil Alexander the Great, Aristotle could have used

All humans are mortal.
Alexander is a human.
Therefore, Alexander is mortal.

That one might have changed the course of history by persuading Alexander the Great to take better care of himself. More likely, it would have gotten Aristotle fired, or worse.

Someday, perhaps soon, as medical technology improves and/or human intelligence migrates outside traditional human bodies, we may need to revisit the proposition "All humans are mortal." For example, the status of this syllogism may be doubtful:

All humans are mortal.
Ray Kurzweil is a human.
Therefore, Ray Kurzweil is mortal.

But if and when that happy day arrives, then people will instead begin with "All pristine humans were mortal," or something of the sort, and proceed as before, but now incorporating finer distinctions. In any case, the mortality of Socrates, Aristotle, and Alexander was never in serious doubt, even well before they actually died, for exactly the reason expressed in the syllogisms, properly understood.

The overarching point, for present purposes, is that broader and firmer foundations make an inference more secure. General assertions can be useful underpinnings for particular ones, even when the latter seem to assert strictly less.

What about Russell's chicken? She reasons:

Every day the Good Farmer feeds me.
Tomorrow is another day.
The Good Farmer will feed me tomorrow.

That looks a lot like the preceding syllogisms! But looks aren't everything. Although this syllogism has the same logical *form*, its effective *content* is quite different. A sufficiently intelligent chicken would notice that the Good Farmer does not deliver chicken feed in precisely the same way or at the same time each day, and also that the Good Farmer does many other things. A sufficiently intelligent chicken would try to make a theory that accounted for more of the Good Farmer's actions, and—if she were a particularly insightful chicken—she would be led to think about the Good Farmer as a self-interested agent with motives. The chicken might also notice that the Good Farmer and his family eat biological products, that crops get harvested, that farm animals mysteriously disappear from time to time, and so forth. At that point, the chicken would come to suspect that a Day of Reckoning will come, when the Good Farmer will *not* behave as the first line assumes. Whereas "All humans are mortal" stands up to hard scrutiny, and forms part of a comprehensive, coherent world-picture, "Every day the Good Farmer feeds me" does not.

In order to take up Russell's challenge, answer Hume, and defend sanity, we must justify our assumption of the uniformity of Nature. We can make that assumption more secure, as we've just discussed, by giving it broader and firmer foundations. Our formulation of uniformity in time as a statement about the symmetry of physical law helps to do this, in several ways. We can draw many consequences from it that are by no means obvious, but that turn out to be true features of the physical world. Most simply, we can repeat delicate measurements at different times, and check whether their results agree. By accessing distant stars and galaxies we look more deeply into the past, because light travels at a constant speed. We can check that their past spectral lines follow the same patterns as those we see today. In this way, we see that the same laws of atomic physics operated then as operate now. And, inspired by Noether, we can check the conservation of energy! That is by no means gilding the lily, because conservation of energy gets a tremendous workout in analyzing elementary particle reactions, which probe very extreme conditions.

Now that all these checks, and others, have been performed with great care and precision, the case for sanity is strong.

To round off this discussion, we should note two additional uniformities of physical law, besides uniformity in time, that are almost equally basic to world-construction: uniformity in space, and uniformity of substance. Uniformity in space, which we've touched on earlier, shares in the laboratory and astronomical tests that check uniformity in time. And, inspired by Noether, we can check it another way—by checking the conservation of momentum! That is by no means gilding the lily, because conservation of momentum gets a tremendous workout in analyzing elementary particle reactions, which probe very extreme conditions.

Finally there is *uniformity of substance*—the observed fact that all electrons (for example) have exactly the same properties. That assumption is implicit in every application of modern atomic physics, electronics, and chemistry. While it is often taken for granted, it is the opposite of obvious.

In human manufacturing, use of interchangeable parts was a revolutionary innovation, and hard work to achieve. Yet long before the innovations of Samuel Colt and Henry Ford, Nature's Artisan anticipated the virtues of interchangeable parts. In today's Core Theory, interchangeability of electrons is a consequence of the fact that all electrons (for example) are minimal excitations—quanta—of a common, world-filling electron fluid, and that the properties of that fluid are uniform in space and time. Thus, within the framework of quantum theory, uniformity of substance does not require separate assumptions. It follows from the uniformity of space and time, and thus, as Emmy Noether taught us, from symmetry.

# QUANTUM BEAUTY IV: IN BEAUTY WE TRUST

## A DODECAHEDRAL PARABLE

The dodecahedron has already appeared several times in our meditation. One of the five Platonic solids, it embodies much geometric symmetry. According to Plato himself, it is the shape of the Universe as a whole. We've seen how Salvador Dalí used dodecahedral symbolism to express a cosmic connection that might otherwise be hard to put on canvas. We've also found a dodecahedron lurking within every one of the infinite variety of buckminsterfullerenes, where its twelve pentagons serve as enablers, allowing the hexagons of graphene to close up into a surface.

Dodecahedra also have a very practical use, as desktop calendars. They are elegantly suited to that use, having twelve sides, all nice and equal, so you can fit a month on each one. On the Internet, you can easily find instructions for making such calendars, using cutout designs of stiff paper or cardboard.

The dodecahedron is a thing of beauty, and by now it's become a familiar friend.

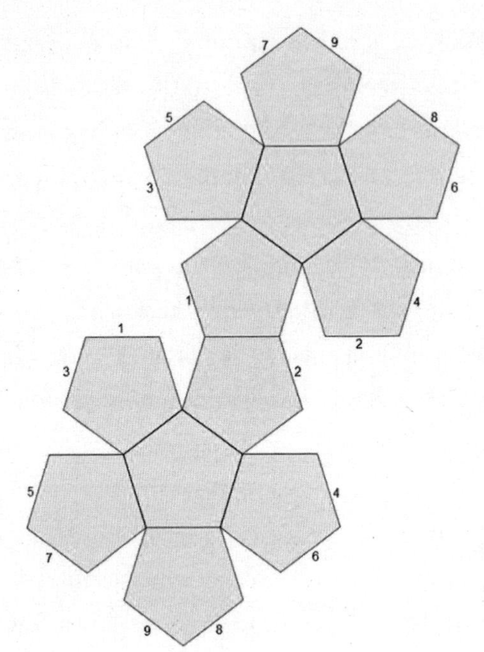


FIGURE 38. THIS LOVELY DESIGN WILL ALLOW YOU TO CONSTRUCT A DODECAHEDRON. YOU'RE MEANT TO TRACE OUT THE PATTERN ON STIFF PAPER OR CARDBOARD, THEN CUT IT OUT AROUND THE BORDER AND FOLD ALONG THE INTERIOR SOLID LINES SO THAT EQUALLY LABELED SIDES COME TOGETHER.

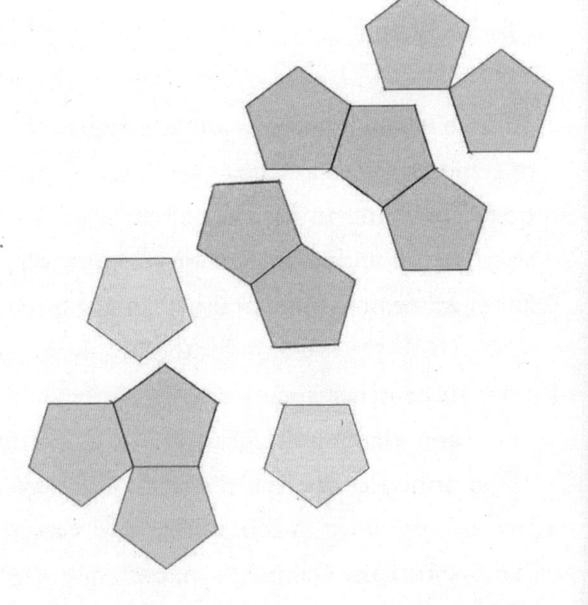

FIGURE 39. WITH THE PATTERN DISASSEMBLED THE IMAGE BECOMES HARDER TO INTERPRET, BUT TELLTALE CLUES TO ITS ORIGIN REMAIN. DODECAHEDRAL LITERACY WILL LEAD US FROM THIS COLLECTION OF PIECES BACK TO FIGURE 38, AND FROM THERE TO OUR PLATONIC DODECAHEDRON.

Now suppose that some playful spirit wants to test our mettle, or to give us the joy of solving a pretty puzzle. She disconnects parts of the design and takes off the labels, producing the enigma of figure 39.

Here it is harder to interpret what we see. Most people, who may not have thought much about dodecahedra recently, wouldn't know what to make of this partial pattern. But those of us who've been meditating on beauty and its embodiments are prepared for the challenge. Twelve identical, regular pentagons, some pairs sharing edges, some triples poised to share a vertex point—it sets bells ringing! We recognize the pattern's hidden potential, and we're ready to make something beautiful of it.

With that triumph in mind, let us now return to our Core Theory. It describes an enormous wealth of facts—hard, quantitative observations about the physical world—using a very compact set of equations. It is, as we discussed, a more than adequate foundation for chemistry, engineering of all kinds, biology (probably), astrophysics, and a lot of cosmology. It is elegant, too. The Core's equations have deep roots in symmetry. Because of that, we can reconstruct the entire Core Theory from a few general instructions about what property spaces different particles have access to, and the (local) symmetry we want those spaces to support. We can specify the necessary data in reasonably simple summarizing images, plates TT and UU.

The Core is a wonderful description of Nature. It would be hard to exaggerate its precision, its power, or its beauty. Yet connoisseurs of ultimate beauty will not be satisfied. Precisely *because* it is so close to Nature's last word, we should hold the Core to the highest possible esthetic standards. When scrutinized in that critical spirit, the Core shows imperfections.

- It contains three mathematically similar forces: the strong, weak, and electromagnetic forces. All are embodiments of a common principle: the local symmetry of property spaces. And gravity is a fourth force. It is based on local symmetry too, albeit of a different sort: local Galilean symmetry. Gravity is also much feebler than the other forces. It would be more satisfying, and beautiful, to have one master symmetry, and one overarching force, supplying a coherent description of Nature. Three (or four) is definitely more than one, so we're not there yet.

- What's worse, even after identifying particles that are "really" the same entity, positioned differently in its space, we find ourselves with six unrelated "fundamental" entities. Six is also definitely more than one.
- We also have the triplication of families, which seems gratuitous.
- We also have the Higgs fluid, which plays a unique and important role in the theory, but appears as yet another independent moving part. The Higgs fluid was introduced to patch things up (which it does), not to beautify them (which it doesn't).

Altogether: It's a kludge, for sure, and a harsh critic might call it a mess.

Might the Artisan, once the Core was rough-hewn, have called it a good week's work, and stopped right there?

Before giving in to that disturbing thought, let's return to the lesson of the dodecahedron. We saw, in that case, how beauty—and, in particular, symmetry—suggests a compelling interpretation of what otherwise might seem a random jumble. Understanding the possible symmetries of objects in space led us to realize that there are only a handful of Platonic solids, and that knowledge allowed us to infer an underlying dodecahedron from partial, distorted evidence.

The Core is based on forms of symmetry more elaborate than the rotations of ordinary three-dimensional space, and on objects (property spaces) less familiar than the dodecahedron. Nevertheless, we can try out the same sort of idea. Might the Core's piecemeal symmetry, and the apparently lopsided and disconnected objects it acts on, be pieces of a larger symmetry acting on a larger object whose connections have been hidden from view?

If we find a positive answer to that mathematical question, it will suggest new theories of physics that might transcend the imperfections of the Core. Yang and Mills showed us, given an assumed symmetry and its action on property spaces, how to construct a corresponding theory of forces and particles. In that construction, symmetries are embodied by their avatars, the gauge particles (e.g., color gluons, weakons, photons), which mediate the forces. Our hypothetical larger symmetry will give us all those Core forces, and more.

Thanks to mathematicians of the late nineteenth and early twentieth centuries—Sophus Lie and his successors—we have a complete inventory of the candidate symmetries and spaces, so we can see whether any of them fit the bill. Just as there are only a few Platonic solids, it turns out that there are only a few candidate larger symmetries that might unify the symmetries of the Core (like the rotations of a dodecahedron), and even fewer reasonable possibilities for property spaces that can link together those of the Core (like the faces of a dodecahedron).

When possibilities are so limited, success is uncertain. If the lopsided, disconnected image in figure 39 (page 296) had been lopsided and disconnected in a different way—say, with three of the pentagons surrounding a triangular hole, or with thirteen altogether, or with variously sized pentagons, or a mixture of pentagons and squares—our attempt to explain it based on hidden symmetry would have failed. Similarly, the lopsided, disconnected structure of the Core has to be lopsided and disconnected in just the right way if it is to fit the pattern of a larger symmetry. So if we do find a pattern that fits, it's unlikely to be a coincidence. It's likely to mean something!

It is therefore gratifying to discover that one of Lie's possible symmetries, acting on a beautiful property space, accommodates reality snugly. The unifying symmetry contains the strong-weak-electromagnetic symmetries of the Core. It can act on a property space that has just the right size and shape to accommodate the known quarks and leptons. And, most important, it contains nothing else. (For experts: The symmetry is based on the group of rotations in ten dimensions, denoted $SO(10)$. The property space is based on the sixteen-dimensional spinor representation of that group. This pattern was discovered by Howard Georgi and Sheldon Glashow.)

Please pause now to take a good look at plates VV and WW, which document that discovery. The discussion that follows here, as a sort of extended caption, spells out the interpretation of the plates. You'll find, in this discussion, all the information you need to appreciate how plates VV and WW encode the content of plates RR and SS, which summarize the Core. The main thread of our meditation relies only on the broad description in the main text. The finer details grace pages 403–4 in our endnotes. I felt it was important to make the details of that marvelous result available for your inspection. You can decide for yourself how deeply you'd like to engage them.

The particles of the Core, within that theory, inhabit six separate, differently shaped property spaces, as described in the previous chapter. Alternatively, we may say they form six distinct entities.

In our unified theory, a larger symmetry connects those property spaces, bringing all the particles into a single entity, or multiplet. This unification of matter echoes the unification we achieved among the disconnected, lopsided pieces of our mystery figure, when we recognized their fit to a dodecahedron. Just as the sides of a dodecahedron are all related through appropriate rotations, here all the particles are related to one another, by mathematical symmetry—and also through concrete physical transformations!

Looking at plates VV and WW, at the top left corner of the page we have a rather abstract table of + and – signs. There are five columns and sixteen rows. The different rows contain every possible distribution of five + and – signs, subject to the constraint that the total number of + signs is even. The top right corner begins the process of unfolding this abstract pattern into physical reality. It has the same structure, but now the columns are interpreted as representing the different strong and weak color charges (and it will emerge that the rows represent the particles of substance). The first three columns represent the three strong color charges red, green, blue, in that order. The last two columns represent the two weak color charges yellow, purple. We translate our previous table of + and – signs into the new format by putting little solid circles of the appropriate color where there are + signs, and little empty circles of the appropriate color where there are – signs.

The solid circles (deriving from the + signs) will be interpreted as one-half units of the charges. Thus a solid red circle represents one-half unit of red color charge, and so forth. (The brilliance of this factor one-half will emerge presently.) The empty circles (deriving from the – signs) will be interpreted as *negative* one-half units of the charges.

Just beneath these two tables there is a simple mathematical formula defining a quantity *Y* as a simple numerical combination of colors. Recall that in the Core census of substances, seen in plates RR and SS, there were funny numbers reflecting the electric charges. Those funny charges were, within the Core Theory, independent of the strong and weak colors, and simply chosen to fit the results of experiments. Momentarily, you'll see how those ugly ducklings of the Core mature into gorgeous swans in our unified theory. For now, just note that I've recorded the values of *Y* we get by applying the formula to our different rows, in the upper middle column of numbers.

The lower left corner is just a copy of the upper right corner, put there for ease of reading. The middle column of numbers is likewise reproduced.

The lower right table is a rewrite of its neighbor on the left, after we use the strong and weak bleaching rules to simplify. Let me walk you through this process for the first row; the others are similar. According to the strong bleaching rule, an equal admixture of red, green, and blue charges has no effect in the strong force. So we can simplify our description of the strong color charges of the particle in the first row, as far as the strong force is concerned, by adding one-half unit to each of red, green, and blue. That operation wipes out the preexisting negative half units for green and blue, while promoting red to a full unit. And in our table in the lower right corner we've depicted the result: a big full red circle, and nothing for green and blue. Turning to the weak charges, we add one-half units to both yellow and purple, appealing to weak bleaching, to get a full yellow and no purple.

And now a miracle is revealed. The list of particles and properties we've arrived at, by spelling out our initial abstract table (in the upper left), exactly matches our census of Core substances (plates RR and SS)! The first row, for example, matches the upper left entry of entity A. The conventional names of the particles are displayed in plates VV and WW, in the final column in the lower right corner of the page, and will help guide your search. It's a joyful exercise, which I highly recommend, to trace through each of the sixteen matches. Before you try it, however, there's one last subtlety that still needs mentioning. The right-handed particles of the Core are represented here through their left-handed anti-particles. So if you see a – sign in front of the name, you must reverse the sign of all the charges (including *Y*) and look for a right-handed match.

That ends our "extended caption."

To sum it up: Upon surveying the entries in our iconic unification plates VV and WW, we find a perfect match to the substance particles of the Core, as captured in *its* iconic synthesis, plates RR and SS! There we studied the world—the Real world—and classified its particles. Here our starting point was quite different. We started with an Ideal—a space of high symmetry, put forward as a candidate for property space—and derived, mathematically, the properties of the particles its local symmetry (Yang-Mills) theory contains. Following those two very different paths, we have arrived at the same destination. The new path is a more unified, principled description. It captures much of what we know about the world of Matter, in a construction of pure Mind. It is a magnificent example of:

Real ↔ Ideal

## REALITY CHECK

If it's right, that is . . .

The mathematics of symmetry has revealed a tantalizing prospect. It has sketched a path that leads from beautiful ideas to the world-governing Core, and beyond. It is a vision reminiscent, in its esthetic inspiration and audacity, of Plato's atomic theory, yet incomparably more sophisticated and precise.

But two serious issues arise as we try to develop that rough sketch into a portrait of reality. One is easy to address. The other is more challenging. It will lure us toward an interesting adventure whose endpoint is not yet clear.

Let's take the easy one first. The expanded theory contains many more gauge (force) particles than the Core, and therefore many more transformative forces. We have, specifically, not only color gluons that change one strong color charge into another, and weakons that change one weak charge into the other, but also mutatrons that change a unit of strong color charge into a unit of weak color charge. (There's no standard name for these particles in the literature, so I made one up. The joke is this: mutatrons produce mutations.) For example, there is a mutatron that transforms a unit of red charge into a unit of purple charge. That operation transforms the first row of plates VV and WW to the fifteenth, as you can check. Thus contact with that particular mutatron will change a red quark into an antielectron. No such process has ever been observed. If mutatrons exist, why haven't we seen their effects?

Luckily, this problem is very similar to a problem we met, and solved, in the theory of the weak force. You may recall that pristine local symmetry predicts that the weakons, like the photon and the color gluons, have zero mass. But if they did, their influence would be much more powerful than what is actually observed. That's the problem the Higgs mechanism solves. By filling space with an appropriate material, theorists got the weakons heavy, and harmonized Real with Ideal. Prior to the actual discovery of the Higgs particle, many physicists were skeptical of that audacious idea,* but now Nature has testified, most eloquently, on its behalf.

An amplified version of the same basic idea will give very large masses to the unwanted mutatrons of the unified theory, and thereby suppress all their unwanted effects. We simply fill the world—or, to say it more humbly (and accurately), we acknowledge that the world *is* filled—with a (selectively) mass-giving material, and proceed.

* I collected on several bets by taking advantage of such misplaced skepticism.

Now let's turn to the more challenging issue. If we are going to have symmetry among the different forces, then those forces must have the same strength. That is an immediate consequence of their presumed equivalence. But oops—they don't. The strong force really is stronger than other forces. The three basic forces are most definitely *not* equal in strength (and gravity is, on the face of it, *hopelessly* feeble in comparison).

(Important but slightly technical interlude: I should pause to explain how the comparison is made. The basic idea is simplicity itself. Each of our gauge theory forces, based as they are on Maxwell-like equations, acts between charged particles. For electromagnetic forces what is relevant is electric charge, for the strong forces it is color charge, and for the weak forces it is weak color charge. For each of our forces there is a unit [quantum] of charge. So to compare the forces, one simply compares how they act between unit charges. In practice, it's a bit more complicated, for two reasons. First: the effect of the weak force is suppressed at distances above $10^{-16}$ cm, and the effect of the strong force is suppressed at distances above $10^{-14}$ cm, for interesting but complicated reasons that we've touched on earlier [the Higgs mechanism and confinement, respectively]. So to make a fair comparison, we should only compare at distances smaller than those. Second: it's not really practical to manipulate particles in space with that level of accuracy. What experimenters actually do to access behavior at small distances is shoot particles at one another, and study the probability that they get deflected through [relatively] large angles. Then we work backward, from the deflections to the forces that cause them. You may recall that this was the strategy Rutherford, Geiger, and Marsden used to explore atomic interiors, circa 1912. The underlying principle hasn't changed, but today, by smashing particles together with much higher energy, we gain access to shorter distances. The comparison of the other forces with gravity is a little trickier. On the one hand, as far as we know there is no basic unit of charge for gravity—it responds to energy. On the other hand, we are using probes that have different energy, to compare forces at different distances. So in assessing the relative power of gravity at those distances, we simply substitute the energy appropriate to that distance, and calculate the gravitational force it exerts. End of technical interlude.)

## *Reimagining Asymptotic Freedom*

Having come this far, however, we should not give up so easily. And indeed, another big lesson from the Core—asymptotic freedom—suggests a solution. In the preceding chapter we saw how important it was, in understanding the strong force, to realize that the strength of that force varies with distance—becoming more powerful at long distances, and feebler at short distances. That variability allowed us to reconcile quark confinement, which indicates a powerful force opposing large separations, with their independence, which indicates a feeble force at small separations.

Asymptotic freedom moves things in the right direction. Because the strength of the strong forces grows weaker at shorter distances, the difference between it and the other forces narrows.

Could it be that they all come together?

To pass from hope to vision, and then from vision to calculation, it will be helpful to reimagine asymptotic freedom, using pictures and concepts that apply generally—beyond the strong force, and even beyond the Core.

Let's give ourselves nimbler and more discerning eyes.

Plate XX shows what we would see, looking at "empty space," if our eyes could resolve time intervals as small as $10^{-24}$ second, and objects or sizes as small as $10^{-14}$ centimeter.

This image, to be more precise, is a snapshot of a typical distribution of the density of energy arising from fluctuations in gluon field strengths. Fluctuations of that kind bubble up spontaneously, everywhere in space, throughout time, as a consequence of quantum mechanics. (They are sometimes referred to as virtual particles, or as zero-point motion.) Spontaneous activity of the gluon fluid is responsible for asymptotic freedom, confinement, and for the bulk of our mass, as we've discussed. Because they appear as central components within calculations that have been checked against reality very accurately and in many ways, the existence of these fluctuations is as certain as anything in science can be. In this calculated image, the most intense concentrations of energy are indicated with the "hottest" colors—red and bright yellow—while less intense regions are assigned less intense yellow, green, and finally light blue. Regions where the energy density is below a cutoff are left empty of color, against a black background. The magnification of this image is approximately $10^{27}$, so the region it depicts is roughly as small, compared with a human, as a human is small compared to the visible Universe. The fluctuations turn over in about $10^{-24}$ second. That time is far smaller, compared to a second, than is a second compared to the time since the Big Bang.

Because QCD has been tested with almost incredible rigor, it is as certain as anything can be in science that this picture accurately depicts something that has happened, is happening, and will continue to happen, all the time everywhere.

But there's more! The gluon fluid is not the only quantum fluid, by any means. We also calculate that the photon (electromagnetic) fluid fluctuates, and that the weakon fluids fluctuate. And so do the fluids associated with creation and destruction of "substance" particles—quarks and leptons. The electron fluid fluctuates, the up quark fluid fluctuates, and so forth. The physical consequences of fluctuations in those other fluids tend to be smaller than the effect of the fluctuations in the gluon fluids, because gluons are numerous (eight of them!) and strongly interacting. But the general principles of quantum theory predict fluctuations in all quantum fluids, and there is overwhelming evidence, from precision measurements, that those fluctuations occur. In correcting our vision, we must take them all into account.

Just as water distorts the way fish see their world, so too the medium of space—in particular, the activity of the quantum fluids that fill it—distorts our perception of the shortest distances. To perceive the underlying fundamentals, we must correct for those distortions. And therein lies our hope. The different forces *appear* to be unequal in strength. But perhaps, once we correct our vision, they will reveal themselves as equal.

## *A Near Miss*

Here's what happens when you carry that program out (figure 40). As you can see, it almost works—the three lines representing the strengths of the different forces *almost* meet in a point. But not quite.

Here is some additional information about figure 40, in case you'd like to understand it in technical detail: To make the result look as simple as possible—three straight lines!—I've had to make two slightly peculiar choices that are indicated in the axis labels. I've made the height of the points represent the inverse strengths, so that the stronger the force, the *lower* it appears. (That seemingly perverse choice also has another impor-

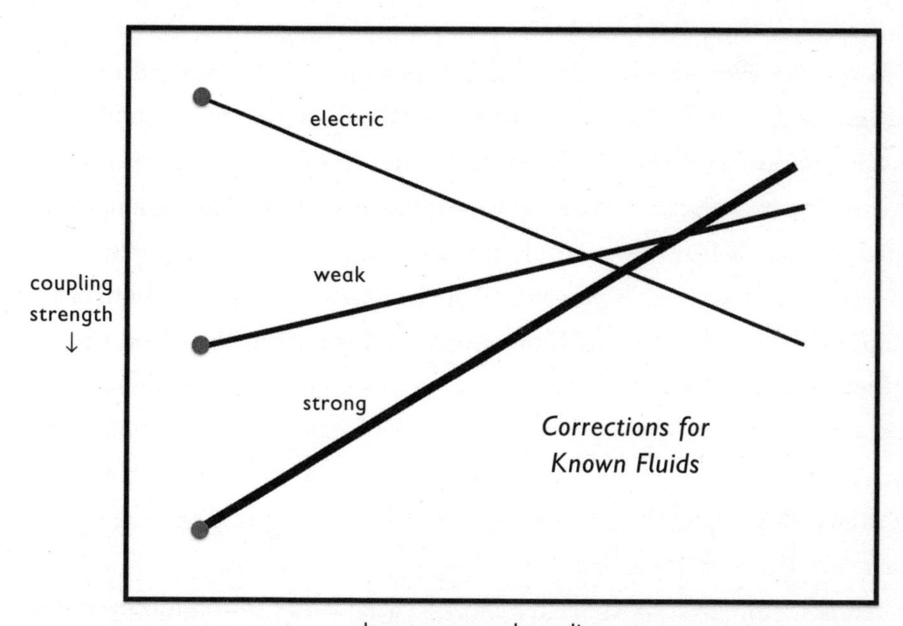


FIGURE 40. ONCE WE CORRECT FOR THE EFFECTS OF KNOWN QUANTUM FLUIDS, WE FIND THAT UNIFICATION IS A NEAR MISS.

tant advantage, which will appear in figure 41 on page 316.) In the horizontal direction, I've used a logarithmic scale. Thus each tick to the right shrinks the distance and increases the energy we need to probe that distance—by a factor of ten! Thus our calculation, despite its modest appearance, carries us way beyond what accelerators currently do. The width of the lines indicates their experimental and theoretical uncertainties.

We were hoping to find that the strengths of the different basic forces become equal when they are measured at short distances or probed at high energy. We take the values we measure at the distances (or energies) made accessible by the most powerful accelerators available, and then use theories and calculations to estimate what we'd find at even shorter distances (or higher energies). In this figure, the base points, representing measurements, appear on the left, highlighted by the big dots. Shorter distances, "accessed" by calculation, extend to the right. You see that it almost works—the three lines *almost* meet in a point—but not quite.

At this juncture we might seek consolation in the ideas of a famous philosopher, Karl Popper. Popper taught that the goal of science was to produce falsifiable theories. We have produced a theory that is not merely *falsifiable*, but *false*. Mission accomplished!

That comfort rings hollow. We have developed a beautiful idea that seemed promising, and it almost worked. Beauty is precious. We must not give up too easily.

And now I'd like to tell you the story of how a couple of my friends and I discovered a possible solution. But first I need to introduce another friend: SUSY.

## INTRODUCING SUSY

Supersymmetry, or SUSY for short, is a new sort of symmetry. Its existence as a mathematical possibility came as a big surprise to physicists. It was first proposed in a mature form in 1974, by Julius Wess and Bruno Zumino.

Symmetry, in general, is Change Without Change. Applied to systems of equations, it is the idea that we can make transformations among the quantities in the equations, without changing the equations' consequences. Supersymmetry is a particular example of that concept, which involves an especially weird kind of transformation.

We've already discussed many examples of physical symmetry. Time translation symmetry involves transforming what we mean by time, by adding or subtracting a constant. Galilean symmetry, the central concept of special relativity, involves transforming the world—i.e., space-time—by adding or subtracting a constant velocity, thus giving it a "boost."

Supersymmetry extends special relativity to allow for a new kind of transformation. It is a quantum version of the Galilean, velocity-giving "boost" transformation. Quantum Galilean transformations, like ordinary Galilean transformations, involve motion, but it is motion into, or back from, strange new dimensions. The new dimensions of supersymmetry are very different from the dimensions of ordinary geometry. We call them *quantum* dimensions.

As we explored earlier, in our discussion of property spaces, *what* can depend on *where*. The same entity, located in different positions within a property space, often manifests itself as several "different" particles. We—or, more to the point, gluons, weakons, and photons—respond to that entity in different ways, depending on where in the property space it is. If you imagine a particle moving through property space—that particle is transformed along the journey from one kind of particle into another kind of particle.

The quantum dimensions of supersymmetry are like that, too. What's new is the radical nature of the transformation that occurs when a particle moves within them.

The Core is divided in two parts, which we've called substance and force (or, poetically, yin and yang). The "substance" sector, comprising quarks and leptons, contains particles that have a certain persistence and grittiness: properties we associate with earthy matter, and substantiality. The precise technical concept, which captures what all those particles have in common, is that they are *fermions*, named for Enrico Fermi.

- Fermions come and go in pairs. As a result, if you've got one, you can't get rid of it altogether. It might turn into another kind of fermion, or into three, or five, together with any number of non-fermions (i.e., bosons—see below), but it cannot dissolve into nothingness, leaving no trace.
- Fermions obey Pauli's exclusion principle. Roughly speaking, this means that two fermions of the same kind don't like to do the same thing. Electrons are fermions, and the exclusion principle for electrons plays a crucial role in the structure of matter. We encountered it when we explored the exuberant world of carbon.

The "force" sector, comprising color gluons, the photon, weakons, and also the Higgs particle and the graviton, contains particles that tend to come and go easily—or, in jargon, to be radiated and absorbed—and that often come in bunches. The precise technical concept, which captures what all those particles have in common, is that they are *bosons* (named for Satyendra Bose):

- Bosons can be created or destroyed singly.
- Bosons obey Bose's "inclusion principle." Roughly speaking, this means that two bosons of the same kind are especially happy to do the same thing. Photons are bosons, and the inclusion principle for photons is what makes lasers possible. When offered the chance, a collection of photons will try to all do the same thing, making a narrow beam of spectrally pure light.

The contrast between substance and force particles—fermions and bosons—is very stark. It took great imagination—and audacity—to conceive that it might be transcended. Yet quantum dimensions accomplish exactly that. When a substance particle steps into a quantum dimension, it becomes a force particle; when a force particle steps into a quantum dimension, it becomes a substance particle. It's a kind of mathematical magic I won't be able to treat with justice here. But I will briefly describe the central weirdness, which is quite entertaining.

We map ordinary dimensions onto ordinary, so-called "real" numbers. We pick a reference point, usually called the origin, and label any point by a (real) number that describes how far you must go to get there from the origin. Real numbers, in a word, are suitable for measuring distances, and labeling continua. They satisfy the multiplication rule

$$xy = yx$$

Quantum dimensions use a different kind of numbers, called Grassmann numbers. They satisfy a different multiplication law,

$$xy = -yx$$

That little minus sign makes a huge difference! Notably, if we put $x = y$, we get $x^2 = -x^2$, and so we conclude $x^2 = 0$. That strange rule encodes, in the physical interpretation of quantum dimensions, Pauli's exclusion principle: you can't put two things in the same (quantum) place.

After those preparations, we're ready to meet SUSY. Supersymmetry is the claim that our world has quantum dimensions, and that transformations exist which interchange ordinary with quantum dimensions (change), without changing the laws of physics (without change).

Supersymmetry, if correct, will be a profound new embodiment of beauty in the world. Because the transformations of supersymmetry turn substance particles into force particles, and vice versa, supersymmetry can explain, based on symmetry, why neither of those things can exist without the other: Both are the same thing, seen from different perspectives. Supersymmetry reconciles apparent opposites, in the spirit of yin-yang.

## *From "Not Wrong" to (Maybe) Right*

Savas Dimopoulos is always enthusiastic about something. In the spring of 1981 it was supersymmetry. Savas was visiting the new Kavli Institute for Theoretical Physics in Santa Barbara, which I had recently joined. We hit it off immediately. Savas was bursting with wild ideas, and I liked to stretch my mind by trying to take them seriously.

Supersymmetry was (and is) a beautiful mathematical theory. The problem with applying supersymmetry is that it is too good for this world. It predicts new particles—lots of them. We have not seen, so far, the particles it predicts. We do not see, for example, particles with the same charge and mass as electrons, yet are bosons instead of fermions.

But supersymmetry wants such particles to exist. When an electron steps into a quantum dimension, it becomes just such a particle.

Based on experience with other forms of symmetry, we have a fallback position called spontaneous symmetry breaking. The fallback is to suppose that the equations for the object of our interest—which, in fundamental physics, is the world as a whole—have the symmetry, but their stable solutions do not.

An ordinary magnet is the classic example of that phenomenon. In the basic equations that describe a lump of lodestone, any direction is equivalent to any other. But the lump forms a magnet, and in a magnet it is no longer true that all directions are equivalent. Each magnet has a polarity, and can be used to make a compass needle. How is such directionality consistent with the nondirectional equations? The point is that there are forces that tend to align the spins of electrons in the magnet *with each other*. In response to those forces, all the electrons must choose a common direction in which to point. The forces—and the equations that describe them—will be equally satisfied with any choice for that direction, but *a choice must be made*. So the stable solutions of the equations have less symmetry than the equations themselves.

Spontaneous symmetry breaking is a strategy for having our supersymmetric cake and eating it too. If we are successful, we can apply beautiful (supersymmetric) equations to describe a less beautiful (asymmetric—or should we say subsupersymmetric?) reality.

Specifically, when an electron steps into the quantum dimension, its *mass* will change. If the new particle it becomes, the so-called selectron, is sufficiently heavy, then it is no wonder that we have failed to observe it. It will be an unstable particle that can only exist briefly, after having been produced at a (very) high-energy accelerator.

At the frontiers of ignorance, applications of spontaneous symmetry breaking involve creative guesswork. You must guess a symmetry that isn't visible in the world, put it into your equations, and show that the world—or, more realistically, some aspect of the world you're trying to explain—pops out of its stable solutions.

Can we use that fallback for supersymmetry? Building world-models with spontaneously broken supersymmetry that are consistent with everything we know turns out to be difficult. I briefly tried my hand at it when supersymmetry was first developed, in the mid-1970s, but after simple attempts failed miserably I gave up. Savas is a much more naturally gifted model-builder, in two crucial respects: he does not insist on simplicity, and he does not give up.

It was an interesting collaboration, reminiscent of *The Odd Couple.* When I identified a particular difficulty (let us call it A) that was not addressed in his model, he would say: "It's not a real problem, I'm sure I can solve it," and the next afternoon he would come in with a more elaborate model that solved difficulty A. But then we would discuss difficulty B, and he would solve that one with a completely different complicated model. To solve both A and B, you had to join the two models, but then new problems arose. Lather, rinse, repeat. Before long, things got incredibly complicated.

Eventually we succeeded in patching everything up, using the method of exhaustion. Anyone (including us) who checked the model for flaws got exhausted following out the complications, before being able to pin down a difficulty. When I tried to write up this work for publication, I found myself inhibited by a certain feeling of embarrassment about the complexity and arbitrariness of what we had come up with.

Savas, as I mentioned, revels in complexity. He was already talking to another colleague, Stuart Raby, about adding supersymmetry to models of force unification, which were complicated to begin with, for other reasons.

I was unenthusiastic about this stacking of speculations. To tell the truth, I wanted to show that it couldn't work, so that I could wash my hands of the mess with a clean conscience. My plan was to find some definite general consequence that wouldn't depend on the patchwork details. It would be wrong, and so finis: good riddance to bad rubbish.

To get oriented and make a definite calculation, I suggested that we start by doing the crudest thing, which was to ignore the whole issue of (spontaneous) symmetry breaking, which was the source of most of the complication and all of the uncertainty. That allowed us to focus on nice, simple, symmetric models, at the price of abandoning realism. We could calculate whether the forces came together in those models. (We were following, without thinking about it, in the footsteps of Pythagoras and Plato. And, of course, heeding the counsel of Father Malley, SJ.)

The result was a big surprise, at least to me. In those early days the relevant measurements were still quite crude, and so the lines in figure 40 (page 307) were thicker, indicating greater uncertainty. The thicker lines *did* overlap. In other words it seemed possible, given the uncertainties, that the strengths of different forces do unify at short distances. That was a tantalizing clue, famous among theorists in the field. What came out of our calculations, and seemed amazing to me, was that the supersymmetric models, though they contained many more fluctuating fluids, also worked! The answers differed—depending on whether you did or did not allow for supersymmetry—but both were consistent with the existing experimental information.

That marked a turning point. We put aside the "not wrong" complicated models that tried to accommodate reality in detail. Instead Savas, Stuart, and I wrote a short paper that, on its face, was clearly unrealistic (i.e., wrong). With supersymmetry unbroken, what we were proposing was too good for this world. Yet it gave a result that was so straightforward and successful that it made the idea of unifying the unifications—joining *force unification:*

strong + weak + electromagnetic

. . . with *supersymmetry unification:*

substance + force

. . . seem (maybe) right. We left the problem of how supersymmetry gets broken for another day.

Sometimes the most important step in understanding *something* is to realize you shouldn't worry about *everything.* It's usually better to be (maybe) right about something than "not wrong" about everything.

## *A Crowning Jewel?*

Figure 41 displays, in a plot, what our calculation revealed.

Supersymmetry introduces new sources of activity into space—new kinds of quantum fluctuations, or virtual particles. We must revisit figure 40 (page 307), to include corrections for their distortions too. Of course, we'll use the best available experimental information so our lines will be similarly thin.

When we do that, it works! The strengths of the different forces—strong, weak, and electromagnetic—come together, with impressive accuracy.

And there's more. So far, we've left the fourth force, gravity, out of our discussion of unification. This was a strategic decision. Unification of the other three forces presents a much riper, simpler problem. The strong, weak, and electromagnetic forces are described by deeply similar theories. Each is an embodiment of local property space symmetry. And while

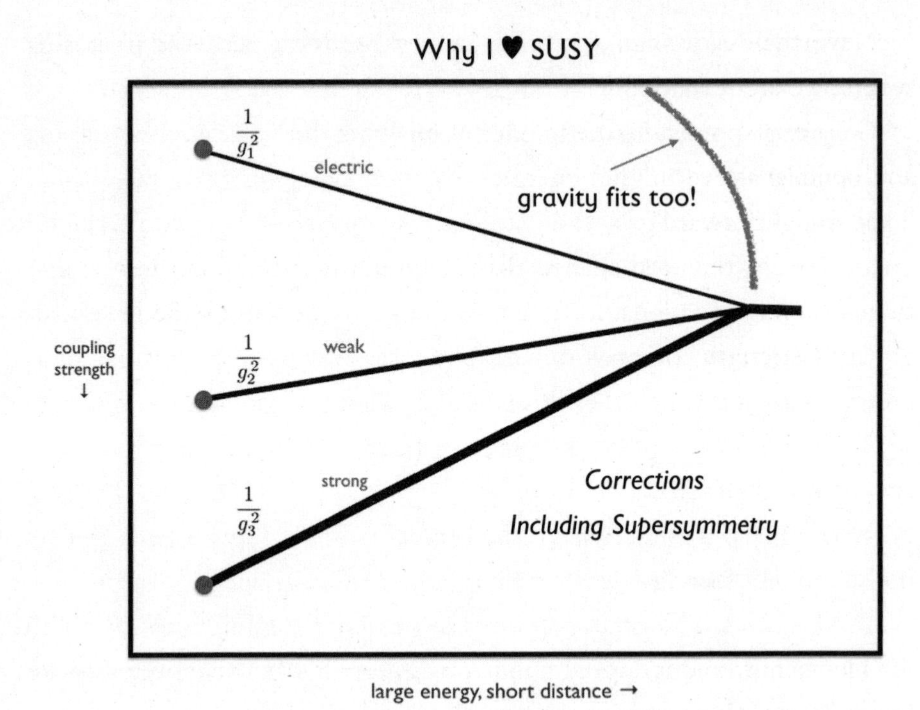


FIGURE 41. AFTER ADDING THE EFFECTS OF NEW QUANTUM FLUIDS REQUIRED BY SUPERSYMMETRY, ACCURATE UNIFICATION OCCURS.

the observed strengths of those forces differ, as indicated by the distribution of the dots in figures 40 and 41, they are not crazily out of proportion. Indeed, they differ by less than a factor of ten.

Gravity is different, on both counts. Its governing theory—Einstein's general theory of relativity—is also an embodiment of local symmetry, as we've discussed, but the symmetry (local Galilean symmetry) is of a different kind. Even more daunting is its *absurd* disparity in strength. Gravity, acting between elementary particles at accessible energies, is much, much, *much* feebler than the other forces. If each "much" were to represent a factor of ten, we'd need forty of them! And so, in figure 41, no circle representing the observed strength of gravity is visible. For that circle lies way, way, *way* outside the visible Universe. It takes about twenty-seven factors of ten to get from the size of our figure to the size of the visible Universe, leaving thirteen overflows.

Nevertheless, we can also bring gravity into the game. And if we persevere, we are rewarded.

Gravity responds *directly* to energy, so as we probe it (with our minds and pencils) at ever higher energies, its strength increases in proportion. That straightforward growth in strength is a much more powerful effect, quantitatively, than the changes in the strengths of the other forces, due to quantum fluctuations. In figure 41, the arc, representing the inverse of gravity's strength, plunges down. It reenters the visible Universe, and pretty nearly joins the other three forces, when they come together.

At the level of coupling strengths, then, we achieve a complete unification among all four basic forces

strong + weak + electromagnetic + gravity

That achievement does not, in itself, represent a fully complete theory of unification. For example, you might have noticed that if we continue following the straight lines of figure 41 to the right, the forces would "un-unify" again! For the strong, weak, and electromagnetic forces, we can flesh out the unification. We can't derive a completely unique theory—there's not enough information for that—but the possible theories have much in common. In particular, they all require new, very heavy particles, like the mutatrons we mentioned earlier. The fluctuations associated with those particles, not included in figure 41, make the couplings stay together, once they have met. (They don't have much effect before then.) When we try to include gravity, the uncertainties are much greater. A major goal of string theory is to explain how gravity unifies with the other forces, but to date that goal has proved elusive.

Despite its limitations, this unification of forces is a great result. It emerges from, and crowns, the search for an answer to our Beautiful Question. It confirms, with impressive precision and accuracy, that beauty, in the specific form of deep symmetry, is embodied in the world.

Or does it?

In order to fulfill our vision, we've had to invoke supersymmetry.

Because there is, as yet, no direct evidence for supersymmetry, that assumption is questionable. (The success of our calculation, to me, is strong *circumstantial* evidence!)

Fortunately, we can test it. If the new particles that supersymmetry predicts are going to do the job we've assigned them, they can't be too heavy. Large masses suppress their quantum fluctuations, and would turn figure 41 back into figure 40. The Large Hadron Collider should soon be able to concentrate enough energy to start producing some of those particles. Within the next five years. I've bet on it.

## IN BEAUTY WE TRUST

*In God we trust: all others pay cash.*

• *Jean Shepherd (book title)*

In beauty we trust, when making our theories, but their "cash value" depends on other factors. Truth is highly desirable, but it is not the only, or even the most important, criterion. Newton's mechanics (centered on conservation of mass) and his theory of colors (centered on conservation of spectral types), for example, are not strictly true, yet they are hugely valuable theories. Fertility—a theory's ability to predict new phenomena, and give us power over Nature—is also a big part of the equation.

Trust in beauty has often, in the past, paid off. Newton's theory of gravity was challenged by the orbit of Uranus, which did not obey its predictions. Urbain Le Verrier, and also John Couch Adams, trusting in the beauty of the theory, were led to propose the existence of a new planet, not yet observed, whose influence might be responsible. Their calculations told astronomers where to look, and led to the discovery of Neptune. Maxwell's great synthesis, as we've seen, predicted new colors of light, invisible to our eyes, but also not yet observed. Trusting in the beauty of the theory, Hertz both produced and observed radio waves. In more recent times, Paul Dirac predicted, through a strange and beautiful equation, the existence of antiparticles, which had not yet been observed, but soon thereafter were. The Core, anchored in symmetry, gave us color gluons, *W* and *Z* particles, the Higgs particle, the charmed quark, and the particles of the third family all as predictions prior to their observation.

But there have been failures too. Plato's theory of atoms and Kepler's model of the Solar System were beautiful theories that, as descriptions of Nature, utterly failed. Another was Kelvin's theory of atoms, which proposed that they are knots of activity in the ether. (Knots come in different forms, and they are not easily undone, so they have, it might seem, the right stuff to make atoms.) Those "failures" were not without fruit: Plato's theory inspired deeper study of geometry and symmetry, Kepler's model inspired his great career in astronomy, and Kelvin's model inspired Peter Tait to develop the theory of mathematical knots, which remains a vibrant subject today—but as theories of the physical world they are hopelessly wrong.

The fate of supersymmetry is not yet decided. Its discovery would, as I've described, reward our trust in beauty as a guide to deep reality. There are good reasons to think that discovery may be imminent, and beautiful reasons to hope so, but it has not yet occurred.

We shall see.

## *A Double Blessing*

According to the story of Doubting Thomas, the Apostle Thomas was skeptical of Jesus's resurrection, withholding his belief in the absence of evidence:

> Except I shall see in his hands the print of the nails, and put my finger into the print of the nails, and thrust my hand into his side, I will not believe.

When Jesus did appear to Thomas, he allowed Thomas to examine his wounds, and Thomas believed. Jesus said:

> Thomas, because thou hast seen Me, thou hast believed: blessed are they that have not seen, and yet have believed.

That story has inspired many works of art, including Caravaggio's *Incredulity of Saint Thomas* (plate YY), which I find resonant. To me, Caravaggio's rendering conveys two profound messages that go beyond the words of the gospel's text. One sees, first, that Jesus does not resist Thomas's inquisitive examination, but rather welcomes it. And one sees that Thomas is fascinated and excited to discover that reality conforms to his deepest hopes. Doubting Thomas is a hero, and a happy man.

Those who believe without seeing are blessed with the joy of certainty. But it is a fragile certainty, and a hollow joy.

Those whose faith is not passive, but engages reality, will receive a second, more fulfilling blessing in the harmony of belief and experience. Blessed are those who believe what they see.

# A BEAUTIFUL ANSWER?

Not all beautiful ideas about deep reality are true. Plato's vision of ideally geometric atoms and Kepler's vision of a geometric Solar System are examples we've discussed. Leonardo's extraordinary *Vitruvian Man* (plate ZZ) alludes to others of a different sort. His drawing suggests that there are fundamental connections between geometry and (ideal) human proportions. This concept connects to philosophical and mystic traditions even older, and much more popular, than the Pythagorean stream we've followed: the idea that the human body reflects the structure of the Universe, and vice versa. Sadly, perhaps, we humans and our bodies don't figure prominently in the world-picture that emerges from scientific investigation.

Nor are all the truths of deep reality beautiful. The Core has many loose ends, and there is little prospect of tying them all up. Even if my dreams of axions, supersymmetry, and unification are fulfilled, the messy (non-) pattern of quark and lepton masses and the conceptually opaque dark energy will remain problematic for the foreseeable future.

Nevertheless, at the conclusion of this meditation, I hope you'll agree that the only fitting answer to its Question

*Does the world embody beautiful ideas?*

. . . is a resounding

Yes!

That answer emerges, with increasing force and clarity, from each preceding page. The most daring hopes of Pythagoras and Plato to find conceptual purity, order, and harmony at the heart of creation have been far exceeded by reality. There really is a Music of the Spheres embodied in atoms and the modern Void, not unrelated to music in the ordinary sense, but adding a strangeness and abundance all its own. The Solar System does not embody Kepler's original vision, but he himself discovered the precision of its dynamical laws, and thus enabled the transcendent beauty of Newtonian celestial mechanics. There really is much more to light even than our wondrous sense of vision reveals, and our imagination—and not only our imagination!—opens new Doors of Perception. The basic forces of Nature embody symmetry, and are implemented by its avatars.

Nor were Leonardo's inspirations entirely mistaken, if we interpret them more broadly. The connection (human body ↔ cosmos) no longer seems central, but its close relative

microcosm ↔ macrocosm

. . . is thriving.

In this meditation we've mainly explored the left-hand side of that pair, but let us now, with plate AAA, look outward. It is a picture of the sky as it would appear to an observer whose eyes sense microwave radiation rather than visible light. Of course, to put the information in a form humans can perceive, some image processing has been done. The intensity of the radiation is encoded as the color of the light, with dark blue corresponding to the lowest intensity, bright red the highest, and a range of colors you can see interpolating between those extremes. Also—a crucial "detail"!—a uniform average has been subtracted away, and the contrast turned up by about ten thousand. The raw image would be a featureless haze; what you see is the pattern of tiny deviations from the average.

The leading interpretation of this picture draws marvelous connections between microcosmos and macrocosmos. The microwave sky is a snapshot of conditions early in the history of the Universe, roughly thirteen billion years ago, and about one hundred thousand years after the Big Bang. Light radiated then is arriving here now, having traveled a very long way. This is the message it brings: Thirteen billion years ago the Universe was almost, but not quite, perfectly uniform. It contained parts-in-ten-thousand deviations from perfect uniformity.

Those deviations from uniformity grew by gravitational instability (denser regions attract matter away from surrounding less dense regions, and the contrast grows). Eventually they gave birth to galaxies, stars, and planets as we know them today. This is all fairly straightforward astrophysics, once one has the seeds. So the big question becomes: How did those seeds arise in the first place?

We need more evidence to be certain, but it seems likely, based on the evidence so far, that they started as quantum fluctuations, similar to the ones on view in plate XX. In present-day conditions quantum fluctuations are significant only at very small distances, but an episode of very, very rapid expansion during the early history of the Universe, through the process known as cosmic inflation, can stretch them to universal proportions.

We humans are poised between Microcosm and Macrocosm, containing one, sensing the other, comprehending both.

## DOWN TO EARTH

As I was completing this book, an unfortunate event occurred that brought me back down to Earth. My laptop computer, which to me is an appendage of my brain, was stolen. I was devastated.

But then a miracle occurred. I had all my data backed up, and within a few days I had a new laptop with everything restored—pictures, words, calculations, music, and so on. All those things had been encoded in numbers—strings of 0s and 1s—so faithfully that they could be made to reappear without noticeable modification. It occurred to me that one could hardly ask for a more tangible, direct, or impressive demonstration of the truth of Pythagoras's vision:

All Things Are Number

It is a beautiful reality, for which I gave—and give—thanks.

## COMPLEMENTARITY AS WISDOM

*Do I contradict myself?*
*Very well, then, I contradict myself,*
*I am large, I contain multitudes.*

• *Walt Whitman,* Leaves of Grass

FIGURE 42. COAT OF ARMS DESIGNED BY NIELS BOHR.

Our exploration of Nature has shown us many new perspectives. It is not easy to reconcile them with everyday experience, or with one another. From his immersion in the quantum world, where contradiction and truth are near neighbors, Niels Bohr drew the lesson of *complementarity:* No one perspective exhausts reality, and different perspectives may be valuable, yet mutually exclusive.

The yin-yang sign is an appropriate symbol for complementarity, and was adopted as such by Niels Bohr. Its two aspects are equal, but different; each contains, and is contained within, the other. Perhaps not coincidentally, Niels Bohr was very happily married.

Once recognized, complementarity is a wisdom we rediscover, and confirm, both in the physical world and beyond. It is a wisdom I embrace, and recommend to you. Let us consider some complementary pairs:

## *Reduction and Abundance*

- The basic building blocks of Nature are few and profoundly simple, their properties fully specified by equations of high symmetry.
- The world of objects is vast, infinitely various, and inexhaustible.

Understanding the basics more deeply cannot undo the richness of experience. It can and does illuminate experience, and empower us to enrich it further.

## *One World, and Many*

- Individual brains are the ultimate repositories of human thought, and they fit comfortably within individual skulls, within individual bodies, here on planet Earth. Most people, most of the time—when they're not philosophizing, or doing astronomy—are concerned with events that occur in a small region around the surface of the Earth. It is where human history—great wars, great art, and billions of absorbing "ordinary" lives—plays out.
- Viewed from even the near afar, Earth is merely a tiny speck of reflected light (figure 43).

Recent developments in cosmology suggest that the part of the Universe we can presently access, even with the most powerful instruments,

FIGURE 43. EARTH, AS SEEN FROM MARS.

is but a small part of a multiverse whose faraway parts might look quite different. Should this be established, it would amplify a theme that has already repeatedly sounded before: each person's experienced "world" is one among billions of such (one per human, at least); Earth is one among several planets of our Sun; our Sun is one among billions of stars in our Milky Way Galaxy; our Galaxy is one among billions in the visible Universe.

The existence of surrounding immensities does not subtract from me, you, or humanity as a whole. It can and does expand our imagination.

## *Object and Person*

- I am, and you are, a collection of quarks, gluons, electrons, and photons.
- I am, and you are, a thinking person.

## *Determined and Free*

- I am, and you are, a material object, subject to the laws of physics.
- I am, and you are, capable of making choices. I am, and you are, responsible for them.

## *Transient and Eternal*

- The state of the world is in flux, and every object within it is subject to change.
- Concepts live outside of time and, because All Things Are Number, liberate us from it.

At the frontiers of physics and cosmology, this complementarity is very much in play. We currently make an uneasy separation between laws of physics and initial conditions, which begs to be transcended. The worldview of any finite observer evolves, but space-time as a whole, which is the most natural arena for world description, does not. The quantum mechanical wave function of a system as a whole can be constant in time, while its parts, regarded separately, experience relative change. (For experts: this occurs routinely, in the energy eigenfunctions of complex systems.) Something of that sort could well be true of the world as a whole. Change Without Change, the great and fruitful principle of symmetry, would then be fully embodied, as Parmenides paradoxically insisted:

> One story, one road, now
> is left: that it is. And on this there are signs
> in plenty that, being, it is ungenerated and indestructible,
> whole, of one kind and unwavering, and complete.

A final complementary pair concludes our meditation:

## *Beautiful and Not Beautiful*

- The physical world embodies beauty.
- The physical world is home to squalor, suffering, and strife.

In neither aspect should we forget the other.

# ACKNOWLEDGMENTS

The seed of this project was planted in 2010, when I accepted an invitation from Darwin College, Cambridge, to lecture on "Quantum Beauty." I would like to thank Christopher Johnson and 3Play Media for preparing an extremely useful transcript of the lecture, and Zoe Leinhardt, Philip Dawid, and especially Lauren Arrington for their suggestions and aid in preparing it as a chapter in the collection *Beauty*.

I would like to thank John Brockman for nourishing the seedling by encouraging me to develop and expand on the ideas expressed there and bringing it to the attention of Penguin Press.

Scott Moyers and Mally Anderson of Penguin Press have been helpful from the start, supplying a wise mixture of enthusiasm, constructive criticism, and creative suggestion that inspired me to write more and to rewrite. Mally has seen the project through its later stages with aplomb, nurturing it into its mature form. I am also grateful to the designers and technical staff at Penguin Press for their professionalism and devotion to the ideal of producing a beautiful product.

Al Shapere made helpful comments on early drafts.

My wife and partner in life, Betsy Devine, read everything and made many suggestions for getting the language more direct and forceful. She also suggested and advocated for "Terms of Art"; without her input, that section would not exist in anything like its present form. Betsy also eased me through the ups and downs that are an inevitable part of such a big undertaking as this book.

I am grateful to my home institution, MIT, for the kind of reliable support that makes an adventure of this kind possible, and to Arizona State University for support during its concluding phase. A crucial part of the book was written during a visit to China, especially during a magical week on the West Lake. The influence of that week can be felt at many places, beginning with the frontispiece. I would like to thank Vincent Liu, Wu Biao, and Hongwei Xiong for arranging the visit.

# TIMELINES

## I: PRE-QUANTUM PHYSICS

**C. 525 BCE** Pythagoras (570–495 BCE) develops numerical laws for geometry and musical harmony.

**C. 369 BCE** Plato's (429–347 BCE) friend Theaetetus develops theory of Platonic solids.

**C. 360 BCE** Plato's dialogue *Timaeus* argues his atomic theory and cosmological speculations.

**C. 300 BCE** Euclid's (323–283 BCE) *Elements* develops geometry as deductive system.

**C. 1400** Filippo Brunelleschi (1377–1446) develops projective geometry as foundation for artistic perspective.

**C. 1500** Leonardo da Vinci (1452–1519) foreshadows fusion of art, engineering, and science.

**1543** Nicolaus Copernicus's (1473–1543) *De Revolutionibus* proposes heliocentric system, based on mathematical esthetics.

**1596** Johannes Kepler's (1571–1630) *Mysterium Cosmographicum* proposes Copernican Solar System model based on Platonic solids. His later work establishes empirical laws of planetary motion.

**1610** Galileo Galilei's (1564–1642) *Sidereus Nuncius* (*Starry Messenger*) announces "mini-Copernican" system of moons orbiting Jupiter, and establishes earthlike nature of our Moon, based on observations using his pioneering telescopes.

**1666** Isaac Newton (1642–1727) develops breakthrough theories in calculus, mechanics, and optics.

**1687** Newton's *Principia* solves earthly and celestial laws of gravity according to mathematical principles.

**1704** Newton's *Opticks* reports his experiments and speculations about the nature of light.

**1831** Michael Faraday's (1791–1867) discovery of magnetic → electric induction.

**1850S–60S** James Clerk Maxwell's (1831–79) publications on color vision commence in **1855**. Major papers on electrodynamics: "On Faraday's Lines of Force," **1855**; "On Physical Lines of Force," **1861**; "A Dynamical Theory of the Electromagnetic Field," **1864**.

**1887** Heinrich Hertz's (1857–94) production and detection of electromagnetic waves verifies Maxwell's induction and lays the theoretical foundations for radio and eventually other telecommunication.

## II: QUANTUM PHYSICS, SYMMETRY, AND THE CORE

**1871** Sophus Lie's (1842–99) thesis introduces concepts of continuous transformations and symmetry, which he develops and refines in later work.

**1899** Ernest Rutherford (1871–1937) identifies nuclear decays with emission of electrons ("beta decay") as a special form of radioactivity, initiating experimental study of the weak force.

**1900** Max Planck (1858–1947) introduces quantization for exchange of energy between matter and light.

**1905** Albert Einstein (1879–1955) introduces the concept that light itself comes in discrete units (quanta = photons).

Einstein's (1879–1955) special relativity and general relativity (**1915**) are powerful physical theories based on assumptions of symmetry. They set the stage for later work on rigid (global) and anamorphic (local) symmetry, respectively.

**1913** Hans Geiger (1882–1945) and Ernest Marsden (1889–1970), following a suggestion by Ernest Rutherford, use scattering experiments to demonstrate the existence of atomic nuclei.

Niels Bohr (1885–1962) introduces successful model of atoms, based on quantum ideas.

**1918** Emmy Noether's (1882–1935) theorem establishes connection between continuous symmetry and laws of conservation.

**1924** Satyendra Bose (1894–1974) introduces concept that photons are an example of what we now call bosons.

**1925** Wolfgang Pauli (1900–58) introduces exclusion principle.

Enrico Fermi (1901–54) and Paul Dirac (1902–84) introduce concept that electrons are an example of what we now call fermions.

Werner Heisenberg (1901–76) introduces modern quantum theory, embedding Bohr's ideas into a mathematically coherent framework.

**1926** Erwin Schrödinger (1887–1961) proposes the Schrödinger equation. This looks very different from, but proves equivalent to, Heisenberg's more abstract proposal.

**1925-1930** Paul Dirac (1902–84) proposes the Dirac equation for electrons, and a quantized version of the Maxwell equations, in a series of brilliant papers. His work establishes quantum electrodynamics (QED) as a rich physical theory.

**1928** Hermann Weyl (1885–1955) shows that the quantum version of Maxwell's theory (quantum electrodynamics, or QED) is an embodiment of anamorphic symmetry.

**1930** Wolfgang Pauli (1900–58) postulates existence of unseen neutrinos, to preserve conservation of energy and momentum in weak decays.

**1931** Eugene Wigner (1902–95) shows the power of rigid symmetry in quantum mechanics.

**1932** Enrico Fermi (1901–54) applies general principles of special relativity and quantum mechanics to weak decays, establishing their validity in this new realm.

**1947-48** Measurements of effects that deviate from the straightforward Dirac theory: a shift in hydrogen energy levels ("Lamb shift") observed by Willis Lamb (1913–2008), and an "anomalous" electron magnetic moment, observed by Polykarp Kusch (1911–93), indicate the importance of including quantum fluctuations.

**1948** Richard Feynman (1918–88), Julian Schwinger (1918–94), and Sin-Itiro Tomonaga (1906–79) show that Dirac's quantum electrodynamics, when solved more accurately, includes quantum fluctuations (virtual particles).

**1950** Freeman Dyson (1923– ) puts preceding work on a firm mathematical footing, and demonstrates its consistency.

**1954** C. N. Yang (1922– ) and Robert Mills (1927–99) combine the ideas of Lie and Maxwell/Weyl to find equations that embody more complex forms of anamorphic symmetry. These Yang-Mills equations are at the heart of our modern Core Theories.

**1956** Frederick Reines (1918–98) and Clyde Cowan (1919–74) observe interactions of neutrinos, establishing their tangible reality.

Tsung-Dao Lee (1926– ) and C. N. Yang propose that the weak force makes a fundamental distinction between left and right ("parity violation"). Experimental confirmation follows shortly thereafter.

**1957** John Bardeen (1908–91), Leon Cooper (1930– ), and J. R. Schrieffer (1931– ) propose the milestone "BCS theory" of superconductivity. Powerful ideas of spontaneous symmetry breaking, including the Higgs mechanism, are implicit in this work.

**1961** Sheldon Glashow (1932– ) proposes anamorphic theory with mixing of weak and electromagnetic forces.

**1961-62** Yoichiro Nambu (1921– ) and Giovanni Jona-Lasinio (1932– ) introduce spontaneous symmetry breaking in a specific theory of fundamental particle interactions. Jeffrey Goldstone (1933– ) simplifies and generalizes their concept.

**1963** Philip Anderson (1923– ) suggests the importance for particle physics of work on equations for massive photons that arose in work by the brothers Fritz (1900–54) and Heinz London (1907–70) in **1935** and by Lev Landau (1908–68) and Vitaly Ginzburg (1916–2009) in **1950**.

**1964** Robert Brout (1928–2011) and François Englert (1932– ); Peter Higgs (1929– ); and Gerald Guralnik (1936–2014), Carl Hagen (1935– ), and Tom Kibble (1932– ) make concrete theoretical models reconciling massive particles with anamorphic symmetry.

Murray Gell-Mann (1929– ) and George Zweig (1937– ) suggest the existence of quarks as primary building blocks of hadrons.

Abdus Salam (1926–96) and John Ward (1924–2000) clarify anamorphic electroweak theory.

**1967** Steven Weinberg (1933– ) incorporates spontaneous symmetry breaking into anamorphic theory, defining mature Core electroweak theory.

**1970** Gerard 't Hooft (1946– ), together with Martinus Veltman (1931– ), puts preceding work on a firm mathematical footing, and demonstrates its consistency.

Jerome Friedman (1930– ), Henry Kendall (1926–99), and Richard Taylor (1929– ) take snapshots of proton interiors, find nearly free quarks plus unknown electrically neutral matter.

**1971** Sheldon Glashow, John Iliopoulos (1940– ), and Luciano Maiani (1941– ) add quarks to anamorophic electroweak theory and predict the existence of charmed quarks.

**1973** David Gross (1941– ), Frank Wilczek (1951– ), and David Politzer (1949– ) establish asymptotically free theories. Gross and Wilczek formulate precise theory of the strong force: quantum chromodynamics (QCD).

**1974** Experimental discovery of heavy quark mesons provides semiquantitative evidence for asymptotic freedom and QCD.

Jogesh Pati (1937– ) and Abdus Salam, and also Howard Georgi (1947– ) and Sheldon Glashow, propose unification of Core Theories.

Howard Georgi, Helen Quinn (1943– ), and Steven Weinberg investigate the relative strength of different forces, using asymptotic freedom.

Julius Wess (1934–2007) and Bruno Zumino (1923–2014) formulate supersymmetry.

**1977** Roberto Peccei (1942– ) and Helen Quinn suggest a new symmetry to solve the "θ problem."

Wilczek discovers the dominant coupling of the Higgs particle to ordinary matter, through color gluons.

**1978** Wilczek and Weinberg point out that Peccei-Quinn symmetry implies the existence of a remarkable new light particle, the axion.

**1981** Savas Dimopoulos (1952– ), Stuart Raby (1947– ), and Wilczek demonstrate quantitative advantages of including supersymmetry in unification.

**1983** Several authors establish axions as candidate to provide astronomical dark matter.

Carlo Rubbia (1934– ) leads experimental team at CERN to observation of weakons (*W* and *Z* particles), establishing anamorphic electroweak theory.

**1990S** Experiments at the Large Electron-Positron Collider show clear jet structure, provide powerful quantitative evidence for asymptotic freedom and QCD.

**2005** Building on ideas of Kenneth Wilson (1936–2013), Alexander Polyakov (1945– ), and Michael Creutz (1944– ), massive computer power is brought to bear on QCD, allowing accurate calculation of hadron masses, including that of protons and neutrons.

**2012** Discovery of Higgs particle at the Large Hadron Collider.

**2020** My bets favoring discovery of supersymmetry at the Large Hadron Collider come due at midnight, December 31.

# TERMS OF ART

This supplement contains definitions and brief commentaries on scientific concepts that may be unfamiliar to general readers, and that are used in this book. In some cases (e.g., energy, symmetry) they describe common words that we are using in special ways—generally with narrower, more specific meanings than in their everyday use. I've tried, to the extent possible, to make this supplement an organic part of the whole, by using themes and examples from the main text in the commentaries. Here also you will find several ideas—including some beautiful ones—that I was tempted to include in the text, but could not fit in smoothly. In many cases I've had to compromise precision and mathematical rigor in favor of brevity and accessibility.

Note on typography: *Italics* are used to state the topic word or phrase of each entry, to mark its major appearances within the entry, and occasionally for emphasis. **Boldface** is used within entries to point out significant use of terms that head other entries.

### ABSORPTION

We say a particle is absorbed when its independent existence ends. Because total energy is conserved, the energy of the particle then takes another form. For example, when a particle of light (photon) hits your retina, it may be absorbed by proteins (rhodopsins) that bend in response. Their bending, in turn, triggers electrical signals that our brains interpret as visual experience.

### ACCELERATION

Velocity is the rate of change of position with time—see **Velocity**—and *acceleration* is the rate of change of velocity with time. One of Newton's great achievements was to point out that the accelerations of bodies are related to the forces acting on them. (He announced this discovery, before fully revealing it, in a memorable anagram, as described in "Newton III.") In early textbooks of classical mechanics, you will usually find Newton's second law of motion: **Force** equals **mass** times acceleration. In the absence of some separate information about forces, of course, that statement is empty. It should be interpreted, really, as a promise that studying acceleration will be rewarding!

Newton supplied some general statements about forces. Notable is his first law of motion, which says that "free" bodies have zero acceleration (i.e., that the velocity of a free body is constant). Implicit in this law is the promise that bodies that are far away from all other bodies are approximately free; or, in other words, that forces decrease with distance.

Newton also developed a detailed theory of one kind of force, the **gravitational** force. It is interesting to observe, in this context, that the gravitational force on a body is proportional to its mass, so that the gravitational acceleration is *independent* of the body's mass! That principle was tested, for the Earth's gravity, in Galileo's famous experiment of dropping objects from the Leaning Tower of Pisa.

In Einstein's theory of gravity, the general theory of **relativity**, the law of motion is expressed directly in terms of acceleration, without separate mention of force.

Acceleration, like velocity, is a **vector** quantity.

### ACCELERATOR

A particle *accelerator* is a machine that produces beams of rapidly moving, highly energetic particles. Historically, and still today, accelerators have been used to reveal fundamental processes in Nature. By studying collisions of the most rapidly moving particles, we glimpse behavior at extremes of high energy, small distances, and short times that are otherwise inaccessible.

### ACTION AT A DISTANCE

*Action at a distance* is a feature of Newton's theory of **gravity**: bodies exert gravitational forces on other bodies, even distant ones, instantaneously through empty space. Newton himself did not like that feature of his theory, but his mathematics led him there. The success of Newton's theory, based on action at a distance, was so complete that that idea was tacitly assumed by early investigators in electricity and magnetism.

Faraday developed an alternative vision, in which electric and magnetic forces are transmitted as pressures by space-filling fluids. Maxwell developed Faraday's intuitions mathematically, and thereby arrived at the fluid (or, alternatively, field) concepts of electromagnetism we use today.

Astrology postulates powerful action at a distance, but there is, to say the least, no rigorous evidence for its validity.

### ALPHA PARTICLE

Early in the experimental study of radioactivity, Ernest Rutherford divided the emitted material into alpha, beta, and gamma rays. They were distinguished by their ability to penetrate matter, their susceptibility to bending by magnetic fields, and other properties. Further studies reveal that alpha rays consist of helium 4 nuclei—that is, they are bound combinations of two protons and two neutrons. We call those nuclei alpha particles.

### AMPERE'S LAW / AMPERE-MAXWELL'S LAW

Ampere's law is now regarded as part of one of **Maxwell's equations**, though historically it was an earlier discovery. Ampere's law, in its original form, states that the **circulation** of the **magnetic field** around a curve is equal to the **flux** of **electric current** through any surface bounded by the curve. To unpack this, see **Circulation**, **Flux**, and **Current**. You will also find plate N helpful.

Maxwell, motivated by considerations of mathematical consistency and beauty, modified Ampere's law by adding an extra term. According to the complete Ampere-Maxwell's law, the circulation of the magnetic field around a curve is equal to the flux of electric current through any surface bounded by the curve *plus* the rate of change in electric flux through that surface.

Maxwell's new term is a sort of dual to **Faraday's law**. Faraday's law says that changing magnetic fields can produce electric fields, while Maxwell's term says that changing electric fields can produce magnetic fields.

### ANALOG

If a quantity can vary smoothly, or as we often say, continuously, we say it is an *analog* quantity. Analog quantities are to be contrasted with **digital** quantities, which can take on only a discrete range of values, and so can vary only discontinuously. In the current foundations of physics, length and duration of time are analog quantities.

An extreme interpretation of the Pythagorean credo All Things Are Number would have it that all quantities are, fundamentally, digital. The impossibility of expressing both the side and the diagonal of a square as multiples of a common unit, and Zeno's paradoxes of motion, gave early warnings of problems with that view.

Digital quantities have great advantages for computation and communication of information, because they allow for correction of small errors. Thus, for example, if you know that the valid results of a computation can be only 1 or 2, and your approximate version of that computation yields 1.0023, you can infer that the correct answer is 1, unless your approximation is very poor.

If the unit of discreteness is sufficiently small, a digital quantity can give a good representation of one that is intrinsically analog. For example, a digital photograph may consist of black dots so finely spaced that to the human eye, with its imperfect resolution, they appear as smoothly varying, in shades of gray that depend on their density.

The mathematical description of analog quantities typically involves **real numbers**, while the simplest digital quantities are described by **natural numbers**.

### ANALYSIS

As used in physics, chemistry, and mathematics, the word "analysis" typically refers to the process of studying things by studying their parts. In this usage "holistic analysis" is oxymoronic; and psychoanalysis is something else again.

Two interesting examples of analysis are the separation of light into spectral colors, and the analysis of functions by studying their variation over small ranges, as in (differential) **calculus**.

### ANALYSIS AND SYNTHESIS

"Analysis and Synthesis" is Newton's phrase for the strategy of reaching complete and powerful understanding of a class of objects by understanding the behavior of their simplest parts precisely (analysis) and then building back up (synthesis). Newton himself applied this strategy with great success to the study of light, to the study of motion, and to the study of mathematical **functions**.

Analysis and Synthesis is a more elegant, appropriate, and historically justified way to express what is commonly called "reductionism," and should always be preferred, outside of polemics.

### ANGULAR MOMENTUM

Angular momentum is, together with **energy** and **momentum** (that is, ordinary or linear momentum), one of the great conserved quantities of classical physics. Each of them has evolved into a pillar of modern physics, as well. Angular momentum is by far the most complicated of those quantities to define and to understand, and you should not expect to grasp its intricacies without substantial effort. The fascinating, often counterintuitive behavior of tops and gyroscopes is a consequence of their angular momentum, for example. Accordingly, our meditation does not lean hard on this concept!

The angular momentum of a body is a measure of its angular motion around a specified center. Quantitatively, it is equal to twice the rate at which a line drawn from the center to the body sweeps out area, multiplied by the body's mass. (This is the nonrelativistic version, accurate for small velocities. **Special relativity** leads to a related, but more complicated, formula.)

Angular momentum has a direction, as well as a magnitude. (Thus it is a **vector** quantity—an **axial vector**, actually.) To define the direction, we first identify the momentary axis of rotation—i.e., the direction perpendicular to the increment of area—and then orient that axis using the right-hand rule. See **Handedness**.

The angular momentum of a system of bodies is the sum of the angular momenta of the bodies separately.

There is a wide variety of circumstances in which angular momentum is conserved. This result is best understood through Noether's general theorem, which connects **conservation laws** to **symmetry**. In that framework, conservation of angular momentum around a center reflects symmetry (i.e., invariance) of the physical laws under transformations that rotate space around that center. In other words, we have conservation of angular momentum when the laws do not depend on any externally specified, fixed direction.

Kepler's second law of planetary motion, according to which a line drawn between a planet and the Sun sweeps out equal areas in equal times, is an instance of the conservation of angular momentum.

In the quantum world, angular momentum continues to be a valid concept, and takes on additional features of great subtlety and beauty. It was the mathematics of the quantum theory of angular momentum that most attracted me to physics when as a student I was considering different career choices. If you'd like to pursue this further, you might consult the "Recommended Reading." Here I will mention only that quantum particles often exhibit an irreducible rotary activity, or **spin**, similar in spirit to their zero-point motion (see **Quantum fluctuation**), or to the spontaneous activity of quantum fluids.

### ANTHROPIC ARGUMENT / ANTHROPIC PRINCIPLE

Roughly speaking, anthropic arguments take the form

> The world must be as it is, in order for me to exist.

Before discussing more sophisticated refinements, let us consider the *anthropic principle* in that basic form.

It is important to distinguish between two aspects of this basic anthropic argument: its truth, and its explanatory power. (See the related discussion in **Consistency**.) Depending on how narrowly one defines "me," there can be a great deal of trivial truth in the anthropic assertion. If "me" is meant to entail a carbon-based life-form with human physiology that has experienced life as I have experienced it (which, among other things, includes reading books about science that make strong, specific assertions about the natural world), well, then, surely the laws of physics as we know them, not to mention many other things about Earth, the history of Europe, the color of my children's eyes, etc., could not be very different before precluding the existence of "me." So the basic anthropic argument, taken literally, is true. But that true argument has very little explanatory power, basically because the existence of "me" is such an all-encompassing assumption—embracing everything that I have or ever will experience—that it leaves nothing left to explain!

More sophisticated versions of the anthropic argument must hinge on looser definitions of "me." We might require that the world, in its basic laws and history, must make possible the emergence of some kind of intelligent or conscious observers, for example. Otherwise it could not be observed—and so to heck with it! But it is very difficult to define what one means by intelligent observers, and also very difficult to assess, once you've decided on a definition, what sort of laws and histories give rise to intelligent observers. I find it hard to imagine that such vague, slippery ideas will lead us to significant explanatory power.

It is noteworthy that the deepest understandings of the world we have achieved, as in our **Core Theory**, involve conceptual principles like **relativity**, **local symmetry**, and the framework of **quantum theory** that are abstract in form and universal in scope. Those principles don't look at all like anthropic arguments! Evidently things are happening in the world that transcend its desire to produce "me."

In general, anthropic arguments, by their nature, shift the focus of discussion from explanation to assumption. Because they compromise explanatory power, as a matter of principle they are best avoided. But in the right, very special circumstances, arguments with an anthropic flavor can be both valid and useful. For what might be interesting examples, see **Dark energy / Dark matter**.

### ANTIMATTER / ANTIPARTICLE

In 1928 Paul Dirac proposed a new equation, which today we call the **Dirac equation**, for the behavior of electrons in quantum mechanics. That work importantly predicted that there should also be a particle, the **positron**, with the same mass and spin as an **electron**, but the opposite charge. This positron, also called the antielectron, is the antiparticle of an electron. Later work revealed that this phenomenon is a far more general result of quantum mechanics and special relativity: for each particle, there is an associated *antiparticle,* which has the same mass and spin but opposite values of the electric charge, and also of weak and strong color charges, and of handedness.

Antielectrons, or positrons, were discovered experimentally in 1932. Antiprotons were first seen in 1955. It would be a major shock to discover a particle that did not have an antiparticle. **Photons** are their own antiparticles. (This is possible because they are electrically neutral, and carry no other charges.)

A particle, when brought together with its antiparticle, can annihilate into "pure energy"—which means, in practice, any of a wide variety of groups of particles and their antiparticles. For example, any particle and its antiparticle can annihilate into two photons or into a **neutrino**-antineutrino pair, although those are usually not the most likely outcomes. The Large Electron-Positron Collider at CERN, which was the predecessor of the **Large Hadron Collider** and occupied the same enormous tunnel, was devoted to the study of annihilation of electrons and positrons accelerated to move rapidly in opposite directions.

Although the term *antiparticle* has clear and precise scientific meaning, the term *antimatter*, although sometimes useful, is a bit more problematic, or perhaps I should say provincial. To understand this usage, we should start with the (equally problematic and provincial) definition of "matter" that it mirrors. In that definition, we declare that some of the particles we're made of, and encounter in everyday life—that is, $u$ and $d$ quarks and electrons—are "matter." We also include, in this definition, their close relatives—thus all kinds of **quarks**, and all kinds of **leptons** ($u, d, c, s, t, b$ quarks, $e, \mu, \tau, \nu_e, \nu_\mu, \nu_\tau$)—as matter, while their antiparticles are called antimatter. Photons do not fall naturally on either side of this divide because they are their own antiparticles. The only thing that distinguishes "matter," in this sense, from "antimatter" is that it is more common, at least in our part of the Universe. If we suddenly changed all the particles in the world into their antiparticles (and simultaneously made a **parity** transformation, interchanging right and left), it would be very hard to tell the difference!

I think the term "antimatter" is more apt to confuse than enlighten, and in this book I have avoided it. When I speak of "matter," without qualification, I mean all forms of matter, including (for example) antiquarks, photons, and **gluons**.

#### ANTISYMMETRIC

We say a quantity is *symmetric* under a transformation, or exhibits **symmetry**, when it does not change under the action of the transformation. We say a quantity is *antisymmetric* under a transformation when it changes sign under the action of the transformation. This notion can be applied to numerical quantities, or to **vectors**, or to **functions**, because in all those cases it makes sense to speak of changing sign.

Examples: The **coordinates** of points on a line are antisymmetric under rotation of the line by 180 degrees around its origin. **Electric charge** is antisymmetric under the transformation that takes particles into their **antiparticles** (see **Antimatter**).

A fundamental characterization of **fermions** states that the quantum mechanical **wave function** describing a system of identical fermions is antisymmetric under the transformations that exchange any two of them.

#### ASYMPTOTIC FREEDOM

The **strong force** between two **quarks** is modified by the ceaseless, spontaneous activity of quantum fluids that permeate space. The force weakens as the quarks approach one another, and strengthens as they separate. This is *asymptotic freedom*.

Asymptotic freedom has many implications and applications that are described at length in the text.

See also **Confinement** and **Renormalization / Renormalization group**.

#### ATOMIC NUMBER

The atomic number of a **nucleus** is the number of protons it contains. The *atomic number* of a nucleus determines its **electric charge**, and thus its influence on **electrons**, and thus its role in the chemistry of atoms or molecules where it figures. Nuclei that have the same atomic number but different numbers of neutrons are said to be **isotopes** of the same chemical element.

Example: Carbon 12 ($C^{12}$) nuclei contain six protons and six neutrons, while carbon 14 ($C^{14}$) nuclei contain six protons and eight neutrons. They have basically the same chemistry—hence both are called "carbon"—but different mass. Carbon 14 nuclei are unstable, and their decays can be used to date biological samples. (When an organism dies it ceases to incorporate new carbon, and hence the ratio of Carbon 14 to Carbon 12 in the corpse gradually decreases. In the atmosphere, Carbon 14 gets renewed by cosmic-ray collisions.)

### AXIAL CURRENT

Axial currents are a special class of currents that do not change sign under a spatial **parity** transformation. Thus axial currents define **fields** of **axial vectors**. I brought in this rather esoteric concept when searching for a reason to introduce the term "**axion**" that could get past the editors of *Physical Review Letters*.

### AXIAL VECTOR

See **Parity**, where this concept appears in its natural context.

### AXION

The axion is a hypothetical particle whose existence would improve the beauty of the **Core Theory**. Axions are also, at present, an excellent candidate to provide the cosmological **dark matter**.

The Core Theory has many virtues, but also exhibits some esthetic shortcomings. Among the latter is the following:

We observe, experimentally, that the laws of physics are, to a very good approximation (though not exactly), **invariant** under a change in the direction of time. Put simply: If you take a movie of any experiment in physics, and run the movie backward, the images you see will still depict events that obey the basic laws of physics. Of course, if you take a movie of everyday life, and run it backward, what you see will *not* look like everyday life. But in the subatomic world, where the basic laws operate most clearly, the distinction disappears. We say, therefore, that the laws of physics are very nearly invariant under reversal of the direction of time, or alternatively that they obey time reversal (*T*) symmetry.

This property of the laws, that they obey *T* **symmetry**, is consistent with the Core Theory, but it is not quite forced by it. There is an interaction among color **gluons** that is consistent with all known general principles including **quantum theory**, **relativity**, and **local symmetry**—and thus, according to the Core, "possible"—but whose existence would violate *T* symmetry.

It is **consistent**, but lame, simply to assert that this interaction does not occur. A more fitting response, championed by Roberto Peccei and Helen Quinn, is to explain this "coincidence" by expanding the Core to support additional symmetry. By doing this in an appropriate way, one can explain the smallness of *T* violation. (Other potential explanations have been advanced, but none has stood the test of time.) This expansion of the Core is not without consequence: as Steven Weinberg and I pointed out, it implies the existence of a new, very light particle with remarkable properties, the *axion*.

Axions have not yet been detected experimentally, but that non-observation is not conclusive, because theory predicts that axions should interact very feebly with ordinary matter, and to date no experiment has achieved the necessary sensitivity. At this writing, several experimental groups around the world are actively working to find evidence for axions, or else to exclude them convincingly.

One can calculate the production of axions during the Big Bang. It emerges, from these calculations, that the Universe is permeated by a gas of axions, a gas that might well supply the cosmological dark matter.

### BARYON. SEE HADRON.

### BOOST

In the primary scientific literature, it is increasingly common to refer to the transformation we make on a system, when we imagine adding or subtracting a constant velocity to the motion of each of its parts, as a *boost*. This term is inspired, I suppose, by the action of booster stages in rockets, which work to impart velocity to their payloads. In this book I refer to these transformations instead as **Galilean transformations**, in homage to Galileo, who memorably emphasized their importance with his elegant thought experiment, wherein he takes us aboard a sailing ship, in an isolated cabin (as quoted in the main text). See **Galilean transformation**.

### BOSON / FERMION

Elementary particles divide into two broad classes: bosons and fermions.

In the **Core Theory**, **photons**, **weakons**, color **gluons**, **gravitons**, and **Higgs particles** are *bosons*. In the text I often refer to these as

**force particles**. Bosons can be created or destroyed singly.

Bosons obey Bose's inclusion principle. Roughly speaking, this means that two bosons of the same kind are especially happy to do the same thing. Photons are bosons, and the inclusion principle for photons is what makes lasers possible. When offered the chance, a collection of photons will try to all do the same thing, making a narrow beam of spectrally pure light.

In the Core Theory, **quarks** and **leptons** are *fermions*. In the text I often refer to these as **substance particles**.

Fermions come and go in pairs. As a result, if you've got one, you can't get rid of it altogether. It might turn into another kind of fermion, or into three, or five, together with any number of non-fermions (i.e., bosons; see above)—but it cannot dissolve into nothingness, leaving no trace.

Fermions obey Pauli's **exclusion principle**. Roughly speaking, this means that two fermions of the same kind don't like to do the same thing. **Electrons** are fermions, and the exclusion principle for electrons plays a crucial role in the structure of matter. It guides, in "Quantum Beauty II," our exploration of the exuberant world of carbon.

### BRANCHING RATIO

When a particle can decay in several alternative ways, we say that it has several decay channels, or decay branches. The relative probability with which a particular decay branch occurs is called its *branching ratio.* Thus if particle $A$ decays into $B + C$ 90 percent of the time, but $D + E$ 10 percent of the time, we say that the branching ratio of $A$ into $B + C$ is 0.90, while its branching ratio into $D + E$ is 0.10.

### BUCKMINSTERFULLERENE / BUCKYBALL

The *buckminsterfullerenes* are a class of pure carbon molecules.

They take the form of quasi-spherical polyhedral surfaces, with each carbon **nucleus** extending chemical bonds to its three nearest neighbors. The faces of the **polyhedron** always include twelve pentagons, plus a variable (generally larger) number of hexagons. The buckminsterfullerene $C_{60}$, containing sixty carbon nuclei, is especially common. Molecules of $C_{60}$ are often called buckyballs, celebrating their uncanny resemblance to (microscopic) soccer balls.

See also **Polygon**.

### CABIBBO ANGLE. SEE FAMILY.

### CALCULUS

The word "calculus" derives from the Latin word for "pebble" or "stone." Its modern use, in mathematics, traces back to the operation of counting, or keeping accounts, with stones (as many people do, even today, using an abacus). We see this origin reflected in the general term "calculation," used for many different procedures and operations to process information.

Mathematics recognizes several kinds of calculus (e.g., propositional calculus, lambda calculus, calculus of variations). But one particular method of processing mathematical information is so important, and made such an impact on the minds of scientists, that when people speak of calculus, without further qualification, it is the default meaning.

Calculus, so understood, is the method of **Analysis and Synthesis**, applied to the study of smoothly varying processes or **functions**. The two branches of calculus, differential calculus and integral calculus, reflect that method. Differential calculus provides concepts and methods for analyzing behavior over very small intervals, while integral calculus provides concepts and methods for synthesizing that local information into global understanding.

An outstanding application of calculus, which Newton had in mind as he developed the subject, is the description of motion. One introduces concepts like **velocity** and **acceleration** to characterize the motion over very brief intervals of time (differential calculus), or, conversely, uses information about velocity and acceleration to construct orbits (integral calculus). In classical mechanics, **force** laws give information about the acceleration of a body. A major challenge in classical physics is how to use that information: to construct the motion of that body, in response to a known acceleration. This is a problem in integral calculus—building up the large from knowledge of the small.

### CELESTIAL MECHANICS

Originally, *celestial mechanics* meant the application of classical mechanics, together with Newton's theory of **gravity**, to the description of motion of major bodies—notably planets, their moons, and comets—in the Solar System. Today the term "celestial mechanics" is used more broadly to describe the application of mechanics to astrophysical bodies, and also to rockets and artificial satellites. Because the relevant laws of physics apply universally, celestial mechanics is really a specialized branch of mechanics, rather than an independent subject.

### CHARMED QUARK

The *charmed quark*, denoted "c," is a member of the second **family** of **substance particles**. Charmed quarks are highly unstable, and they play a very minor role in the present-day natural world. Charmed quarks were discovered in 1974, and their experimental investigation was instrumental in establishing the **Core Theory**.

### CIRCULATION

**Vector fields**, whatever their true nature, can be regarded mathematically as representing the flow of an ordinary fluid such as air or water. (The imagined flow has, at every point, its velocity proportional to the value of the vector field at that point.) In this model, the *circulation* of the vector field at a point is a measure of the fluid's angular motion. Thus, for example, the circulation of the atmosphere is especially large around curves that encircle a tornado's center.

Let us define this more precisely. Imagine that our curve forms the center of an imaginary, narrow cylindrical tube, and compute the amount of air that is transported around the tube, per unit time, divided by the cross-sectional area of the tube. (Flow of air into or out of our imaginary tube is simply ignored.) Then you will have computed the circulation of the flow around the curve.

Exploiting the flow analogy—that is, regarding the electric field as a velocity field—we can likewise define the circulation of the **electric field** around a curve, or of the **magnetic field** around a curve. Those quantities appear as star players in **Maxwell's equations**. See **Ampere's law / Ampere-Maxwell's law** and **Faraday's law**.

Here I'd like to add a personal coda, combining hero worship and esthetics. The pioneering papers of Faraday and Maxwell, where the field concepts of electromagnetism first appeared, are largely written in terms of word definitions and mental images close to what I've given you here for *circulation,* and below for **flux**, rather than in conventional mathematical equations. To keep such complicated pictures clearly in mind, and to make connections among them, was an astonishing feat of visual imagination that I have found inspiring, and beautiful, to re-imagine. Picturing the equations brings them into a domain of experience humans are well prepared to enjoy.

### COLOR (OF LIGHT) / SPECTRAL COLOR

It is important, in thinking about light, to distinguish between physical color and perceptual color.

*Spectral color* is a physical concept, independent of human perception. In principle, it can be defined and explored entirely using physical tools—lenses, prisms, photographic plates, etc. We can produce light of any pure spectral color by passing a beam of white light through a prism and selecting a small portion of the emergent "rainbow," as discussed in the main text. We now understand that pure spectral colors correspond to **electromagnetic waves** that **oscillate** periodically with a definite **frequency**. The different pure spectral colors correspond precisely to different frequencies. According to Maxwell's (well-tested) theory one can have electromagnetic waves with any possible frequency, so that the pure spectral colors form a continuum. The human eye is sensitive only to electromagnetic waves within a narrow range of frequencies; but it is often natural to speak of "light" in a more general way that includes electromagnetic waves in the form of radio waves, microwaves, infrared, ultraviolet, X-rays, and gamma rays. The complete range of possibilities constitutes the **electromagnetic spectrum**.

Spectral colors are analogous to **pure tones** in music. Indeed, pure tones are also oscillations—in this case, sound waves—with definite frequency. White light, in this analogy, corresponds to a cacophony of **tones**, which inspires the term "white noise."

The concept of *perceptual color* involves a mixture of physics and psychology. Our richest color experiences, as in fine art, are extremely complex, and involve high-level brain processes that are poorly understood. Some basic facts about the early stages of vision are well established, however, and they already highlight a vast gap between the analysis of light that is possible, according to basic physical principles, and the analysis our sense of color provides. Most profoundly: Whereas the pure spectral colors form a continuum, and a complete analysis of incoming light would provide the intensity of each kind, human eyes extract only three averages over these intensities.

For much more on these topics, which are central to our meditation, you should consult the main text!

### COLOR CHARGE / STRONG COLOR CHARGE / WEAK COLOR CHARGE

Our **Core Theories** of the **weak** and **strong forces** build on ideas first developed in **electrodynamics**. In particular, they feature variants of electric charge called *color charges.* The charges, in all cases, are conserved quantities that control the behavior of photon-like particles—**photons** for **electric charge**, color **gluons** for *strong color charges,* **weakons** for *weak color charges.*

There are three strong color charges. In the text, they are referred to as red, green, and blue. Eight color gluons respond to, and induce transformations among, those charges.

There are two weak color charges. In the text, they are referred to as yellow and purple.

Needless to say—but said nonetheless—"color," as used in the context of color charge, is a completely different concept from "color" as used in the context of light.

### COMPLEMENTARY / COMPLEMENTARITY

We say two ways of regarding the same thing are *complementary* when each is valid and coherent on its own, but they cannot both be used at the same time, because each interferes with the other. This is a common situation in **quantum theory**. For example, one can choose to measure the position of a particle, or alternatively to measure its **momentum**—but one cannot choose to do both simultaneously, because either measurement interferes with the other. Partly inspired by examples like that, but also by his wider experience of life, Niels Bohr suggested that it is wise to apply the concept of *complementarity* much more broadly as an imaginative method for engaging difficult problems and reconciling apparent contradictions. This broader concept of complementarity, which I find to be valuable and liberating, is best conveyed through examples. You will find several in our concluding postscript, "A Beautiful Answer?"

### COMPLEX DIMENSION

Ordinary ("real") **dimensions** are naturally described by supplying numbers—**coordinates**—that are **real numbers**. Thus the position of a point on a computer screen is specified by two real coordinates, representing vertical and horizontal position, while a point in ordinary space is specified by three coordinates. In many mathematical and physical contexts, it is natural to consider spaces where the coordinates are **complex numbers**. In that case we say we have a complex space, and that the required number of coordinates is the number of complex dimensions in that space. Because a complex number can be specified by two real numbers—specifically, by the magnitude of its real and imaginary parts—a complex space can also be considered as a real space (with extra structure). So regarded, it has a number of real dimensions equal to twice its number of complex dimensions.

### COMPLEX NUMBER

The imaginary unit, denoted $i$, is a quantity which, when multiplied by itself, gives the result $-1$. Thus, in an equation, $i^2 = -1$. *Complex numbers* are numbers $z$ of the form $z = x + iy$, where $x$ and $y$ are **real numbers**; $x$ is called the real part of $z$, and $y$ its imaginary part.

Complex numbers can be added, subtracted, multiplied, and divided, much like real numbers.

Complex numbers were introduced into mathematics so that general equations involving sums and powers—so-called polynomial equations—might have solutions. Thus, for example, the equation $z^2 = -4$ has no solution in real numbers, but it is solved by $z = 2i$ (and by $z = -2i$). It can be proved that complex numbers as we have defined them are fully adequate to this task. (This result, the so-called fundamental theorem of algebra, is not at all obvious, and its proof is a major highlight of mathematics.)

As the term *imaginary* (and its explicit contrast with *real*) suggests, human mathematicians had great difficulty coming to terms with these sorts of numbers. Their "existence" seemed somehow dubious. But a few adventurous souls wisely heeded the advice of Father Jim Malley—"It is more blessed to ask forgiveness than permission"—and used them. Familiarity, and continued success, eventually brought complex numbers into very high esteem. Nineteenth-century mathematics was largely an exploration of the dazzling perspectives that use of complex numbers brings to **calculus** and geometry.

In the twentieth century the process of introducing new kinds of *objects* by listing a set of desirable *properties*, and declaring them to be embodied, which was so successful for complex numbers, became a standard operating procedure. Emmy Noether was a major force in advancing this style of thought. Were Plato informed of these developments he might feel vindicated, considering that mathematicians had fully embraced his philosophy, and discovered the joy of Ideals.

(Indulgent digression, to be read as poetry: Indeed *ideals*, called by that name, are an important class of mathematical object. Perhaps Emmy Noether's masterpiece in pure mathematics, comparable in depth and significance to the conservation theorem we celebrated in our main text, is the concept of a Noetherian ring. What is a Noetherian ring? It is a ring in which any chain of ever larger *ideals* eventually terminates. End of indulgent digression.)

Another useful way to represent a complex number is to write it as $z = r \cos \theta + ir \sin \theta$, where $r$ is a positive real number, or zero, and $\theta$ is an angle; $r$ is called the magnitude of the complex number, and $\theta$ is called its phase. Thus either $(x, y)$ or $(r, \theta)$ can serve as **coordinates** for the complex number.

In **quantum theory**, complex numbers are **ubiquitous**.

Complex numbers are God's numbers.

### CONFINEMENT

The basic ingredients in **quantum chromodynamics (QCD)**, our theory of the **strong force**, are **quarks** and **gluons**. There is overwhelming evidence (partially described in "Quantum Beauty III") that this theory is correct. But neither quarks nor gluons are seen as individual particles. They are found only as constituents of more complicated objects: **hadrons**. In describing this situation, we say that quarks and gluons are *confined*.

We might imagine attempting to liberate (that is, "de-confine") a quark from a proton either gradually, by pulling the proton apart with tweezers, or by bombarding the proton with energetic particles, thus smashing it (that is, the proton) into its constituent parts. Each of those attempts fails in an interesting—and I'd say *beautiful*—way.

If we try to do it slowly, we find that there is an irresistible **force** pulling the quark back in.

If we do it rapidly, we get **jets**.

For more on these topics, see "Quantum Beauty III," especially part 2.

### CONSERVATION LAW / CONSERVED QUANTITY

We say a quantity is *conserved* if its value does not change with time. A *conservation law* is a statement that some quantity is conserved. Many of our most basic insights about the world can be expressed as conservation laws. Emmy Noether proved an important theorem, described at length in the text, making a close connection between conservation laws and statements of **symmetry**, or invariance.

Examples: Conservation of **energy**, conservation of **momentum**, conservation of **angular momentum**, and conservation of **electric charge** are conservation laws; energy, momentum, angular momentum, and electric charge are conserved quantities.

The phrase "conservation of energy" deserves special comment because its scientific usage differs from its common use. We are often urged to conserve energy, for example, by turning off electric lights at night, or turning down our thermostats, or walking rather than driving. Does the world really need our help to enforce its basic laws? The point here is that when we are urged to conserve energy, we are really being urged to maintain energy in forms we can later use to do useful work, rather than allowing it to flow into forms that are useless (heat) or harmful (chemical reactions liberating toxins). The concept of *free energy* in thermodynamics captures some of this distinction. Free energy, which is the generally useful kind, is not conserved. It tends to diminish, or as we often say dissipate, over time.

### CONSISTENCY / CONTRADICTION

We say a system consisting of assumptions and observations is *consistent* if it cannot be used to derive a contradiction. We say we have a contradiction when both a statement and its denial are true.

In purely speculative theories that make no claims about concrete physical phenomena, observations cannot lead to contradictions. Such immunity from contradiction makes those theories consistent, but it does not make them good. Newton expressed that opinion forcefully in his *Principia:*

> Whatever is not deduced from the phenomena must be called a hypothesis; and hypotheses, whether metaphysical or physical, or based on occult qualities, or mechanical, have no place in experimental philosophy.

In assessing the value of physical theories, we should consider not only their consistency, but also their power and their economy. For more on these topics, see **Falsifiable / Powerful** and **Economy (of ideas)**.

### CONTINUOUS GROUP. SEE GROUP.

### CONTINUOUS SYMMETRY

If a structure admits a continuous range of transformations that leave it unchanged, or **invariant**—in other words, if our structure admits a smooth range of *symmetry transformations,* we say that we have a *continuous symmetry* of the structure, or that the structure admits a *continuous group* of symmetry transformations.

Example: A circle can be rotated through any angle around its center, while remaining the same circle. Thus the circle is invariant under a smooth range of rotations. An equilateral triangle, on the other hand, is invariant only under rotations around its center by whole-number multiples of 120 degrees. An equilateral triangle therefore admits discrete, but not continuous, symmetry.

See also **Analog** and **Digital.**

### COORDINATE

When we use sets of numbers to specify the points of a space, we call those numbers *coordinates.*

The introduction of coordinates connects "left brain" concepts of calculation and quantity to "right brain" concepts of shape and form. While the underlying psychology is murky in detail, there is no doubt that the method of coordinates helps our brains' diverse modules to communicate with one another, and to pool their strengths.

The simplest, most basic example of the use of coordinates is the description of a straight line using **real numbers.** To set this up, we need three steps:

- Pick a point on the line. (Any point will do.) This chosen point will be called the origin.
- Choose a length. (Meters, centimeters, inches, feet, furlongs, light-years, etc., are possible choices.) This chosen length is called the unit of length. To be definite, let's choose meters.
- Choose a direction on the line. (There are just two possibilities.) This chosen direction is called the positive direction.

And now, to determine the coordinate of a point $P$, we measure the distance, in meters, between $P$ and the origin. This is a positive real number. If the direction from the origin to $P$ is the positive direction, that number is $P$'s coordinate. If the direction from the origin to $P$ is opposite to the positive direction, then minus that number is $P$'s coordinate. The coordinate of the origin itself is zero.

In this way, we establish a perfect correspondence between real numbers and points on a line: each point has a unique real-number coordinate, and each real number is the coordinate of a unique point.

In a similar way, we can specify the points of a plane using pairs of real numbers, or the points in a model of three-dimensional space using triples of real numbers. We say those numbers are the *coordinates* of the points. We can also use **complex numbers** as coordinates to describe a plane. Indeed, the representation $z = x + iy$ encodes two real numbers $x$, $y$—and thus a point in a plane—in the single complex number $z$.

Of course, if we have only part of a line, we can still use real numbers to specify its points, but not all real numbers will be represented, and similarly in the other cases.

Experience with maps shows us how, by appropriate **projection**, we can represent curved surfaces on a plane (e.g., a flat piece of paper). In this way, we can use coordinates to specify points on curved surfaces.

The basic idea of coordinates admits many variations and generalizations:

- We can use more numbers! While it is difficult to visualize more than three dimensions, working with quintuple, or larger, sets of real numbers is not much more difficult from working with triples. Thus higher-dimensional spaces come within our intellectual grasp. See **Dimension**.
- We can reverse the procedure! Coordinates are introduced in order to allow us to describe geometric objects using sets of real numbers. On the other hand, in human color perception, we find that any perceived **color** can be matched, essentially uniquely, using mixtures of three basic colors, say red, green, and blue. Different intensities of red, green, and blue are described by three positive real numbers, and each such combination of intensities corresponds to a different perceived color. We can interpret these triples as the coordinates of a three-dimensional **property space**, the *space* of perceived colors. There are many examples of this general sort. Spaces based on **color charges** play a central role in our **Core Theory**.
- We can define what we mean by curved three- (or more) dimensional spaces! Again, these concepts are difficult to visualize directly. But the procedures we use to represent distances on maps, where we represent surfaces on planes, can be expressed algebraically, using a **metric**, and then easily generalized.
- We can define space-time, including time on the same footing as space! To do this, we need only regard the *date* of an event, together with the *place* it occurs, as an additional coordinate. (It's amusing to note that negative numbers make a disguised appearance in dates BCE. The year 5 BCE could, and probably should, be called the year minus five, and written −5 CE.) In **general relativity** we combine this idea with the previous one to define curved space-times.
- We can use different kinds of numbers! Coordinates based on complex numbers are widely used in quantum theory, and coordinates based on **Grassmann numbers** have enabled us to formulate the promising idea of **supersymmetry**.

### CORE THEORY

As used in this book, *Core Theory* refers to the reigning theories of the **strong**, **weak**, **electromagnetic**, and **gravitational forces**, embodying the principles of **quantum theory** and **local symmetry** (including **general relativity**, the local version of **Galilean symmetry**).

The Core Theory, or its sub-theory excluding gravity, is often called the Standard Model. For reasons explained in the text, I feel a better name is called for.

(Why would anyone even consider excluding gravity in defining the Core? It is often said that there is a fundamental conflict between quantum mechanics and general relativity, and it is sometimes claimed that that conflict entails a paralyzing crisis in physics. Both statements are exaggerations, and the second is positively misleading. For example, astrophysicists routinely combine general relativity and quantum mechanics in their work without encountering serious difficulties.

One can bring general relativity into the equations of the Core in a unique, compelling way, using the same deep principle—local symmetry—that we use to get the other forces. The rules of quantum theory continue to apply.

The Core Theory, thus defined, fails to give convincing answers to thought experiments about certain aspects of black hole physics, and its equations become singular and unusable when we extrapolate to the origin of the cosmological Big Bang, so it is not a Theory of Everything. We knew that already, thanks to the **family**, **dark energy**, and **dark matter** problems, among others. Nevertheless it is a coherent, **falsifiable**, **powerful**, and **economical** theory. It is entirely appropriate to include general relativity as part of the Core, and I have done so.)

### COSMIC RAY

When we speak of "seeing" the cosmos—stars, nebulae, galaxies, as so forth—what we usually have in mind is sampling some of the electromagnetic radiation those objects rain down on Earth. See **color (of Light)**. In the language of quantum theory, we might say we see it through **photons**. Photons travel freely through the vast empty regions of space, and we know how to orchestrate them, using lenses, into images of their sources. By "empty" here we mean devoid of **normal matter**. Because normal matter is, basically, the kind of matter that disturbs photons, there's an element of circularity in the definition—but the point remains that there are such regions. As we discuss in **Vacuum**, space that is "empty" in this sense still contains **dark energy**, often **dark matter**, one or more **Higgs fields**, and a ceaseless churning of spontaneous quantum activity (see **Quantum fluctuation**).

The celestial objects emit, besides photons, other particles: **electrons**, **positrons**, **protons**, and a variety of heavier atomic **nuclei**, notably iron. Some of these particles have enormous energy—much larger than has been achieved at the **Large Hadron Collider**, for example, and some of them find their way to Earth. Those other particles, and also the most energetic photons (gamma rays), are what we call *cosmic rays*. Cosmic rays that are electrically charged particles follow curved paths because they are deflected by galactic magnetic fields. That makes it difficult to infer their origin.

In the trailblazing era of high-energy physics before the advent of powerful **accelerators** and colliders, cosmic rays were the best available source of highly energetic particles. Several fundamental discoveries, including the existence of positrons, muons ($\mu$), and pions ($\pi$), were made by studying cosmic rays. It is possible that close encounters between dark matter particles cause them to annihilate into energetic sprays, which could be the source of interesting cosmic rays. Several experiments are currently under way, exploring that possibility.

### CURRENT

Electric current is a measure of the movement of **electric charge** from one place to another. The simplest, idealized case of an electric *current* is associated with one moving **electron**. Then we have an electric current equal to the electric charge of the electron times its **velocity** at the momentary position of the electron, and zero elsewhere. If the electron's velocity stays constant, the current is constant in magnitude, but its location moves with the electron.

In a situation where we have many electrons together with other electrically charged particles, the total electric current is the sum of the electric currents due to each particle (the charge times the velocity, in all cases) separately. The value of this fundamental, "microscopic" electric current is defined at any point in space and at any time. In other words, the electric current is a **vector field**.

The magnitude of the microscopic electric current, so defined, is strictly zero where there are no moving electrically charged particles, and varies erratically in space and time. It is usually convenient, in practical applications, to average quantities over spatial regions that contain many electrons. In this way we define averaged electric currents, which vary smoothly in space and time. In common discussions of the electric currents within electric circuits or electrical appliances, such averaging is taken for granted.

In a similar way, we have currents associated with the transfer of other kinds of charge, such as the two weak **color charges** of the **weak force**, or the three strong color charges of the **strong force**. We also have (substituting, in the definition, "mass" for "charge") *mass currents* associated with the transfer of mass, *energy currents* associated with the transfer of energy, and so forth. In ordinary language we use the word "current" most frequently to describe flows of water, and have in mind a mass current.

### DARK ENERGY / DARK MATTER

The **Core Theory** gives us a detailed, profound understanding of essentially all the matter we find on Earth and nearby. This "normal" or "ordinary" matter is composed from *u* and *d* **quarks**, color **gluons**, **photons**, and **electrons**, plus a relatively sparse flow of **neutrinos**. Astronomical observations reveal, however, that the Universe as a whole contains other kinds of matter that indeed contribute most of its total mass. The nature of this additional stuff is at present unknown in detail, but we can organize the known facts in a simple and suggestive way.

- **Normal matter** contributes about 5 percent of the total mass of the Universe. It is distributed very unevenly into galaxies (which break down further into gas clouds, stars, and planets) separated by large regions nearly devoid of normal matter.
- *Dark matter* contributes about 27 percent of the total mass of the Universe. It is also clumpy, but less so than normal matter. Astronomers commonly say that galaxies are surrounded by more diffuse dark matter haloes, but in view of their relative mass it would be more appropriate to say that galaxies are concentrated impurities within dark matter clouds. Dark matter interacts very feebly with ordinary matter, including light. Thus it is not dark in the conventional sense but, rather, transparent.
- *Dark energy* contributes about 68 percent of the total mass of the Universe. It is distributed homogeneously, as if it were a universal mass density associated to space itself. There is evidence that this density has also been constant in time, over many billions of years. Like dark matter, dark energy interacts very feebly with ordinary matter, and is transparent rather than dark.

Dark matter and dark energy, and their distribution in space, are inferred from observations of normal matter. We find, in many astrophysical and cosmological contexts, that we can account for the motion of normal matter using the known laws of physics (i.e., the Core) only if we assume that there are additional sources of mass besides normal matter. In other words, the motion we calculate for normal matter, under the influence of its own gravity, does not agree with its observed motion.

This discrepancy might, in principle, be due to a failure of general relativity, but despite many attempts no attractive alternative has emerged (even allowing, here, a very low bar for "attractive").

On the other hand, ideas for improving the Core have suggested, on quite independent grounds, the existence of new forms of matter that could account for the dark matter. Both **axions** and new particles suggested by **supersymmetry** theories could fit the bill: they are sufficiently stable and they interact feebly with normal matter. Furthermore, they are calculated to have been produced during the Big Bang in roughly the required abundance, and to clump in the observed patterns. These possibilities are currently the subject of very active experimental investigation.

The dark energy has the properties expected of Einstein's "cosmological term," and also of the energy densities associated with the **Higgs field**, the spontaneous activity of **quantum fluids**, and several other more or less plausible sources. It is possible that several of these effects make independent contributions to the dark energy, with some adding to, and others subtracting from, the total. In contrast to the situation for dark matter, existing theoretical ideas about dark energy are vague and hard to **falsify**.

It is worth mentioning, here, that the modern dark matter / dark energy problem has two remarkable historical precedents. Painstaking work in **celestial mechanics**, based on Newton's theory of **gravity**, had by the mid-nineteenth century revealed two small discrepancies between calculation and observation. One concerned the motion of Uranus; the other concerned the motion of Mercury. The difficulty with Uranus was resolved by a form of "dark matter." Urbain Le Verrier, and also John Couch Adams, suggested that its discrepant acceleration was caused by the gravitational force from a new, hitherto unknown planet whose position they could calculate. The required planet—Neptune—was duly found! The difficulty with Mercury was resolved when Einstein's **general relativity** supplanted Newton's theory of gravity. The new theory, put forward for entirely different and profound reasons, makes slightly different predictions for the **orbit** of Mercury, and its predictions agree with observation.

Arguments with an **anthropic** flavor have been applied to both the dark energy and the dark matter problems. The structure of the arguments is similar, in both cases:

- The part of the Universe we can presently observe is only one portion of a larger structure, sometimes called the *multiverse*. (Note that as time goes on the region of space accessible to observation expands, due to the finite speed of light.)
- Physical conditions in other, distant parts of the multiverse can be different. In particular, the density of dark energy, or of dark matter, can vary.
- In regions where the density of dark energy, or of dark matter, is drastically different from what we observe in our Universe, intelligent life cannot emerge.
- Therefore the only values of those densities that can be observed are close to those we do observe.

The second and third steps are at present controversial, so these ideas remain speculative. But as our knowledge of fundamental laws, and our abilities to assess their consequences, improve, it is logically possible that they will be generally accepted. Should that occur, it seems to me that this line of reasoning will be compelling. In that case, we will have discovered that major features of the world we observe—that is, the dark energy and/or dark matter densities—are determined not by abstract principles of **dynamics** or **symmetry**, but by *selection*, in the style of biology.

### DIGITAL

If a quantity cannot vary smoothly, we say it is a *digital* quantity. For more on this concept, see the entry for **Analog**.

### DIMENSION

Intuitively, a *dimension* is a possible direction of motion. Thus we say that a line, or a curve, has one dimension. A plane, or a surface, has two dimensions because it requires motion in two independent directions—we might say "horizontal" and "vertical," or "north-south" and "east-west," for example—to reach any point from any other. Everyday space, or a solid body, has three dimensions.

A more flexible concept of "space," and of dimension, arises naturally from the introduction of coordinates. You should refer, at this point, to **Coordinate** for a discussion of that concept. The dimension of a space described by coordinates is simply the number of coordinates it requires. That concept, when applied to simple, smooth geometric objects, agrees with the preceding intuitive idea.

Mathematicians have generalized these more or less intuitive concepts of dimensions in many ways. Two notable generalizations are **complex dimensions** and fractional, or *fractal*, dimensions. Complex dimensions add more coordinates, but coordinates that are **complex numbers**. Fractional dimensions can arise when one considers objects that contain extremely rich local structure and are far from smooth: see **Fractals**. In recent years physicists have introduced, in connection with **supersymmetry**, the concept of **quantum dimensions**. The coordinates of quantum dimensions are **Grassmann numbers**.

There is another, quite different scientific use of the word "dimension." In that usage, we refer to the units in which a quantity is measured as its "dimensions." Thus in this sense area has dimensions of length squared, while **velocity** has dimensions of length divided by time, **force** has dimensions of mass times length divided by time squared, and so forth. In this book, to avoid possible confusion, I have avoided using "dimension" in that sense.

### DIRAC EQUATION

In 1928 Paul Dirac (1902–84) proposed a **dynamical equation** describing the behavior of **electrons** in **quantum mechanics**, which today we call the *Dirac equation*. The Dirac equation improves on **Schrödinger**'s earlier electron equation in much the same way that Einstein's equations for mechanics improve on Newton's. In both cases, the new equations are consistent with the **special relativity** theory, while the simpler ones they replace are not. (And in both cases, the new equations reproduce the predictions of the old ones, when describing the behavior of bodies that move much slower than the speed of light.)

The Dirac equation has additional solutions, besides those that represent electrons in different states of motion (and spin). These solutions describe particles with the same mass as electrons, but the opposite electric charge. The new particles are called antielectrons, or **positrons**. Positrons were discovered experimentally in 1932 by Carl Anderson, through study of **cosmic rays**. See also **Antimatter**.

The Dirac equation, with suitable (relatively minor) adaptations, describes not only the behavior of electrons, but also the behavior of other fundamental particles with **spin** ½, including all **quarks** and **leptons**—in other words, the **substance particles** of the main text. With slightly more significant modifications, it also describes the behavior of spin-½ **hadrons**, including **protons** and **neutrons**.

### DYNAMICAL LAW / DYNAMICAL EQUATION

Dynamical laws are laws that specify how quantities change in time. Dynamical laws are formulated in *dynamical equations*.

Example: Newton's second law of motion specifies the **acceleration** of bodies, which is how their **velocities** change in time.

Counterexample: **Conservation laws**, by contrast, state that quantities do not change in time.

The basic laws of our **Core Theory** are dynamical laws, but they imply conservation laws for a few special quantities.

Second counterexample: Within the Core Theory, there are a number of so-called *free parameters*. These are quantities that appear within the equations, whose values are not fixed by any general principle, but rather taken from experiment. They are tacitly assumed to be constant in time.

Possible counter-counterexample: The central idea of **axion** physics is that one of these parameters, the so-called $\theta$ parameter, obeys a dynamical equation within a larger theory. In that larger theory, the "coincidence" that $\theta$ is observed to be very small becomes a consequence of solving a *dynamical equation*. More generally, one may hope that other free parameters of the Core will someday be determined by solving dynamical equations within more powerful theories.

See also **Initial conditions**.

### ECONOMY (OF IDEAS)

We say that an explanation, or more generally a theory, is *economical* if it assumes little and explains a lot.

While it does not involve the exchange of goods and services, this concept is not altogether disconnected from economics proper. There we would say that intelligent use of limited resources to create valuable products is an economical use of those resources, which is a similar idea.

It seems reasonable, intuitively, to prefer economical explanations over their opposites—explanations that invoke many assumptions to explain a limited range of facts or observations. This intuition gets support from Bayesian statistics, which assure us that the more economical of two explanations is more likely to be correct when both fit the same data equally well.

### ELECTRIC CHARGE

In modern physics, and specifically in our **Core Theory**, *electric charge* is a primary property of matter, one that cannot be explained in terms of anything simpler. Electric charge is a discrete (i.e., **digital**), **conserved** quantity to which **electromagnetic fields** respond.

The simplest manifestation of electric charge is its ability to generate **forces**. According to Coulomb's law, two electrically charged particles feel an electric force proportional to the product of their electric charges (and inversely proportional to the square of their separation). When the charges are of the same sign the force between them is repulsive, but it is attractive when the signs are opposite. Thus there is a repulsive electric force between two **protons**, or between two **electrons**, but an attractive electric force between a proton and an electron.

Electric charge comes in whole-number multiples of the proton's charge. Electrons, relative to protons, carry equal and opposite electric charge. (Theoretically, **quarks** carry charges that are fractions of the proton's charge. However, quarks appear not as individual particles, but only within **hadrons**. The electric charges of hadrons are always whole-number multiples of the proton's charge.)

### ELECTRIC CURRENT. SEE CURRENT.

### ELECTRIC FIELD / ELECTRIC FLUID

The value of the *electric field*, at any point, is defined as the ratio of the electric **force** felt by a charged particle positioned at that point, divided by its **electric charge**. Force is a **vector** quantity, so the electric field is a **vector field**.

That definition is widely used in molecular biology, chemistry, electrical engineering, and other applications. But in applications to fundamental physics, where **quantum fluctuations** are important, it becomes problematic because both forces and positions fluctuate. It can be salvaged as an *approximate* notion by doing some averaging over time and space.

In fundamental physics a different approach, which bypasses these problems, has proved more useful. We do not insist that the concepts we use correspond, at every stage, to observable quantities. We do want all observable quantities to appear somewhere in the equations, but we might—and we do—find it convenient to include other things, besides! (See, in particular, **Renormalization**.)

In that spirit, I define *electric fluid* to be the space-filling, active *thing* that appears in **Maxwell's equations**.

The necessity to distinguish between electric field and electric fluid becomes crystal clear if we consider how we should interpret the statement "The electric field in intergalactic space vanishes." That statement makes sense (and is approximately correct), given our definition of the electric field in terms of the forces it generates, on average. On the other hand, it would be quite wrong to say that the quantum-mechanical entity which appears in Maxwell's equations, and exhibits spontaneous activity, vanishes anywhere. So the usual terminology, which does not distinguish between those two concepts—the entity itself, and its average value—is fundamentally flawed. (That flaw doesn't seem to bother most physicists much, but it bothers me!) We solve the problem by calling the entity itself the *electric fluid,* and its averaged value the *electric field.*

(That said, when there is no danger of confusion, I will sometimes use "electric field" both for the entity and for its averaged value. Bad habits die hard.)

See also **Quantum fluid**.

### ELECTRICITY

"Electricity" is a broad term used to refer to a wide range of phenomena associated with the influence and behavior of **electric charges**.

### ELECTRODYNAMICS / ELECTROMAGNETISM

These two terms are used interchangeably to mean the body of science that concerns **electricity**, **magnetism**, and the relations between them.

Since the work of Faraday and Maxwell, people have realized that electricity and magnetism are inseparably connected. Magnetic fields that change in time produce electric fields, according to **Faraday's law**, and electric fields that change in time produce magnetic fields, according to **Maxwell's law** (see **Maxwell's term**). **Electromagnetic waves**, arising from the interplay of those laws, contain both electric and magnetic fields.

In the **special relativity** theory, we learn that **Galilean transformations** transform electric and magnetic fields (and fluids) into one another.

### ELECTROMAGNETIC FLUID / ELECTROMAGNETIC FIELD

Because electric and magnetic fluids have great influence over each other, it is convenient, and appropriate, to treat them together as a unified whole. *Electromagnetic fluid* is simply the fluid whose two components are the electric and magnetic fluids. The electromagnetic field at any point is its average value at that point.

### ELECTROMAGNETIC SPECTRUM. SEE COLOR (OF LIGHT).

### ELECTROMAGNETIC WAVE

When we combine **Faraday's law**, through which changing magnetic fields produce electric fields, with **Maxwell's law**, through which changing electric fields produce magnetic fields, we find the possibility of self-sustaining activity in those fields. That self-sustaining activity takes the form of **transverse waves**, which move through space at the speed of light. We call these waves *electromagnetic waves.*

Maxwell discovered the possibility of electromagnetic waves and calculated their speed. Finding that it matched the speed of light, he proposed that light consists of electromagnetic waves. This remains, today, our fundamental description of light.

Visible light corresponds to only a small band within the **electromagnetic spectrum**, which encompasses electromagnetic waves with all possible **wavelengths**. Today we understand not only light, but also radio waves, **microwaves**, infrared radiation, ultraviolet radiation, X-rays, and gamma rays all as electromagnetic waves that differ in wavelength and **frequency**.

### ELECTRON

Electrons are a constituent of **normal matter**. They were first clearly identified in 1897 by J. J. Thomson.

In the **Core Theory**, *electrons* are **elementary particles**, and one defines them through the equations they satisfy.

In normal matter, electrons carry all the negative **electric charge**. Although they contribute only a very small fraction of the mass of normal matter, electrons play a dominant role in chemistry and the structure of materials. Controlled manipulation of electrons—in other words *electronics,* broadly defined—is the foundation for much of modern technological civilization.

### ELECTRON FLUID

The *electron fluid* is a world-filling, active **quantum fluid** or **medium**. According to **quantum theory**, as used in our **Core Theory** description of the world, **electrons** and their **antiparticles**—antielectrons, or **positrons**—are disturbances in the electron fluid. They are similar, in this description, to waves in water, which if left unimpeded can hold together and move (or, we sometimes say, "propagate") over long times and distances.

There are fluids of the same sort associated with every species of elementary particle. (In the physics literature, they are often called "**quantum fields.**") These space-filling fluids coexist—the presence of one does not crowd out any of the others. The **dynamical equations** of the Core Theory describe how they influence one another.

Like many ideas we now use to describe matter, our modern understanding of electrons builds on concepts that first arose in the study of **electromagnetism** and light. The electron fluid is deeply similar to the **electromagnetic fluid**, and we now view electrons as minimal disturbances in the electron fluid. They are its **quanta**, similar to **photons** in the electromagnetic fluid.

### ELEMENTARY PARTICLE

We say a particle is *elementary* if it obeys simple equations. In the **Core Theory**, **quarks**, **leptons**, **photons**, **weakons**, color **gluons**, **gravitons**, and the **Higgs particle** are elementary particles.

**Protons** and **neutrons** were once thought—or rather, hoped—to be elementary particles, but further investigation revealed that they do not obey simple equations. Similarly, atoms and molecules are not elementary particles. In all those cases, we now understand that the objects—protons, neutrons, atoms, and molecules—are composite structures, built up from simpler bits. They are built up, in fact, from a few of the Core's *elementary particles* (namely *u* and *d* quarks, color gluons, **electrons**, and photons).

### ELLIPSE

An *ellipse* is a planar geometric figure that looks like a stretched-out circle. Ellipses are usually defined as follows: Pick two points $A$ and $B$, and a distance $d$ greater than the distance between $A$ and $B$. Then the collection of all points $P$ with the property that the distance between $A$ and $P$, plus the distance between $B$ and $P$, is equal to $d$, is an ellipse. $A$ and $B$, which are not themselves on the ellipse, are called its focal points, or foci.

Circles are special cases of ellipses, which arise when $A$ and $B$ coincide. When $d$ is much larger than the distance between $A$ and $B$, the ellipse will be nearly circular. As $d$ decreases and becomes only slightly more than the distance between $A$ and $B$, the ellipse becomes an oval tightly enclosing the line interval connecting $A$ and $B$; in the limiting case, as $d$ becomes equal to the distance between $A$ and $B$, the ellipse degenerates to that line interval.

Ellipses can be defined in several other very different-looking, but mathematically equivalent, ways. This, my favorite, is perhaps the simplest to visualize: Draw a circle on a rubber sheet, and then stretch the sheet uniformly in any chosen direction. The circle then deforms into an ellipse; all ellipses can be obtained in this way.

Ellipses were studied in amazing depth by ancient Greek geometers for their own sake, because they seemed beautiful. Many centuries later, Kepler discovered, through careful study of Tycho Brahe's astronomical observations, that orbits of planets around the Sun form ellipses, with the Sun at one focus. Though Kepler himself was initially disappointed to abandon the "perfect" circular shape for **orbits**, in retrospect, this embodiment of Greek geometry appears as a near miraculous instance of Ideal becoming Real.

Kepler's laws of planetary motion guided Newton to his theories of mechanics and gravity. In Newton's framework, we discover that the orbits of planets around the Sun are only approximately elliptical, as they are distorted by the gravitational influence of other planets. Ultimate beauty, in this case, resides in the **dynamical laws** themselves, rather than in their solutions. For more extended discussion, see "Newton III."

### ENERGY / KINETIC ENERGY / MASS ENERGY / ENERGY OF MOTION / POTENTIAL ENERGY / FIELD ENERGY

*Energy* is, together with **momentum** and **angular momentum**, one of the great **conserved quantities** of classical physics. Each of them has evolved into a pillar of modern physics, as well.

Practical discussions of energy introduce many categories, such as wind energy, chemical energy, heat energy, and others, that arise from packaging the fundamental forms of energy in different ways. But even when expressed in terms of fundamentals, energy comes in several different forms. Here we'll be looking at the concept of energy from the perspective of fundamentals.

Total energy, which is the conserved quantity, is the sum of several terms: kinetic energy, mass energy, potential energy, and field energy. Those different terms refer to aspects of reality that seem, on the face of it, quite different. Much of the power of the concept of energy, in applications, comes precisely from its ability to describe, and relate, several different aspects of reality.

Kinetic energy was the first form of energy to be discussed, historically, and its importance is the easiest to grasp intuitively. Qualitatively, kinetic energy is motion. Often, in designing machines, we want to get things moving. Because moving things have kinetic energy, often a major goal in power engineering is to transform other forms of energy into kinetic energy.

**Quantitatively**, in Newtonian mechanics, the kinetic energy of a particle is equal to half its **mass** times the square of the magnitude of its **velocity**. In Einstein's modification of mechanics, to satisfy the **special relativity** theory, *energy of motion* gets tied up with a new form of energy, *mass energy*, to which we now turn.

In Newtonian mechanics one has two separate **conservation laws**: conservation of mass, and conservation of *energy*. Special relativity requires radical revision of the concept of mass. As part of that revision, one must abandon the conservation of mass. Conservation of energy continues to hold, but with a significantly different definition of what energy is. Though I've never seen it presented in quite this way, the logic of the concept *mass energy* emerges most clearly, if one considers it as a way of reconciling the nonrelativistic and relativistic concepts of energy. The next three paragraphs spell this out.

For momentum and angular momentum, passage from relativistic to Newtonian definitions is smooth. The relativistic expressions for those quantities become equal to the Newtonian expressions, approximately, when the velocities of all the bodies involved are much smaller than the speed of light. For energy, on the other hand, that smooth passage is not completely straightforward. It works only if we include a new contribution to the usual Newtonian definition of energy. This new contribution is *mass energy.*

The mass energy of a body is equal to its mass times the square of the speed of light. When we symbolize, as is usual, the speed of light as $c$, the corresponding formula is perhaps the most famous equation in science:

$$E_{\text{mass}} = mc^2$$

I've attached the subscript "mass" here to emphasize that this is just one form of energy among many. The total mass energy, when we have several bodies, is simply the sum of their separate mass energies. Thus the total mass energy is just the total mass multiplied by the square of the speed of light. The total "corrected" Newtonian energy is the classic Newtonian energy, kinetic plus potential, which you find defined in textbooks (and also below!), *plus* the mass energy. It is this corrected Newtonian energy, not the classic Newtonian energy, that emerges smoothly from relativistic mechanics.

To the extent that total mass is conserved, the corrected and classic Newtonian energies differ by a constant amount (and both are conserved). The corrected energy works more generally, however. It covers some cases, such as **nuclear** reactions, where the conservation of mass is *not* a good approximation. In those cases, the mass energy at the beginning of the process is not equal to the mass energy at the end. Total energy is conserved, so the difference between those mass energies must appear in other forms. That is what is meant when one speaks of conversion of mass into energy, or of energy into mass. Or rather, it should be. The concept has been a potent source of misunderstanding and confusion in the literature of popular science. I hope I've helped to clarify it here.

For accurate work, and in applications where there are several particles moving with near-light speeds, one must use the full relativistic formulas for *energy of motion,* and separating it into mass energy and kinetic energy is artificial.

For the benefit of readers who have some algebra at their command, here is that formula:

$$E_{\text{motion}} = \frac{mc^2}{\sqrt{1 - \frac{v^2}{c^2}}}$$

When the magnitude of the velocity is much less than the speed of light, $v << c$, this becomes approximately equal to the sum of the mass energy $mc^2$ and the Newtonian kinetic energy $\frac{1}{2}mv^2$, as we described in words before. As the magnitude of the velocity approaches the speed of light, the energy of motion grows without bound.

*Potential energy,* qualitatively, is energy of position, or distance. For example, potential energy of a stone near the surface of Earth can be stored up by lifting the stone, or released by dropping it. As the stone falls after being dropped, its velocity, and therefore its kinetic energy, increases. To maintain conservation of energy, then, we must have a decrease in its potential energy.

The concept of potential energy can be extended to cover many more general cases. When bodies exert forces on one another, the potential energy associated with their interaction is a function of their distance. Potential energy—energy of distance—is a natural concept within theories based on **action at a distance**, such as Newton's theory of gravity. Like those theories, it continues to be useful in many applications, where it provides a good enough, user-friendly approximation. But in fundamental physics, since the revolution initiated by Faraday and Maxwell, force-transmitting **fields** replace action at a distance. *Field energy* replaces potential energy.

Field energy is stored throughout space wherever there are nonvanishing fields. The density of field energy associated with an electric field at a point in space, for example, is proportional to the *square* of the magnitude of the electric field at that point.

The possibility of replacing the concept of *distance-dependent* potential energy with the concept of *locally defined* field energy is both profound and very pretty. Consider the potential energy between a positively (electrically) charged and a negatively charged particle. For the same sorts of reasons that we discussed in connection with a stone near Earth, there is potential energy associated with the distance between those particles. In the Faraday-Maxwell picture, the same quantity of energy arises in quite a different way. Both of our charges generate electric fields, and the total electric field is the sum of their contributions. The energy density associated with that total electric field is its square, and so contains not only the squares of each field separately, but also a cross-term reflecting their simultaneous presence. (If that idea is unfamiliar, let's step back momentarily. The square of $1 + 1 = 2$ [that is, $2 \times 2 = 4$] is not equal to the square of 1 twice over; i.e., 2. There is an extra contribution, or cross-term, as the two independent contributions to the sum meet each other in its square. Algebraically, more generally, we have $(a + b)^2 = a^2 + b^2 + 2ab$, with the cross-term $2ab$.) The cross-terms that appear in the total field's energy density will depend on the relative geometry of the two fields from which it is composed, which in turn depends on the relative distance between the particles. When you add up the total energy density, taking contributions from all space, to get the total *field energy,* you find that the contribution of these cross-terms exactly matches the *potential energy* of the old theory, and can replace it.

In this example field energy is merely a different—and more complicated!—way of getting to the same answer as potential energy. But in a more complete account of physics the fundamental laws are formulated *locally,* and they lead naturally to field energy. Potential energy is an approximate, *emergent* concept that is useful in some problems, but inadequate in others.

The **conservation** of energy, and ultimately energy itself, is best understood through Noether's general theorem, which connects conservation laws to **symmetry**. In that framework, conservation of energy reflects symmetry (i.e., **invariance**) of the physical laws under **time translation**—that is, under a transformation that advances (or retards) all events through a common interval of time. In other words, we have conservation of energy when the laws do not depend on any externally specified, fixed time.

In the quantum world, energy takes on additional features of great subtlety and beauty. Especially notable is the **Planck-Einstein relation**, which relates the energy of photons to their **color**. When combined with Bohr's ideas, this connection allows us to decode the message of **spectra**. The colors of an atom's spectral light encode the energies of its **stationary states**, providing a visible Music of the Spheres.

### EXCLUSION PRINCIPLE / PAULI EXCLUSION PRINCIPLE

The *Pauli exclusion principle,* in its original form, states that no two **electrons** can share the same quantum state. This principle applies to all **fermions**: no two identical fermions can share the same quantum state. The reluctance of electrons, or fermions in general, to do the same thing results in an effective repulsion between them. This repulsion is a purely quantum-mechanical effect that supplements more conventional forces, such as electrical forces.

The exclusion principle is fundamental to understanding atoms, because it prevents the electrons in an atom from piling up near its nucleus, despite the latter's powerful electrical attraction. Outer electrons, being remote from the nucleus, are open to influences from other nearby atoms. In this way, the exclusion principle opens the door to chemistry.

### FALSIFIABLE / POWERFUL

When a proposition (or theory) can be compared with empirical observations, and thus potentially disproved, we say it is *falsifiable.* Sir Karl Popper (1902–94) advocated falsifiability as a criterion for distinguishing science from other human endeavors. Though it is bracing, I do not think that Popper's falsifiability criterion adequately reflects scientific practice, because we are often more concerned with empowering good ideas than with pruning away bad ones.

Falsifiability is more appropriate as a (partial) criterion for the *maturity* and *fruitfulness* of theories, as opposed to their status as science or nonscience. In that context, *falsifiability* should be considered together with *power.* Theories that make many successful predictions but also occasionally fail (for example, practical meteorology), or whose predictions are in some cases inherently statistical and therefore not easily falsifiable (for example, **quantum theory**) may nevertheless be of great value, and should qualify as scientific in any reasonable definition of that term.

One should not regard a powerful but imperfect theory as simply false, but rather—until proved otherwise—as a promising platform for improvement. Newtonian (nonrelativistic) mechanics, classical (non-quantum) **electromagnetism**, and many lesser theories have been falsified, yet we revere them, for good reasons:

- They continue to be useful, due to their predictive power and relative simplicity.
- The theories that superseded them lean heavily on their conceptual structure.
- Within those later theories, they survive as approximations, valid in limiting cases.

See also **Consistency / Contradiction, Economy (of ideas).**

### FAMILY

The **substance particles** of the **Core Theory**—i.e., the **quarks** and **leptons**—feature a peculiar threefold repetition. We say they form three *families.* Each family features sixteen particles that fall into the same patterns of **strong, weak,** and **electromagnetic charges.**

Alternatively, using the geometric language of "Quantum Beauty III," we can say that each of the three families features six entities, which occupy identical **property spaces** in every case.

Transitions associated with the **weak force** that change a unit of yellow **weak charge** into a unit of purple weak charge will change a (left-handed) $u$ quark into a (left-handed) $d$ quark, as we discussed in the main text. There I alluded to complications, and here I will be more specific. The complication is that the *weak color transitions can be accompanied with family transitions.* Thus in addition to $u \rightarrow d$, we also have $u \rightarrow s$ and $u \rightarrow b$. To describe the relative likelihoods of these transitions, we need to introduce additional numbers into the **Core Theory.** The Cabibbo angle, for example, gives a measure of the likelihood of the second, relative to the first. There are many additional transitions among quarks to consider (e.g., $c \rightarrow d$), and even more when we bring in leptons. To describe them all, within the Core Theory, we need to introduce about a dozen new numbers. The values of these "mixing angles" have been measured experimentally, but there is no compelling theory explaining why they have the values they do.

For that matter, there is no compelling theory explaining why Nature indulges in her threefold family repetition at all.

### FARADAY'S LAW

This law states that the **circulation** of the **electric field** around a curve is equal to minus the rate of change of the **flux** of the **magnetic field** through any surface the curve bounds. *Faraday's law* is enshrined as one of **Maxwell's equations.**

### FERMION.
### SEE BOSON / FERMION.

### FIELD / FLUID

In introducing the concept *field*, it is probably best to proceed through examples.

- In describing weather, it is useful to consider the value of temperature at many points in space, at various times. The totality of those values defines the temperature field.
- In describing motion within a body of water, it is useful to consider the value of the **velocity** of the water at many points in space, at various times. The totality of those values defines the velocity field.

- In describing **electric** phenomena, it is useful to consider what forces would be exerted on an electrically **charged** particle, were one present, at different points in space, at different times. This leads us to define, after dividing by the magnitude of the charge, the **electric field**.

In general, we say we have a "*field* of type $X$" when we have values of $X$ at different positions and times. Put another way, a field of type $X$ gives a quantity of type $X$ as a **function** of space and time.

The term *fluid*, as used in this book, refers to any of a variety of things that fill space and exhibit activity. Examples include **electric fluid**, **magnetic fluid**, **gluon fluid**, and **Higgs fluid**. For more on the subtle but important conceptual distinction between field and fluid, see especially the entry **Electric field / Electric fluid**.

See also **Medium**.

#### FIELD ENERGY. SEE ENERGY.

#### FLAVOR

There are six distinct kinds, or *flavors*, of **quarks**: *u* (up), *d* (down), *s* (strange), *c* (charm), *b* (bottom), and *t* (top), in ascending order of mass. Each inhabits the same three-dimensional color **property space**, and (therefore) all of them behave in the same way, as regards the **strong force**. Quarks *u*, *c*, and *t* have **electric charge** equal to ⅔ that of a proton, while *d*, *s*, and *b* have electric charge equal to −⅓ of a proton. They have different, slightly complicated behaviors with regard to the weak interaction—see **Family**.

The deep significance of this proliferation of quark types, if any, is at present unclear. Among the quarks, only *u* and *d* play large roles in the present-day natural world because they appear prominently within **protons** and **neutrons**.

There is a parallel proliferation of leptons; here too people speak of different lepton *flavors*.

#### FLUID. SEE FIELD / FLUID.

#### FLUX

Vector fields, whatever their true nature, can be regarded mathematically as representing the flow of an ordinary fluid, such as air or water. The mathematically imagined flow has, at every point, its velocity proportional to the value of the actual vector field at that point. In this model, the *flux* through a surface is simply the rate at which fluid is being transported through the surface (up to an overall sign, which we shall presently discuss). This definition of flux makes sense whether or not the surface has a boundary.

Thus if we consider a flowing river, and draw a surface that is impacted head-on by the flow, there will be significant flux through that surface. On the other hand, there will not be significant flux through surfaces that present small profiles to the flow.

At this point you should consult, if you have not done so previously, **Circulation**! For I will now fill in one remaining subtlety that involves the relationship between those two concepts. With that, you will have seen everything you need for an honest understanding of what **Maxwell's equations** are, entirely in terms of geometric concepts and images.

In two of Maxwell's equations, we are asked to consider a surface bounded by a curve, and to compare the circulation of one thing around the curve with the flux of something else through the surface. (In **Faraday's law**, we relate the circulation of electric field to the flux of magnetic field; and in **Ampere-Maxwell's law**, we relate the circulation of magnetic field to the fluxes of electric current and electric field.)

In order to calculate the circulation for use in these equations, we need to be definite about the direction we go around the curve. There are two possible choices, and the answers they give for the circulation differ in sign. In order for Maxwell's equations to remain the same, regardless of which choice we make, we need to make sure that the sign of the flux through the surface also changes when we change the direction of its bounding curve (and thus the sign of the circulation).

For this purpose, we use a simple *right-hand rule:* If the fingers of your right hand follow the direction of the curve, then we regard the transport of fluid, in the definition of flux, as positive when it is moving in the direction of your thumb, and negative in the opposite direction. If we follow this rule, then changing the direction of the curve will change the sign of both the circulation and the flux, and so the relationship between circulation and flux will stay the same.

In two others of Maxwell's equations—the electric and magnetic Gauss's laws—we consider flux out of a closed surface. In that case, we consider the flux as positive when it transports fluid from the inside of the surface to the outside, and negative in the opposite case.

### FORCE

In physics, and in our meditation, the term *force* is used in two distinct ways.

In Newtonian mechanics, force is a measure of one body's influence on another. The *force* a body projects is its ability to produce acceleration in other bodies. See **Acceleration**.

In another common but less precise usage, we speak of *forces* of Nature, meaning mechanisms through which Nature acts. In our **Core Theory**, we identify four basic forces of Nature: gravity, electromagnetism, and the strong and weak forces. It is also common to speak here of *interactions,* in place of forces (and thus of the electromagnetic interaction, the strong interaction, and so forth). I have chosen to use "force" consistently, because it is more forceful.

### FORCE PARTICLE

"Force particle" is an informal phrase that I use to refer, collectively, to the fundamental particles of the **Core Theory** that are **bosons**: the **photon**, the **weakon**, the color **gluon**, the **graviton**, and the **Higgs particle**. It is meant to be user-friendly and convey some rough idea of those particles' role in Nature.

### FRACTALS

Fractals are geometric objects that have structure on all scales. Thus when you magnify a complicated fractal image, to zoom in on its details, you find that each detail is as complicated as the original whole—indeed, in many fractals, a magnified part is *identical* to the whole!

Fractals come in many sizes and shapes. There is no single, strict definition that applies to all the objects people have described as "fractals." Instead there is a vast zoo of interesting examples that embody this broad concept of inexhaustible inner structure.

Because the small parts of a fractal are as complex as its whole, the method of **Analysis and Synthesis**, and its classic mathematical realization, **calculus**, lose most of their power. Different ideas, based on recursion and self-similarity, come into play. (Here, I will leave it at that. Although these ideas are fascinating, they are only loosely connected to our main themes.)

Very complex fractals can be constructed by following simple rules through many steps. That procedure is wonderfully adapted to computer graphics. It has led to the production of stunning images and enabled new forms of visual art.

### FREQUENCY

If we have a process that repeats itself in time, its **period** is the time between repeats, and its *frequency* is the number one divided by the period, or equivalently the inverse period. Thus a high-frequency process is a process that repeats itself very frequently. Frequencies are measured in inverse seconds, a unit also called the hertz, in honor of Heinrich Hertz, the discoverer of **electromagnetic** radiation.

Examples: If a process repeats itself every two seconds, its frequency is ½ hertz. If a process repeats itself twice a second—that is, once every half second—then its frequency is 2 hertz. Young, healthy humans are able to hear vibrations in air, or sound waves, when the vibrations' frequencies lie roughly between 20 and 20,000 hertz. Human eyes are sensitive to **electromagnetic waves** with frequencies between $4 \times 10^{14}$ and $8 \times 10^{14}$ hertz—quite a rapid rate of oscillation!

### FUNCTION

When some quantity varies with time, we say it is a *function* of time. More generally, we say that a quantity $y$ is a *function* of some other quantity $x$ when each value of $x$ determines a value of $y$. We write $y(x)$ for the value of $y$, determined by $x$.

Examples:

- The temperature in Boston is a function of time.

- The temperature at the surface of Earth, more generally, is a function of position on the surface *and* time. It is, in other words, a function of space-time.

See also **Field**.

### GALILEAN TRANSFORMATION / GALILEAN SYMMETRY / GALILEAN INVARIANCE

A *Galilean transformation* is the sort of transformation we make on a system when we imagine adding or subtracting a constant velocity to the motion of each of its parts. Galileo, as quoted in the main text, described a beautiful thought experiment which makes it plausible that Galilean transformations leave laws of physics unchanged, or **invariant**: If you're belowdecks in a closed cabin within a ship, in calm weather, you can't tell, from your experience within the cabin, how fast the ship is progressing. The hypothesis that the laws of physics are invariant under Galilean transformations, or alternatively that they exhibit Galilean symmetry, is one of the pillars of **special relativity**. See also **Boost**.

### GAUGE PARTICLE

In order to implement **local** (i.e., gauge) **symmetry**, one must introduce appropriate **fluids** whose properties are precisely tailored to fit that idea. In the **Core Theory**, **gravitational**, **strong**, **weak**, and **electromagnetic fluids** are introduced for this reason. The minimal units, or quanta, of these fluids—**gravitons**, color **gluons**, **weakons**, and **photons**—are therefore referred to as *gauge particles*. The term is bland, but it encodes a profound and beautiful fact: these particles, which mediate the basic forces of Nature, are *embodiments of symmetry*.

### GAUGE SYMMETRY

This is another term for **local symmetry**.

### GAUSS'S LAW

There are actually two Gauss laws, of very similar form.

The Gauss's law for the **electric field**, or electric Gauss's law, states that the **flux** of electric field through any closed surface is equal to the amount of **electric charge** the surface encloses.

The Gauss's law for the **magnetic field**, or magnetic Gauss's law, states that the flux of magnetic field through any closed surface is equal to zero. Alternatively, we can say that this flux is equal to the amount of magnetic charge the surface contains, and that one does not find magnetic charge in Nature.

These Gauss laws are enshrined as two of the **Maxwell's equations**.

### GENERAL COVARIANCE

This is Einstein's original term for **local Galilean symmetry**, the foundational principle of **general relativity**.

### GENERAL RELATIVITY

The *general theory of relativity* is Einstein's theory of **gravity**.

John Wheeler described the essence of general relativity this way:

Matter tells space-time how to curve.
Space-time tells matter how to move.

The main text contains an extensive explanation (and critique!) of that capsule summary.

The "general" in "general relativity" is Einstein's coinage, positioning the new theory in relation to his earlier **special relativity**. In our meditation we express that relationship in a different and more systematic language that was developed in describing the other forces. Special relativity considers **Galilean transformations**, and general relativity considers more *general* transformations, which amount to allowing the use of different Galilean transformations at different places in space-time. In our language, general relativity is based on ***local* symmetry**, while special relativity is based on nonlocal, or alternatively (and better), ***rigid* symmetry**.

### GEODESIC

On a curved surface, there may not be any straight lines, but *geodesics* are the nearest substitute. A geodesic curve has the property of providing the shortest path between any two of its nearby points. We must restrict to "nearby" points because after a long excursion a geodesic may loop close to its earlier parts, and there may be shorter paths that short-circuit the long trajectory.

Example: The geodesics on a sphere are

its *great circles*, obtained by cutting the sphere with a plane through its center. Thus the equator is a great circle, and a geodesic, on Earth, as are lines of longitude. Polar routes for air travel approximate geodesics to save on fuel.

The concept of geodesic, so defined, is not restricted to surfaces. We can find geodesics in curved spaces of higher dimension, and—with an appropriate definition of distance—in space-time.

### GLUON / COLOR GLUON

Gluons are minimal units, or **quanta**, of the **gluon fluid**.

### GLUON FLUID / GLUON FIELD

The *gluon fluid* is the active, space-filling entity responsible for the **strong force**. The *gluon field* at a point is a measure of the influence of the gluon fluid at that point, averaged over some appropriate small volume of space and interval of time.

### GRAPHENE

*Graphene* is a chemical substance made entirely from carbon. In graphene, the carbon atoms form a two-dimensional sheet, with their **nuclei** arranged in a honeycomb pattern. Graphene has remarkable mechanical and electrical properties.

### GRASSMANN NUMBER

These are numbers that satisfy the **antisymmetric** multiplication rule

$$xy = -yx$$

*Grassmann numbers* appear, in supersymmetry, as the coordinates of quantum dimensions.

### GRAVITON

Gravitons are the smallest units, or **quanta**, of disturbance in the **gravitational fluid**, also called the **metric fluid**. Thus gravitons are, to **gravity**, what **photons** are to **electromagnetism**. Individual gravitons are predicted to interact extremely feebly with **normal matter**, and there is little prospect of observing them as individual objects directly. Potentially detectable gravity waves are built up from vast numbers of gravitons.

### GRAVITY

As it acts between elementary particles, *gravity* is by far the feeblest of the four **Core Theory** forces. But the other three **forces** respond to **charges** that can have either sign, and tend to cancel when many particles are brought together. Gravity, on the other hand, responds primarily to **energy**, and does not cancel, but rather acquires enhanced power when many particles are brought together. In **celestial mechanics**, gravity is the dominant force.

Under almost all circumstances gravity leads to attraction between bodies. **Dark energy** is exceptional in this regard. I've suggested two references where you can find more information on this topic in the endnotes. Here I will simply note three consequences for the present, future, and past of the Universe as a whole:

- At present the gravity of **normal matter**, together with **dark matter**, dominates the gravitational effect of dark energy in our neighborhood, up to the scale of our Galaxy and a bit beyond. On cosmological scales, however, normal matter and dark matter are distributed patchily, while the dark matter, though much less dense in our neighborhood, is everywhere, and its effect accumulates. As a result, the gravity of dark energy, which is essentially repulsive, dominates the evolution of the Universe as a whole. The expansion of the Universe, which one might have expected to be slowed by gravitational attraction, is actually speeding up.
- Straightforward extrapolation of present-day cosmology to the far future suggests that after hundreds of billions of years our Galaxy, having merged with Andromeda and perhaps a few other nearby dwarf galaxies, will form an isolated island, with the rest of normal (and dark) matter in the Universe having expanded away so far, and so rapidly, that it is no longer accessible to observation, due to the limiting speed of light.

  Of course, this is a very bold extrapolation, considering how much scientists' views on cosmology have changed over much shorter time scales. It's been less than *one hundred* years since the expansion of the Universe was discovered!

- For most of the thirteen billion or so years since the Big Bang, the gravity of normal matter and dark matter dominated the effect of dark energy, even on cosmological scales. It is only in the past two billion or so years, as those forms of matter got diluted by expansion of the Universe, with the density of dark energy remaining constant, that the latter came to dominate. There are, however, good reasons to suspect that in the *very* early history of the Universe things were different, and dark energy dominated, with its repulsive gravity leading to a period of rapid cosmological *inflation*.

Newton's theory of gravity was an epochal event in the history of human thought. By providing an accurate account of many aspects of celestial motion, based on a few precisely stated mathematical principles, it set new standards for scientific accuracy and ambition. In the early twentieth century, however, Newton's theory was superseded by Einstein's **general relativity**, which remains foundational today.

### GROUP (OF TRANSFORMATIONS) / CONTINUOUS GROUP / LIE GROUP

It is often useful to consider the transformations that leave some structure as a whole unchanged, or **invariant**, while generally moving its parts—in other words, the **symmetry transformations**, or simply **symmetries**, of that structure—not only individually, but as a collective. Such collections of symmetry transformations are called *groups* of transformations.

Groups of transformations come in many varieties. Some allow continuous variation, while others are discrete, for example. (See **Continuous symmetry**.) But all groups of transformations share a few important features:

- We can combine two symmetry transformations, by performing one and then the other. The combined operation also leaves our structure invariant, so it defines a new symmetry operation.
- For each symmetry operation there is an opposite, or (as it is usually called) *inverse*, transformation. If the original transformation changes $x$ into $x'$, its inverse changes $x'$ into $x$.
- If we combine a transformation with its inverse (in either order), following our first rule, the result is the trivial *identity transformation*, which "changes" every $x$ into itself.

The Norwegian mathematician Sophus Lie, beginning late in the nineteenth century, made profound studies of groups of transformations that allow smooth variation, and can be studied using the methods of **calculus**. In his honor, these smooth symmetry groups are called *Lie groups*. The symmetry groups of circles, spheres, and their higher-dimensional generalizations, consisting of all the transformations we can get by repeatedly combining rotations around all possible axes through all possible angles, are Lie groups.

Those groups of rotations, as well as other Lie groups, find many applications in modern quantum physics. Most notably, the symmetry groups of **property spaces** based on different sorts of **charge** that are the cornerstone of our **Core Theories** of **strong**, **weak**, and **electromagnetic forces** are Lie groups—as are the larger *symmetry groups* we contemplate in attempting to unify those theories. See also **Local symmetry**.

### HADRON

Because they are subject to the **strong force**, **quarks**, antiquarks, and **gluons** can bind into a wide variety of small objects. *Hadron* is the generic term for objects of this kind. **Protons** and **neutrons** are examples of hadrons, as are atomic **nuclei**. All other known hadrons are highly unstable, with lifetimes ranging from a few nanoseconds (few × $10^{-9}$ seconds) to much smaller times.

Most hadrons can be understood semi-quantitatively in the framework of the **quark model**. (See, if necessary, **Quantitative**.) According to the quark model, hadrons fall into two broad classes: **baryons** and **mesons**. Baryons (the class including protons and neutrons) are bound states containing three quarks, while mesons are bound states containing a quark and an antiquark. (We also have antibaryons, based on three antiquarks. See **Antimatter.**) In a more accurate picture, based on **quantum chromodynamics (QCD)**, those two basic body plans should be considered as skeletons that get fleshed out with gluons and additional quark-antiquark pairs.

It is widely anticipated that there are hadrons that fall outside the quark model body plans altogether, such as "glue balls," wherein gluons dominate over quarks and antiquarks. This is an area of ongoing research.

See also **Quantum chromodynamics (QCD)**, and the extended discussion in "Quantum Beauty III," part 2.

### HANDEDNESS. SEE PARITY.

### HARMONY

We say that musical tones are in *harmony,* or are harmonious, when they sound well together. The origins of this psychological phenomenon in physiology are at present obscure; in the text, a candidate theory is broadly sketched. The musical notion of harmony is often extended, following the lead of Pythagoras, to a more general conception of "things that go well together."

### HIGGS FIELD / HIGGS FLUID

The *Higgs fluid* is a space-filling, active entity that appears in the equations of the **Core Theory**. The *Higgs field* is a measure of the average influence that the Higgs fluid has on other particles. See **Field / Fluid**, **Electric field / Electric fluid**, and also the extended discussion in "Quantum Beauty III," part 3.

### HIGGS MECHANISM

We would like to use the beautiful equations of **local symmetry** to describe the **weak force**. But those equations, applied to empty space, suggest that the **quanta** of the weak force **fluid**—the **weakons**—are particles with zero mass, like **photons**. In fact, the weakons have masses several tens of times larger than the proton's mass. The *Higgs mechanism* allows us to keep the beautiful equations while respecting reality. The central idea in the Higgs mechanism is that space is permeated by a field—the **Higgs field**—which modifies the behavior particles would otherwise exhibit.

According to the *Higgs mechanism,* we live inside a **superconductor** for **currents** of weak charges.

See **Higgs field / Higgs fluid, Higgs particle / Higgs boson**, and also the extended discussion in "Quantum Beauty III," part 3.

### HIGGS PARTICLE / HIGGS BOSON

These terms are used interchangeably, to refer to the minimal unit, or **quantum**, of **Higgs fluid**.

See **Higgs field / Higgs fluid**, and also the extended discussion in "Quantum Beauty III," part 3.

### HYPERCHARGE

The average **electric charge** within each entity of the **Core Theory** is called its *hypercharge.* (These *entities* are defined in "Quantum Beauty III," part 4.)

There is a complicated relationship among the **weak force**, **hypercharge**, and **electromagnetism** that I have glossed over in the main text. It would require several pages of dry exposition to explain, and the explanation does not greatly illuminate our main themes. I've suggested two references where you can find more information on this topic on pages 403–4 in the endnotes.

### INFINITESIMAL

"Infinitesimal," linguistically, is a condensed expression of "infinitely small."

In today's physics and mathematics, we define quantities like **velocity** and **acceleration** by limiting processes. Thus, to define the velocity of a particle, we consider its displacement $\Delta x$ during a small amount of time $\Delta t$, take the ratio $\Delta x/\Delta t$, and consider its limiting value as the interval $\Delta t$ is taken smaller and smaller. That limiting value, by definition, is the velocity.

In the early days of calculus the pioneers did not have firm foundations or clear definitions. They were guided, instead, by intuition and guesswork. Leibniz, in particular, was fond of the idea that instead of taking a limit, one could consider an "infinitely small" change in time $\delta t$, and the corresponding displacement $\delta x$, and take the ratio of these *infinitesimals.* However neither Leibniz nor his disciples ever made that idea precise. It lay dormant and basically forgotten for many decades, until twentieth-century mathematicians showed that it could be made **rigorous** in several ways.

The idea of the infinitesimal is similar in spirit—though opposite in direction!—to the idea that guides us to the **points at infinity** of **projective geometry**. In both cases, we replace limiting *procedures* by accomplished *objects.*

Infinitesimals provide a new way to embody the Ideal. They have not yet played any significant role in describing the physical world, but they are a beautiful idea, and they deserve to do so.

### INITIAL CONDITIONS

The basic laws of physics, as presently understood, are **dynamical equations**. They specify, in other words, how the state of the world at one time is related to its state at other times. They do not, however, tell us what we should assume for our starting point. Thus we must supply *initial conditions* to get our description started.

### INTENSITY (OF LIGHT)

The *intensity* of light is a precise concept that corresponds to the perceived quality of brightness. The *intensity* of a light beam incident on a surface is the amount of energy that beam brings to that surface, per unit time and per unit area. This definition allows us to generalize the concept of intensity to all parts of the **electromagnetic spectrum**, such as radio, infrared, ultraviolet, and X-rays.

### INVARIANCE

We say something is *invariant* under a transformation when making the transformation doesn't change it.

Examples:

- The distance between objects is invariant if you move all the objects in the same direction by the same amount (invariance of distance under **translation**).
- The shape of a circle is invariant if you rotate it around its center (invariance of a circle under rotation).
- The speed with which light beams progress is invariant if you move at any constant velocity. Thus we say that the speed of light is invariant under **Galilean transformations**, or equivalently **boosts**, that transform between views from platforms moving at different velocities.

The third of these examples describes the key assumption of Einstein's **special relativity**.

### ISOTOPE

Nuclei that have the same number of protons, but different numbers of neutrons, are said to be *isotopes.* Nuclei that are isotopes have the same amount of **electric charge**, and lead to very nearly the same chemical behavior, though they differ significantly in **mass.**

### JET (OF PARTICLES)

In the aftermath of collisions at modern high-energy accelerators, including notably the **Large Hadron Collider**, one frequently observes that streams of energetic **hadrons**, all moving in very nearly the same direction, emerge. Such streams are called *jets.*

Jets have a remarkable interpretation, deriving from **quantum chromodynamics (QCD)** and **asymptotic freedom**, as follows. We can describe the initial "bang" of our collision, involving the **strong force**, directly in terms of **quarks**, antiquarks, and **gluons**. But as those particles emerge from the initial fireball, they come into equilibrium with the ever-present spontaneous activity of QCD's **quantum fluids**—their **quantum fluctuations**, or **virtual particles**—and, in the process, produce swarms of hadrons. Because **energy** and **momentum** are **conserved**, the swarms inherit those properties from the quarks, antiquarks, and gluons that triggered them. Thus an energetic quark, say, will produce a swarm of hadrons moving, as a whole, in the direction of the quark's momentum, and sharing its energy—a *jet*! It is not a great stretch to say that when we view jets, we are glimpsing the reality of quarks, antiquarks, and gluons, which cannot exist as free particles. See **Confinement**.

### KINETIC ENERGY. SEE ENERGY.

### LARGE HADRON COLLIDER

The **Large Hadron Collider**, or LHC, is a project housed at the CERN laboratory near Geneva. The major goal of the project is to probe fundamental processes at higher energies, and thereby, in effect at smaller distances and times, than have ever been accessed before.

This is accomplished in the following way. **Protons** are accelerated to acquire very high **energy of motion**, and are organized into two narrow beams. The beams are stored within a gigantic underground ring, twenty-seven kilometers in circumference, where they circulate in opposite directions, with powerful magnets guiding their paths. (The ring must be large, and the magnets powerful, because it is difficult to deflect such energetic protons from straight-line motion!) At a few observation points, the beams are made to cross. Close encounters between high-energy protons moving in opposite directions result in "collisions" that concentrate an enormous amount of energy in a very small space, recreating extreme conditions last seen during the earliest moments of the Big Bang. Huge, sophisticated "detectors"—instruments measuring tens of meters in all three dimensions, and crammed full of cutting-edge electronic technology—extract physical information from the residues of those collisions, which is then analyzed by big teams of highly trained scientists, assisted by a worldwide network of powerful computers.

The LHC is our civilization's more than worthy addition to the pyramids of Egypt, the Roman aqueducts, the Great Wall of China, and the cathedrals of Europe, all awesome monuments to collective effort and technological achievement by humans.

In July 2012 scientists working the LHC announced the discovery of the Higgs boson. For more on this, see "Quantum Beauty III," part 3. In future operations, at higher energies, attractive ideas about **unification** of forces and **supersymmetry**, described in "Quantum Beauty IV," will be tested.

### LEPTON

The **electron** $e$ and its **neutrino** $\nu_e$, together with their relatives the muon $\mu$ and its neutrino $\nu_\mu$, and the $\tau$ particle and its neutrino $\nu_\tau$, are called, collectively, *leptons.* Their antiparticles are *antileptons.*

### LIE GROUP. SEE GROUP.

### LINE OF FORCE

Under the influence of a bar magnet, iron filings supported by a sheet of paper will form curved lines extending from one pole of the magnet to the other, as pictured in figure 20. This pretty phenomenon, and others of a similar sort, inspired Faraday's imagination. He was led to envision that the lines had an independent, prior existence, and were *revealed,* rather than *created,* by the iron filings. These intuitions guided him to new experimental discoveries. They were developed into precisely formulated mathematical ideas by Maxwell. Modern physics, with its space-filling **fluids**, emerged from these ideas. They supplanted **action at a distance** as a model for fundamental understanding.

### LOCAL SYMMETRY

We say a symmetry is *local* when it allows its **transformations** to be made independently at different places and different times.

Local symmetry is, together with **quantum theory**, the basis of the **Core Theories** of all four forces that summarize our present knowledge of the basic laws of Nature. It is also, together with **supersymmetry** (and in the framework of quantum theory), the basis of an attractive attempt to unify and improve the Core, as described in "Quantum Beauty IV."

Local symmetry is to conventional (that is, **rigid**) symmetry as anamorphic art is to conventional perspective.

Local symmetry is a major focus of our meditation, and dominates its later parts.

### MAGNETISM / MAGNETIC FIELD / MAGNETIC FLUID

"Magnetism" is a broad term, used to refer to a wide range of phenomena associated with the forces that **electric currents** exert on one another, and of their interactions with a few special magnetic substances that exert similar forces. Magnetic substances, which often involve iron ores, are used to make the familiar "magnets" used for compass needles, refrigerator-note-holder-uppers, and many other purposes.

A technical discussion of precisely how the magnetic field is defined, and the forces it leads to, would be broadly similar to our discussion in **Electric field / Electric fluid**, but the details are considerably more complicated and fussy. I've suggested two easily available references, where you can find more information on this topic, on page 404 in the endnotes.

### MASS

The scientific concept of *mass* has evolved over time, and the word is presently used in several closely related but not entirely consistent ways. Here I'll describe the three most important ones.

1. The earliest reasonably precise, scientific use of the concept *mass* occurs in Newtonian mechanics. There, mass is taken as a primary property of matter, which can never be created or destroyed, nor explained in terms of anything simpler. Mass measures a body's inertia, or resistance to **acceleration**. A body with large mass will tend to maintain a constant velocity unless subjected to large outside influences (**forces**). This concept of mass becomes quantitative in Newton's second law of motion, which says that the acceleration of a body is equal to the force acting upon it, divided by its mass. Newton's concept of mass is still very widely used, and still called "mass," because Newtonian mechanics, although not exact, is often an adequate approximation, and it is easier to use than the more accurate relativistic mechanics.

2. In Einstein's modification of mechanics, to make it consistent with special relativity, mass is a different concept. In relativistic mechanics, mass is a property of individual particles, but mass can be created or destroyed when particles interact with one another. Mass, in the relativistic theory, is a measure of a particle's contribution to **mass energy** and governs its **energy of motion**. Mass is a property of particles, but it is not a well-defined (**conserved**) property of the world as a whole.

Each of the **elementary particles** of our **Core Theory** has a definite mass, but it is very far from true that the sum of particle masses entering a collision is equal to the sum of particle masses leaving it. In collisions between high-energy **electrons** and **positrons**, one commonly finds that the total mass of the particles leaving the collision is hundreds of thousands times larger than the total mass of the particles that entered it.

It is not mass, but rather **energy**, that is conserved in relativistic mechanics. I like to summarize the status of mass and energy in relativistic mechanics in the epigram:

Particles have mass, the world has energy.

3. In cosmology, one speaks of the fraction of mass in the Universe due to various kinds of stuff: **normal matter** (5 percent), **dark matter** (27 percent), **dark energy** (68 percent). This is a sloppy use of the term "mass." (Dark energy, in particular, does not have mass in either of its more usual senses, as defined above.) But it is very widespread both in scientific and in popular literature, so we're stuck with it. What it means is the following: Using **general relativity**, we can relate the rate at which the rate of expansion of the Universe changes with time—roughly speaking, its acceleration—to the average density of energy within it. We can divide that average density of energy by the square of the speed of light to get something measured in units of mass density. The percentages quoted above are the relative fractions of that "something," contributed by the various stuff.

Because mass is not conserved, fundamentally, we can hope to explain it in terms of something simpler. There is, in fact, an extremely beautiful explanation of the *origin* of most of the mass of **normal matter** that emerges from **quantum chromodynamics (QCD)**. The important building blocks of **protons**—up and down **quarks**, and color **gluons**—all have masses much smaller than the mass of a proton, so the proton's mass must have some other source.

A key step toward understanding the origin of the proton's mass is to understand properly what a proton *is*. What is a proton? From the perspective of modern fundamentals, a proton is a stable, localized *pattern of disturbance* in quark and gluon **fluids**. Such a pattern can move along—**Galilean symmetry** assures us of that—and if we view it from afar (compared with its size) it will look like a particle. There is gluon **field energy** associated with that disturbance, and **energy of motion** of the confined quarks. If we call the energy of a stationary disturbance $\varepsilon$, then $\varepsilon/c^2$ will be the mass of the particle we interpret it to be—that is, the mass of a proton. And that is, overwhelmingly, the origin of *your* mass. It is Mass Without Mass, arising from embodied energy.

### MASS ENERGY. SEE ENERGY.

### MAXWELL'S EQUATIONS

The *Maxwell equations* are a system of four equations that express relations among **electric fields**, **magnetic fields**, and the distributions of **electric charge** and **electric current** in space. They are discussed extensively in the text and the endnotes.

See also **Ampere's law / Ampere-Maxwell's law**, **Faraday's law**, and **Gauss's law**, where the four Maxwell equations are explained individually.

### MAXWELL TERM / MAXWELL'S LAW

In order to reconcile inconsistencies among the **dynamical laws** for **electric** and **magnetic fields**, as they were known, Maxwell suggested that there must be an additional effect. The new effect, which I've called *Maxwell's law,* is that electric fields which change in time induce (i.e., "create") magnetic fields. This is a sort of dual to **Faraday's law**, which states that magnetic fields that change in time induce electric fields. Maxwell's law supplements another way of inducing magnetic fields, by electric currents (**Ampere's law**). The complete equation, which results from adding the new *Maxwell term* to Ampere's law, is known as **Ampere-Maxwell's law**.

### MEDIUM / MEDIA

A *medium,* for us, is something that fills space. *Media* is, for us, simply the plural of *medium*.

Thus *medium* can be used interchangeably with **fluid**. At the margins, "fluid" suggests a material whose parts can change places with one another, as in flows of air or water, while "medium" suggests something more tangible that may vibrate but has structural integrity, like glass or Jell-O. But the media, or fluids, that according to our Core Theory make up the most basic world-substance, such as the **gluon fluid** and the **electron fluid**, are so different from air, water, glass, Jell-O, or any other everyday fluid or medium that it seems silly to insist on either metaphor exclusively.

### MESON. SEE HADRON.

### METRIC / METRIC FLUID

We say a space has a metric when it is possible to say what the distance is between two very nearby points. The *metric* itself is the secret sauce that turns a collection of points into a structure that has size and shape.

Let us suppose, to get things started, that we know how to measure the distance between two nearby points in ordinary space, say by using little rulers. Then we can also measure distances between nearby points on any reasonably smooth surface, using the same rulers. The restriction to small rulers and nearby points is important here, because if we have a curved surface and long flat rulers, then the rulers may be a poor fit to the surface over large distances, and we won't know how to lay them down.

Now let's consider representing our surface using an ordinary, flat paper map. We can certainly do that, in many ways, simply by setting up a correspondence between the two sets of points—the points on the surface, and the points on the map. We put Prague here, New Delhi there, and so forth, taking care to put points that are neighbors in reality at neighboring places on the map. There is great freedom in how this is done, and in atlases one can find many quite different representations of the same region.

Without further specification, however, a map does not tell us how far apart the points it represents are in reality, back at the surface. The *metric,* an addition to the map, supplies that information. To be a bit more precise, the metric is a **function** of positions on the map—i.e., it assigns a "thing," or value, to each point on the map. At each point, the value of the metric is a gadget that tells you, for each direction you can proceed from that point, the scale you should use on little rulers, so that the distance you measure between nearby points on the map is the same as the distance between the points they represent, back at the surface.

Having seen what it takes to turn a flat plane (our map) into a surface with size and shape, we can get creative with the idea, and build on it, or play variations on the theme. To reach the concept of metric most important for physics, we need to do two things.

First, we switch focus from the problem of surface measurement that led us to introduce the metric concept, to the metric concept itself. Thus we call any gadget that tells us the scales we should assign to little rulers a metric on our map, whether that gadget came from a surface or not. (In taking this step, we are following the path by which Bernhard Riemann [1826–66] generalized the work of his professor, Carl Gauss [1777–1855].) In other words, we let this concept of a metric take on a life of its own.

Second, we add some **dimensions**. There's nothing to stop us from adding the same sort of scale-specifying gadget to points in all of three-dimensional space, not just the points on a flat piece of paper. Carrying this thought further, we can use the method of **coordinates** to represent three-dimensional space and time as a combined four-dimensional space-time, and consider adding a metric gadget to that. In this way, we have found a very flexible procedure that can represent—or, we might say, define—what we should mean by a curved three-dimensional space, or a curved space-time, in an "obviously correct" way that generalizes what we do for surfaces, where our intuition is clear.

So much regarding the *mathematical* concept of metric. It is a conceptual gadget that fills space (or space-time)—a conceptual **field**. Other fields include **electric fields**, **magnetic fields**, and the **field** of **velocities** in a body of water. In those cases and many others, we find that the fields are important elements of reality. They dance to the music of **dynamical equations**, influenced by matter, and in turn affect the behavior of matter. We may say, loosely but with justice, that they have a *physical* existence. Einstein, in his theory of **general relativity**, postulated that the *metric of space-time* is, like those others, a physical entity, with a life of its own. We call it the *metric fluid*, or also the **gravitational fluid**, in view of the role it plays in **general relativity**.

There are many variations and generalizations of the concept "metric" as described in this entry that are useful in different applications. What they have in common is that they deal with some kind of distance. The version described above is the one that is currently most useful in physics, and that figures in our meditation.

Not all spaces have an obvious notion of distance. Or, alternatively, a space may suggest several different possibilities for defining distance. In such cases, we can either do without a metric or experiment with different **complementary** possibilities. The three-dimensional space of color perceptions is an interesting example in this regard.

Is it possible to define, in a precise and quantitative way, the distance between different perceived colors? Several serious thinkers have wrestled with that question, notably including Erwin Schrödinger (of **Schrödinger equation** fame). They came up with several different answers. Each of those answers is internally consistent, but as yet none has proved enormously useful, nor clearly superior to the others.

### MICROWAVE / MICROWAVE BACKGROUND RADIATION

**Electromagnetic waves** with **wavelengths** between roughly a millimeter and a meter are called *microwave radiation*.

Early in its history, the matter in our Universe was so hot and dense that atoms could not hold together. The plasma of protons, helium **nuclei**, and **electrons** glowed white-hot, and the Universe was filled with light. As the Universe expanded and cooled, eventually atoms could hold together, and as a result, rather suddenly, the Universe became transparent to light and other forms of electromagnetic radiation, as it remains today. The ambient light continued to pervade the Universe, but as expansion continued it got stretched to longer wavelengths.

Today, most of that light has been shifted into the microwave part of the **electromagnetic spectrum**. It has become the *microwave background radiation.*

The microwave background radiation was discovered experimentally by Arno Penzias and Robert Wilson in 1964, and ever since it has inspired intense study. In view of its origin, the microwave background radiation gives us access to clean information about conditions in the very early Universe.

### MOMENTUM

Momentum is, together with **energy** and **angular momentum**, one of the great **conserved** quantities of classical physics. Each of them has evolved into a pillar of modern physics as well.

The *momentum* of a body is a measure of its rate of motion. Quantitatively, it is equal to its **mass** times its **velocity**. (This is the nonrelativistic version, accurate for small velocities. **Special relativity** leads to a related but more complicated formula.)

Momentum has a direction, as well as a magnitude. Thus it is a **vector** quantity.

The momentum of a system of bodies is the sum of the momenta of the bodies separately.

There are a wide variety of circumstances in which momentum is conserved. This result is best understood through Noether's general theorem, which connects **conservation laws** to **symmetry**. In that framework, conservation of momentum reflects symmetry (i.e., invariance) of the physical laws under **spatial translations**—that is, under the transformations that move everything in the system of interest through a common displacement. In other words, we have conservation of momentum when the laws governing our system do not depend on any externally specified, fixed position.

In the quantum world momentum continues to be a valid concept, and takes on additional features of great subtlety and beauty.

### MULTIVERSE. SEE UNIVERSE / VISIBLE UNIVERSE / MULTIVERSE.

### MUTATRON

In theories that unify the strong and weak forces, there are particles that induce transformations between strong and weak colors. We (or, more accurately, I) call those hypothetical particles *mutatrons.*

### NANOTUBE

Nanotubes are a class of molecules made entirely from carbon. As the name suggests, they take the form of tubes, and they can extend indefinitely in one dimension. *Nanotubes* come in many sizes and shapes, and they have remarkable mechanical and electrical properties. For example, some classes of nanotubes are extremely strong in their long direction. Fibers made from such nanotubes can be lightweight, but stronger than steel. For a more extended discussion, and pictures, see "Quantum Beauty II."

### NATURAL FREQUENCY / RESONANT FREQUENCY

Many objects, especially stiff ones, prefer to vibrate in a few special patterns. These are called their natural *modes of vibration.* In each natural mode, the object goes through a cycle of shape changes that repeats after a fixed interval of time. That interval is called the **period** of the mode, and one divided by that period is called the **frequency** of the mode. The frequencies of the natural modes of a body are called its natural frequencies. Because vibrations of bodies in air set up sound waves, we can hear the natural frequencies of bodies as the **pure tones** they emit.

Examples:

- Tuning forks are designed to have a single audible natural frequency.
- Gongs typically have several natural frequencies, as do bells. One may hear different combinations of tones as the gong or bell vibrates, depending on where or how it was struck. This is because different hits, by setting up different **initial conditions**, excite the natural modes with different relative strengths.

The natural frequencies of a body are also called its *resonant frequencies.*

These phenomena of musical instruments and sound have close parallels in atoms and light. The natural modes of an instrument are analogous to the **stationary states** of an atom, and the tonal palette of a musical instrument is analogous to the **spectrum** of an atom. These parallels are not only metaphorical, but extend to the equations describing these systems, which are very similar. In the spectra of atoms, a very real, visual Music of the Spheres is on display.

### NATURAL NUMBER

The numbers 1, 2, 3, . . .—the numbers that naturally arise from the act of counting—are called *natural numbers*. They are the sorts of numbers Pythagoras most approved of. Natural numbers form a discrete series. They should be contrasted with **real numbers**.

### NEUTRINO

Each of the three electrically charged **leptons**—electron $e$, muon $\mu$, tauon (or simply tau) $\tau$—has an associated *neutrino*. The neutrinos, written $\nu_e$, $\nu_\mu$, $\nu_\tau$, are electrically neutral particles. Left-**handed** neutrinos carry a unit of yellow **weak charge**, but zero **electric charge** and no strong **color charge**. Thus neutrinos participate in the **weak force**, but not in the **electromagnetic** or **strong** forces. As a result, the interactions of neutrinos with **normal matter** are exceedingly feeble. A dramatic illustration: Every second about 65 billion neutrinos, emitted in the course of the weak transitions that power our Sun, stream through each aligned square centimeter at Earth. Yet we generally feel no effect from those neutrinos, and very elaborate detectors are needed to detect that stream at all.

Neutrinos can be calculated to have been produced in the Big Bang in considerable abundance. The resulting cosmological gas has so far escaped detection, simply because neutrinos interact so feebly. At one time neutrinos were thought to be a good candidate to supply the **dark matter**, but this idea has not survived, basically because we now know they are too light for that purpose.

Many other interesting facts have been discovered about neutrinos. I've suggested two easily available references, where you can find more information on this topic, on page 404 in the endnotes.

### NEUTRON

Neutrons, together with protons, are the building blocks of atomic **nuclei**. *Neutrons* have zero **electric charge**, but weigh about the same as protons. Most of the mass of **normal matter** arises from the mass of its constituent **protons** and neutrons. Neutrons were once thought to be fundamental **elementary particles**, but today we know that they are complicated objects, built up from more elementary **quarks** and **gluons**.

### NORMAL MATTER

"Normal matter" is a convenient term I use to refer to the kind of matter that is made from **quarks**, color **gluons**, **electrons**, and **photons**. Normal matter is the dominant form of matter on Earth and in its immediate environment. It is the kind of matter we are made from, and that we study in chemistry, biology, materials science, all forms of engineering, and almost all of astrophysics. *Normal matter* should be distinguished from **dark energy** and **dark matter**.

### NUCLEON

Nucleon refers to the particles that build up atomic nuclei. *Nucleon* simply means "proton or neutron."

### NUCLEUS

Every atom has a very small center, or *nucleus*, that contains all of its positive charge and almost all of its mass. As described in "Quantum Beauty III," study of atomic nuclei revealed the existence of two new forces of Nature, the **strong** and **weak** forces, and led, over the course of the twentieth century, to our marvelous **Core Theory**.

### ORBIT / ORBITAL

The concept of the *orbit* of a planet revolving around our Sun, or of an artificial satellite revolving around Earth, is widely understood, and needs no special comment here. It is, basically, the sequence of positions that the body occupies over time, collected into a curve.

An *orbital,* as used in quantum physics and chemistry, is the wave function of a **stationary state**. We say an **electron** "occupies an orbital" when the state of that electron is described by the **wave function** associated with that orbital. The term "orbital" is a vestige of Bohr's atomic model, in which stationary states were associated with selected classical orbits.

### OSCILLATION

We call a physical process that goes through many cycles of identical behavior, repeating after a fixed interval of time, an *oscillation.* The vibrations of plucked strings or struck tuning forks, in music, are examples of oscillations.

### PAULI EXCLUSION PRINCIPLE. SEE EXCLUSION PRINCIPLE.

### PARITY / PARITY TRANSFORMATION / PARITY VIOLATION / HANDEDNESS

There are several places in mathematics, and in the formulation of physical laws, where one finds appeals to use your right hand (or, much more rarely, your left hand).

In most cases, these "right-hand rules" are mere conventions. The corresponding left-hand rule could also be used, and would amount merely to a renaming of things. Take, for instance, the way we assign a direction in space to rotation around an axis. If an object rotates around an axis, we can assign a direction to the axis using a right-hand rule, as follows. Imagine our spinning object as an ice skater. The axis around which she rotates is the straight line passing between her head and her toes. That line picks out an orientation in space, and so almost defines a direction, but to complete the definition we need one last step: we must decide between "up" and "down." The usual right-hand rule for breaking that ambiguity is to say that if her rotation brings her right hand toward her abdomen, we choose up—i.e., the direction from her toes to her head—while if the rotation brings her right hand toward her back, we choose down. Clearly, if we interchanged right with left and, at the same time, in the rule, interchanged up and down, the resulting "left-hand rule" would be completely equivalent.

Here are two more examples of how that rule gets used:

- The motion of the hands of a clock amounts to rotation about an axis perpendicular to the clock. If you're looking down at the face of the clock, then applying the right-hand rule to the "clockwise" rotation of the hands gives you the direction down.
- In order to advance a standard screw into place, we need to make it rotate around its axis. If we are looking down on the screw, then we must turn it clockwise to make it advance downward. This works because a standard, good screw is grooved to harmonize with the preceding right-hand rule, and so we call it a right-hand screw. The other kind, bad screws, are left-handed.

In all these cases, we could equally well replace right with left, to describe exactly the same situations. We need only interchange, in the definitions, "clockwise" and "counterclockwise," and "good" and "bad."

Similarly, in textbooks of physics you will find many right-hand rules describing how to calculate the direction of **magnetic fields**, and the **forces** that magnetic fields generate. But if you changed right to left, and at the same time changed the *definition* of the direction of the **magnetic field** by simple reversal, then nothing in the laws of physics would change.

Physicists believed, until 1956, that all appearances of right and left in physics are mere conventions—that is, agreements about how to define things, adopted for convenience. Conventions can be very useful. It's very important, for instance, to be able to tell screw manufacturers which way to cut their grooves. But they're not fundamental. One might have made different agreements!

Another way of stating this assumption that locates it nicely within the mainstream of deep thinking about fundamentals is as an assumption of **symmetry**. We say a set of equations has *parity* symmetry, or is invariant under *parity transformations,* if you can interchange left and right, and make appropriate changes in definitions, without changing the content of those equations.

(Slightly technical elaboration, and fun

exercise: "Interchange left and right" needs some spelling out, here, because left and right are properties of objects [for instance, hands] located in space, and we can't just change all left hands into right hands [and all left-handed screws into right-handed screws, and so forth] without making changes in space itself, so that the transformed objects can continue to fit together! The simplest way to do it is to pick one point *O*—an origin—as reference, and to transform every other point to its antipode with respect to *O*. That is, you move any point *P* into the point diametrically opposite, when viewed from *O*.)

When we make our parity transformation, reflecting points to their antipodes, it is natural for **vectors** to change their direction. For example, it is a nice exercise to visualize that the vector drawn between two points *A* and *B* points in the direction opposite to the direction of the vector drawn between their antipodes $-A$ and $-B$.

Here's a fun exercise with that idea: use the thumb and first two fingers of the right hand to point in three perpendicular directions, do the same with your left, and line them up so that each pair of corresponding fingers points opposite. By means of this exercise, you embody a parity transformation: your fingers indicate directions, and reversing all three directions interchanges left and right hands!

In 1956 Tsung-Dao (T. D.) Lee (1926– ) and Chen-Ning (Frank) Yang (1922– ), after analyzing some puzzling experiments, suggested that although most appearances of "hands" in physics, including the right-hand rules that have confused generations of students of magnetism, are merely conventional, the **weak force** is different and really does make a distinction between left and right. In other words, they suggested that parity symmetry is not strictly correct. Alternatively, to make it short and sweet, they suggested *parity violation*. Their suggestion was soon confirmed experimentally, and that breakthrough led to much better understanding of the weak force.

Today we recognize that parity violation is a key feature of the weak force, and an essential part of formulating its **Core Theory**. The weak force makes a big distinction between left and right, which can't be defined away!

To state the distinction precisely, we must introduce the *handedness* of a particle. When a particle with **spin** is also moving, there are two directions associated with it: the direction associated with its spin, which we defined earlier, and the direction of its velocity. In defining the direction associated with spin, we used a right-hand rule. Accordingly, if the direction of a particle's spin, defined that way, is the same as the direction of its velocity, we say that it is a right-handed particle. On the other hand, if the direction of a particle's spin is opposite to the direction of its velocity, we say it is a left-handed particle.

With that preparation, we are ready to describe how the weak force violates parity: left-handed quarks and left-handed leptons, and right-handed antiquarks and antileptons, participate in the weak force, but particles with the opposite handedness do not.

Finally, let me make good on the promise made in the entry for **axial vector** by defining that term. Above, we saw that vectors associated with going from one position to another change direction in response to a parity transformation. Vectors that transform this way are called natural vectors, or polar vectors. Not all vectors behave that way, however! Vectors whose definition involved a right-hand rule will change direction twice over when we make a parity transformation: once because they are vectors, and again because they were defined by the "wrong" rule, as far as the transformed system is concerned (because right has changed to left!). Vectors whose direction does *not* change when we make a parity transformation are called unnatural vectors, or axial vectors. In physics, the magnetic field is a field of axial vectors.

### PERIOD / PERIODIC

A *periodic* process is one that repeats. Usually the term refers to repetition in time, though in the scientific literature it is also often used for repetition in space. The *period* of a process that is periodic in time is the amount of time that passes between repeats. See also **Frequency**.

**PERIODIC TABLE**

The *periodic table* of chemical elements is an informative geometric arrangement of the list of chemical elements, in which the columns contain elements with similar chemical properties. Within each column, the atomic numbers and atomic weights increase as we go from top to bottom; and within each row, the atomic numbers and atomic weights increase as we go from left to right. In the most straightforward version of the periodic table, the atomic number increases by one with each rightward step, and also as we go from the rightmost entry in a row to the leftmost entry in the row below. (There are many variants. A common one is to snip out the rare earths and the actinides into separate sub-tables.) Quantum mechanics explains the structure of the periodic table theoretically, as an application of the **Schrödinger equation**. It is a glorious example of

$$\text{Ideal} \rightarrow \text{Real}$$

Within that explanation, the quantum theory of **angular momentum** and the **Pauli exclusion principle** play dominant roles.

**PERIOD (OF OSCILLATION). SEE FREQUENCY.**

**PERSPECTIVE. SEE PROJECTIVE GEOMETRY / PERSPECTIVE.**

**PHOTON / PLANCK-EINSTEIN RELATION**

A *photon* is a minimal disturbance in the **electromagnetic fluid**.

Classically, according to **Maxwell's equations**, the **energy** of an **electromagnetic wave** can be arbitrarily small. In **quantum theory** that is not the case. The energy comes in discrete units, or **quanta**. Because these units cannot be broken down further they have the sort of integrity we associate with particles, and in some circumstances it is helpful to think about them that way. In that sense, photons are particles of light.

(The quantum-mechanical description of photons does not strictly conform to the classical idea of a wave, or to the classical idea of a particle. Those ideas, taken from everyday experience with large bodies, have no reason to be an adequate description of what goes on in the unfamiliar domain of very small bodies, and they aren't. Either picture can be useful, but neither by itself does full justice to reality. See **Complementarity**.)

For pure spectral colors, there is a simple quantitative relationship between the unit of energy—i.e., the energy of a single photon—and the **frequency** associated with **electromagnetic waves** of that color. It was proposed theoretically by Planck and Einstein in the early days of the twentieth century and is called the *Planck-Einstein relation*. The Planck-Einstein relation has survived, since then, without substantial modification. It also supports an important application that is central to our meditation and helps us to answer our Question.

Here it is: The energy in a photon is equal to the frequency of light it represents, times **Planck's constant**.

And here is how we use it: When an atom emits or absorbs a photon, it makes a transition between two stationary states. Because energy is conserved in this process, the energy of the photon is related to the difference in energy between those two stationary states. Thus the spectrum of an atom encodes the energies of its possible states.

For more on this remarkable connection, see **Spectra**.

**PLANCK-EINSTEIN RELATION. SEE PHOTON.**

**PLANCK'S CONSTANT / REDUCED PLANCK'S CONSTANT**

In 1900 Max Planck (1858–1947), in the course of studying how the **electromagnetic fluid** comes into equilibrium with hot gases, found it necessary to assume that the transfer of energy between matter and electromagnetic radiation did not take place in arbitrarily small amounts, but only in **quantized** units. He was led, in fact, to postulate that the unit of energy transfer is proportional to the frequency of light. The proportionality constant that appears in this relation—the **Planck-**

**Einstein relation**—is now known as *Planck's constant.*

Einstein proposed that the **Planck-Einstein relation** applies to the **electromagnetic fluid** itself, not only to its exchange of energy with atoms. In his atomic model, Bohr proposed rules for determining the **stationary states** of an electron in the hydrogen atom, in which Planck's constant appears centrally. The success of Bohr's rules in accounting for the **spectrum** of hydrogen brought Planck's constant into the description of substance as well as light.

In modern **quantum theory**, Planck's constant has become **ubiquitous**. An important example is its appearance in the description of particle **spin**. Many kinds of particles, including **electrons**, **protons**, **neutrinos**, and **neutrons**, have "**spin ½**." As described in the entry **Spin**, this means that they display spontaneous rotary motion. Planck's constant appears in the **quantitative** description of that motion. The rotary motion involves, specifically, a quantity of angular momentum equal to one-half of the *reduced Planck's constant,* which is simply Planck's constant divided by $2\pi$.

### PLATONIC SOLID / PLATONIC SURFACE

A *Platonic solid* is a **polyhedron** whose faces are all copies of the same regular **polygon**, and whose faces meet in the same way at all vertices. There are precisely five different (finite) Platonic solids: tetrahedron, octahedron, icosahedron, cube, and dodecahedron. They are described at length in the text.

The mathematical construction of those solids, and the proof that they are the only ones possible, is the climax of Euclid's *Elements.*

The surfaces of the Platonic solids are in many ways more fundamental than the solids they surround. I've referred to them as *Platonic surfaces.*

The Platonic solids have inspired admiration from mathematicians, scientists, and mystics for many centuries.

### POINT AT INFINITY / VANISHING POINT

If we stand vertically over a flat plane, and view two parallel lines in the plane as they extend away from us, we will see that they seem to be converging as they approach the horizon. If we imagine painting what we see, or if we **project**, geometrically, these lines onto a canvas, it is natural to add the limiting point, where they do meet, as an element of the image. This is the **point at infinity**, or **vanishing point**. We picture, amplify, and meditate on the implications of that construction in the main text.

### POLARIZATION (OF LIGHT). SEE TRANSVERSE WAVE / POLARIZATION (OF LIGHT).

### POLYGON / REGULAR POLYGON

A *polygon* is a figure obtained by connecting a sequence of points in a plane by line segments, so as to form a closed loop. Triangles and rectangles are familiar examples of polygons. The polygon's defining points, where its sides meet, are called its vertices.

A *regular polygon* is a polygon whose sides are all equal in length, and meet at equal angles at all its vertices. Equilateral triangles are regular polygons with three sides, squares are regular polygons with four sides, and so forth.

### POLYHEDRON

A *polyhedron* is a three-dimensional solid with flat **polygonal** faces, straight edges where faces meet, and sharp vertices where edges meet.

### POSITRON

"Positron" is another word for antielectron, the **antiparticle** of an **electron**.

### POTENTIAL ENERGY. SEE ENERGY.

### PRESSURE

This concept arises when we discuss forces in the context of continuous media (as opposed to particles). Each part of the continuum exerts forces on its neighboring parts, through the surfaces that divide them. (These surfaces are introduced as objects of thought, and need not be material discontinuities.) *Pressure* is defined, in this context, as the force per unit area.

### PROBABILITY CLOUD

In classical mechanics particles occupy, at each time, some definite position in space. In quantum mechanics, the description of a particle is quite different. The particle does not occupy a definite position at each time; instead, it is assigned a *probability cloud* that extends over all space. The shape of a probability cloud may change over time, though in some important cases it does not. See **Stationary state.**

As the name suggests, we can visualize the probability cloud as an extended object, which has some non-negative—that is, positive or zero—density at each point. The density of the probability cloud at a point represents the relative probability for finding the particle at that point. Thus the particle is more likely to be found where the density of its probability cloud is high, and it is less likely to be found where the cloud's density is low.

Quantum mechanics does not give equations for probability clouds directly. Rather, probability clouds are calculated by squaring wave functions, which satisfy Schrödinger's equation. See **Wave function**, **Schrödinger equation**.

### PROJECTION

This word is used very flexibly in mathematics and physics. It has not just one but several different precise technical definitions, within various subfields. In all cases, a *projection* is a mapping from one space to another, by which information about the first space is presented in a new form. Often (but not always) some information is lost in the process. In this book I have used the word "projection" informally, rather than with technical precision, in several closely related ways:

- Projection of shadows, in the metaphor of Plato's Cave. Here the shadows produce two-dimensional, colorless versions of the objects they represent, and much information is lost.
- Projection by our eyes, in vision. Our retinas receive a two-dimensional version of the three-dimensional world. Focusing by the eye's lens allows production of images that (in perfect vision) focus all the light emerging from a point in an object to a very small area in the retina, thus preserving useful spatial information.

As we discussed extensively in "Maxwell II," there is much more information available within the incoming **electromagnetic signal** we call light than our eyes extract.

Human vision performs a **projection** of the infinite-dimensional space of **spectral** color intensities onto a three-dimensional space of *perceived* color, and discards information about **polarization.**

- Geometric projection: of Platonic surfaces onto circumscribed spheres, by continuing lines from the center to the surface; of light rays onto canvas, in geometrically accurate painting (the artistically inspired science of **perspective**); and of surfaces, such as bits of terrain or the whole surface of Earth, onto flat planes of paper, in maps.
- **Color** projection, in color **property space**. Here, for example, we projected the three-dimension color property space whose coordinates are **intensities** $R$, $G$, $B$ of red, green, and blue onto a two-dimensional property space, simply by dropping the $B$ coordinate.

### PROJECTIVE GEOMETRY / PERSPECTIVE

*Projective geometry* is a wide-ranging branch of mathematics, with close connections to the artistic study of *perspective.* Its central concern is to understand the relationships among the images we receive from an object when we view it from different vantage points (in other words, from different perspectives). What do those images have in common? How can one use the information in one such image to construct others? Those are the sorts of questions addressed in projective geometry. Projective geometry offers us an appealing realization of deep ideas including **transformation**, **symmetry**, **invariance**, **relativity**, and **complementarity**, as explained in the main text.

### PROPERTY SPACE

In human color perception, we find that any perceived color can be matched, essentially uniquely, using mixtures of three basic colors, say red, green, and blue. Different **intensities** of red, green, and blue are described by three positive **real numbers**, and each such combination of intensities corresponds to a different perceived color. We can interpret these triples as the **coordinates** of a three-dimensional *property space,* the *space* of perceived colors.

There are many examples of a similar kind, where we use numbers to encode properties, and regard sets of numbers as coordinates, to define a property space. Property spaces based on **color charges** play a central role in our **Core Theories**.

### PROTON

*Protons,* together with neutrons, are the building blocks of atomic **nuclei**. Protons have the opposite **electric charge** to **electrons**, and weigh roughly two thousand times as much. Most of the mass of **normal matter** arises from the mass of protons and neutrons within it. Protons were once thought to be fundamental particles, but today we know that they are complicated objects, built up from more elementary quarks and gluons.

### PURE TONE. SEE TONE / PURE TONE.

### PYTHAGOREAN THEOREM

The *Pythagorean theorem* was an early, striking discovery in geometry. The Pythagorean theorem states that the squares of the lengths of the two shorter sides of a right triangle add up to the square of the length of its longest side (the hypotenuse). It is discussed extensively, and pictured, in the main text.

### QUALITATIVE / QUANTITATIVE

We say a concept, theory, understanding, or measurement is *quantitative* when it is expressed using numbers. Otherwise it is *qualitative.* The "numbers" used in a quantitative description can be **natural numbers**, **real numbers**, **complex numbers**, or others, depending on the application.

We also speak of *semiquantitative* concepts, theories, understandings, or measurements, when these things are expressed using numbers, but without complete precision or consistency. Thus, for example, we might find that different practitioners, using the same semiquantitative physical theory, derive different predictions depending on how they fill in the theory's vague details.

The word "qualitative" can also be used for emphasis, in the following way. In saying that an idea or phenomenon is *qualitatively* new, we mean that it is not merely an elaboration or intensification of what came before, but fundamentally different, so that the two cannot be compared quantitatively. For example, the **wave functions** of **quantum theory** differ qualitatively from the **orbits** of classical physics, which they supersede.

### QUANTIZATION

This is used in three distinct senses: a general sense, a closely related special sense, and a piece of jargon.

General: When we map or, as we say, **project** a continuous quantity onto a discrete one, we say we have *quantized* that quantity. In other words, the process of quantization converts an **analog** quantity into a **digital** representation of it. Quantization, in this sense, is a very common practice in modern engineering and information processing because digital quantities are easier to communicate, and to keep accurate, than analog quantities. (See **Analog** and **Digital** for further discussion.) With a few esoteric exceptions, modern computers work with digital information exclusively, so readings of analog signals, such as light **intensities**, are quantized before they are read in. The act of quantizing something is called *quantization.*

An important result of quantum mechanics is that it *quantizes,* in the preceding sense, many quantities that were, in classical physics, continuous. (This is an act of Nature, or of the Artisan, not of any human engineer!) Examples:

- **Energy** in an electromagnetic wave. See **Photon**.
- **Energy** in an atom. According to classical mechanics, a negatively charged electron can orbit around a positively charged proton in many slightly different orbits, allowing a continuous range of energies. In quantum mechanics, the

allowed orbits are discretely different, i.e., *quantized,* as therefore are the allowed energies. See **Stationary state, Spectra (atomic, molecular, and other)**, and the extended discussion, with pictures, in "Quantum Beauty I."

- **Elementary particles**, in general. See **Quantum (unit of matter) / Quanta**.

Jargon: Physicists often call the process of applying quantum mechanics to a physical system "quantizing" that system. This is a significantly different use of the word, apt to cause confusion. Professionals, speaking among themselves, can use it safely, but in this book I have avoided it.

### QUANTUM (UNIT OF MATTER) / QUANTA

The objects we commonly call **elementary particles** are understood, according to our **Core Theory**, to be disturbances in **quantum fluids**. Thus **photons** are disturbances in the **electromagnetic fluid**, **electrons** are disturbances in the **electron fluid**, **gluons** are disturbances in the **gluon fluid**, **Higgs particles** are disturbances in the **Higgs fluid**, and so forth. If we treat the motion of those fluids according to the rules of classical physics, we find that a continuous gradation in their energy is allowed. But when we treat them according to the rules of **quantum theory**, we find that the allowed disturbances come in irreducible units—namely, the sort of thing we recognize as elementary particles!

See especially **Photon** for more on the **quantum** of **electromagnetic field**—the original "light-quanta" of Planck and Einstein.

### QUANTUM CHROMODYNAMICS (QCD)

*Quantum chromodynamics,* or *QCD,* is our **Core Theory** of the **strong force**.

QCD introduces many new ideas into the description of Nature, including **quarks**, **color charge**, color **gluons**, **asymptotic freedom**, **confinement**, and **jets**.

QCD provides, within its domain, a clear, positive answer to our motivating Question: Does the world embody beautiful ideas? For QCD embodies the gorgeous principle of **local symmetry**, in the uniquely rich context of strong **color charge property space**.

### QUANTUM DIMENSION

*Quantum dimensions* are **dimensions** whose **coordinates** are **Grassmann numbers**. Quantum dimensions are the soul of **supersymmetry**.

### QUANTUM DOT

Physicists are developing refined techniques to sculpt very small material structures that span only a few atoms on a side. These structures are called *quantum dots.* Quantum dots are, in effect, designer molecules.

### QUANTUM ELECTRODYNAMICS (QED)

*Quantum electrodynamics,* or *QED,* is our **Core Theory** of **electromagnetism**.

QED is based on **Maxwell's equations**, unchanged in form, but interpreted according to the rules of **quantum theory**. Thus disturbances in the **electromagnetic fluid** come in discrete units, or **quanta—photons**—and the fluid comes to exhibit spontaneous activity—**quantum fluctuations**.

QED provides a firm and complete foundation, as Paul Dirac said, for "all of chemistry and most of physics."

### QUANTUM FLUCTUATION / VIRTUAL PARTICLE / VACUUM POLARIZATION / ZERO-POINT MOTION

In the theory of **quantum fluids**, which is the foundation for our deepest understanding of Nature, we come to regard the particles we observe in a new way. They are minimal disturbances, or **quanta**, in quantum fluids. Thus **photons** are quanta of the **electromagnetic fluid**, **electrons** are quanta of the **electron fluid**, and so forth.

There's more to those fluids, however, than the particles they support, just as there's more to water than the waves it supports. In particular, the fluids have spontaneous activity: *quantum fluctuations.* Because the spontaneous activity and the disturbances in a quantum fluid that we recognize as particles are closely related—they're two aspects of the same fluid!—it is common to refer to the spontaneous activity as being composed of *virtual particles.* Thus virtual particles are a mind game we play on ourselves, to represent activity as objects. They are imagery.

The spontaneous activity of quantum fluid can be influenced by the presence of particles, and vice versa. Thus the properties of particles get modified by feedback from quantum fluids: a particle's presence affects the fluids' activity, and that activity feeds back to affect the particle. This feedback loop is called *vacuum polarization*. We can form a nice simple image of this effect, using the concept of virtual particles. The virtual particles form a gas filling space, and the properties of any real particle are affected by the buffeting it receives from that gas.

*Zero-point motion* is yet another way to refer to the spontaneous activity of quantum fluids. The phrase "zero-point motion" emphasizes that this is activity, or motion, that is present, even when all sources of energy have been removed, and so even at the absolute zero of temperature.

Particles, being disturbances within fluids that have spontaneous activity, inherit that spontaneity. They too exhibit zero-point motion. In experiments designed to detect small effects, like gravity waves or the cosmic **axion** background, through their influence on **normal matter**, this provides a complication—a source of background "noise" as your instrument wiggles and shakes. This quantum noise, which arises from fundamental physics, can't be removed by cooling your instrument to low temperatures, or by isolating it. The best you can do is to understand what you're up against, and try to work around it.

The effects of quantum fluctuations on the observed behavior of particles—i.e., *vacuum polarization*—are central to our understanding of Nature's deep workings. **Asymptotic freedom** is an aspect of vacuum polarization, and the quantitative aspects of **unification** of forces rely on it. Much of "Quantum Beauty III" and "Quantum Beauty IV" revolves around those ideas.

See also **Renormalization / Renormalization group**.

### QUANTUM FLUID / QUANTUM FIELD

In **quantum theory**, the properties of **fluids**, or **fields**, are significantly different from the properties of media we encounter in pre-quantum, classical physics. Most notably:

- *Quantum fluids* exhibit spontaneous activity, even in the absence of external influences or "causes." See **Quantum fluctuation / Virtual particle / Vacuum polarization / Zero-point motion**.
- Disturbances, or excitations, in *quantum fluids* cannot be arbitrarily small, but come in minimal units, or **quanta**.

Quantum fluids are the primary ingredients from which our **Core Theory** is constructed.

### QUANTUM JUMP / QUANTUM LEAP

See **Stationary state**, where these concepts are discussed in their natural context. Here I will note only that *quantum leaps* are actually very small leaps. Thus if someone boasts of making a "quantum leap in thinking," and he knows what he is talking about, then he is making a very modest claim.

### QUANTUM THEORY / QUANTUM MECHANICS

A great discovery of the early twentieth century is that the laws of physics used to describe large bodies, as epitomized in Newtonian mechanics and Maxwellian electrodynamics, are inadequate to describe atoms and their **nuclei**. To describe the behavior of matter on atomic and subatomic scales turned out to require not merely adding to what was known before, but the construction of a radically different framework in which many ideas thought to be secure had to be abandoned. The general term *quantum theory*, or *quantum mechanics*, refers to the new framework. It was mostly in place by the late 1930s. Since then our techniques for dealing with the mathematical challenges quantum theory poses have vastly improved—see **Renormalization**—and we have achieved, through our **Core Theories**, powerful, detailed understanding of the main forces of Nature. But those developments have taken place *within* the framework of quantum theory.

Many physical theories can be stated as reasonably specific statements about the physical world. **Special relativity**, for example, is basically the dual assertion of **Galilean symmetry** together with **invariance** of the speed of light. Each of our **Core Theories** is basically an assertion of **local symmetry**, together with specifics of how the associated **symmetry transformations** act on space-time and matter.

Quantum theory, as presently understood, is not like that. Quantum theory is not a specific hypothesis but a web of closely intertwined ideas. I do not mean to suggest quantum theory is vague—it is not. With rare and usually temporary exceptions, when faced with any concrete physical problem all competent practitioners of quantum mechanics will agree about what it means to address that problem using quantum theory. But few if any, I think, would be able to say precisely what assumptions they have made to get there.

While precise definition is elusive, it is possible, and illuminating, here to highlight a few qualitatively new themes that quantum theory brings in to our description of the physical world:

- In the description of matter, the fundamental objects are not particles occupying positions in space, or even **fields** (like **electric fields**) that fill all space with numbers or vectors, but **wave functions**. Wave functions assign **complex numbers** called **amplitudes** to the possible configurations of the object they describe.

  Thus the wave function of a single particle assigns an amplitude to all possible positions of the particle—in other words, to every point in space. The wave function of a pair of particles assigns an amplitude to pairs of points in space—or, alternatively, to points in the six-dimensional space of pair positions. The wave function of an electric field is an object of mind-boggling vastness. Because it assigns an amplitude to every possible value of the electric field as a whole, the wave function of the electric field is a function of (vector) functions!
- Any valid physical question about a physical system can be answered by consulting its wave function. But the relation between question and answer is not straightforward. Both the way that wave functions answer questions, and the answers they give, have surprising, not to say weird, features.

To be concrete, let's consider this first in the relatively simple context of a single particle. (Here we partially recapitulate a discussion from the main text.) To pose questions, we must perform specific experiments that probe the wave function in different ways. We can perform, for example, experiments that measure the particle's position, or experiments that measure the particle's **momentum**. Those experiments address the questions: Where is the particle? How fast is it moving?

How does the wave function answer those questions? First it does some processing, and then it gives you odds.

For the position question, the processing is fairly simple. We take the value, or amplitude, of the wave function—a complex number, recall—and square its magnitude. That gives us, for each possible position, a positive number, or zero. That number is the probability for finding the particle at that position. (Strictly speaking it is the probability density, but let us not multiply complications.)

For the momentum question, the processing is considerably more complicated. To find out the probability to observe some momentum, you must first perform a weighted average of the wave function—the exact way to do the weighting depends on what momentum you're interested in—and then square that average.

Three major points are:

- You get probabilities, not definite answers.
- You don't get access to the wave function itself, but only a peek at processed versions of it.
- Answering different questions may require processing the wave function in different ways.

Each of those three points raises big issues.

The first raises the issue of *determinism*. Is calculating probabilities really the best we can do?

The second raises the issue of *many worlds*. What does the full wave function describe, when we're not peeking? Does it represent a gigantic expansion of reality, or is it just a mind tool, no more real than a dream?

The third raises the issue of *complementarity*. Answering different questions may require different ways of processing the wave function that are mutually incompatible. In that case it is impossible, according to quantum theory, to answer both questions at the same time. You can't do it, even though each question on its own may be perfectly legitimate, and have an informative answer. Precisely this situation arises for our position question and momentum question, where it is known as Heisenberg's uncertainty principle: You can't measure both the position and the momentum of a particle at the same time. If someone were to figure out how to do that, experimentally, he'd have disproved quantum theory because quantum theory says it can't be done. Einstein tried repeatedly to devise experiments of that kind, but never succeeded, and eventually conceded defeat.

Each of these issues is fascinating, and the first two have got a lot of attention. To me, however, the third seems especially well-grounded and meaningful. Complementarity, as an aspect of reality and a lesson in wisdom, looms large in our meditation.

Although I've illustrated these issues for single particles, they all persist, in spades, when we ask questions of more complicated systems.

- Because the wave function gives us odds, rather than unique answers, we will get different answers if we pose the same question repeatedly to the same wave function. This is closely related to the intuitive idea, which I'm very fond of and invoke frequently, that quantum objects exhibit spontaneous activity. See **Quantum fluctuation.**
- Many quantities that are continuous according to classical physics become discrete in quantum theory. See **Photon** and **Spectra**.
- Last, but by no means least: Although quantum theory leads, in general, to probabilistic answers, it makes many predictions that are perfectly definite, too. For example, quantum mechanics underlies theories that predict the spectrum of hydrogen, the strength and electrical conductivity of nanotubes, and the masses and properties of hadrons, all with astonishing precision. These are perfectly definite quantities, not probabilities. Their elucidations are highlights in the recent history of our Question, as discussed in "Quantum Beauty I," "II," and "III."

### QUARK

The concept of *quarks* was introduced by Murray Gell-Mann and by George Zweig, independently, in 1964. They introduced the essential ingredients of the **quark model**, which brought order to the zoology of **hadrons**. A continuous line of development connects their pioneering work to the modern concept of quarks, which appear prominently among the **substance particles** of our **Core Theory**.

### QUARK CONFINEMENT. SEE CONFINEMENT.

### QUARK MODEL

The *quark model* is a semiquantitative model of **hadrons**. It played an important role, historically, in organizing facts about the **strong force**. For more on the quark model, see "Quantum Beauty III," part 2.

### REAL NUMBER

Intuitively, *real numbers* are numbers that allow smooth variation. Similarly to how **natural numbers** naturally arise in the act of counting, *real* numbers really arise in the act of measuring length.

Lengths can be divided very finely. Because there is not an *obvious* limit to that process of division, mathematicians assume, as a working hypothesis, that there is no limit. How is that hypothesis reflected in numbers? Because each succeeding number in a decimal fraction, as we proceed to the right, corresponds to a smaller division of quantity, the implication is that we should allow decimals that continue forever.

Newton was tremendously impressed by infinite decimals, which were in his lifetime a recent invention. They were the direct inspiration for his work in infinite series, and in calculus:

> I am amazed that it has occurred to no one . . . to fit the doctrine recently established for decimal numbers to variables, especially since the way is then open to more striking consequences. For since this doctrine in species has the same relationship to Algebra that the doctrine of decimal numbers has to common Arithmetic,

its operations of Addition, Subtraction, Multiplication, Division, and Root extraction may be easily learnt from the latter's.

In other words, Newton considered his central innovation to be that he allowed himself to use, in place of concrete numbers, the indeterminate *x* of algebra in decimal-like expansions. The deepest achievements of genius often seem to emerge, as here, from a childlike simplicity and spirit of play.

"Decimals that continue forever" is an excellent description of real numbers, and corresponds to how most mathematicians, and essentially all physicists, usually think about them. But it is not a **rigorous** definition. The challenge, in making the definition precise, is to capture the guiding idea of something that "goes on forever," using sentences that do not go on forever. It is actually quite difficult to provide a rigorous definition of real numbers. It was accomplished only in the late nineteenth century, though people had been using real numbers for hundreds of years previously.

In modern physics, because of the discovery of atoms and the weirdness of **quantum theory**, the correctness of the hypothesis that there is no limit to the division of length is far from obvious. Nevertheless, real numbers continue to provide the intellectual material out of which our Core Theories are forged. Why? It seems deeply mysterious, at least to me. See, in this regard, **Infinitesimal**.

### REDUCTIONISM

A pejorative term for "Analysis and Synthesis." See **Analysis and Synthesis**.

### RELATIVITY

In physics, *relativity* usually refers to one or both of Einstein's theories that include that word, i.e., the **special** or **general relativity** theories. Which of these is intended should be clear from the context.

In our meditation, we have emphasized that both theories of relativity are, in their essence, statements of **symmetry**, in our precise sense. That is, they are statements that we can make **transformations** on the quantities that appear in the laws of physics, without changing the content of those laws—Change Without Change. The word "relativity" emphasizes the aspect of "change," but leaves the aspect "without change"—that is, **invariance**, the complement to relativity—unsaid. This quirk has had the unfortunate effect of leading some people to infer, and even to assert, absurdities like "Einstein taught us that everything is relative." He didn't, and it isn't.

### RENORMALIZATION / RENORMALIZATION GROUP

**Quantum fluids** are the primary ingredients of our **Core Theory**. They exhibit spontaneous activity, or **quantum fluctuations**, that generally become increasingly violent at short distances. These ceaseless fluctuations permeate space and modify the behavior that matter would have in their absence. Calculating those modifications is called *renormalization*.

When we study the properties of particles more closely, going to higher energies or shorter distances or, as we say, using finer resolution, we become less sensitive to effects of the more gradual, gentle quantum fluctuations. We get closer to seeing the "bare" particles. The *renormalization group* is a mathematical technique for making quantitative connections between the properties of a particle when it is viewed at different resolutions.

**Asymptotic freedom** in the strong force, and the quantitative study of unification discussed in "Quantum Beauty IV," are examples of putting the renormalization group to good use.

### RIGID SYMMETRY

We say a symmetry of physical law is *rigid* if it requires one to make the same transformation everywhere (and every when) in space-time. **Local symmetry**, by contrast, allows transformations that vary over space-time.

### RIGOR

We say an argument is *rigorous,* or has *rigor,* when it is both precisely stated and difficult to challenge. We say that a concept is rigorous if its meaning is precisely formulated, and thus suitable for use in rigorous arguments.

Rigor itself is not a rigorous concept because "difficult to challenge" is a bit fuzzy. (How difficult?) For example, there is overwhelming evidence, based on solutions of the equations of **quantum chromodynamics (QCD)** using computers, that the theory produces the phenomenon of **quark confinement**, and correctly predicts the **spectrum** of **hadrons**. (In other words: the computations correctly predict what strongly interacting particles exist, their masses, and their other properties; quarks are not among those particles.) But mathematicians generally do not regard this conclusion as rigorous.

### SCHRÖDINGER EQUATION

The *Schrödinger equation* was proposed by Erwin Schrödinger (1887–1961) in 1925. It is a **dynamical equation** that determines how the **wave functions** of **electrons**, or of other particles, change in time.

The Schrödinger equation is approximate, in two important ways. First, it is based on nonrelativistic (Newtonian) mechanics, rather than on Einstein's relativistic mechanics. Paul Dirac, in 1928, provided another equation for electron wave functions that obeys the assumptions of **special relativity** (see **Dirac equation**). Second, it does not contain the effects of **quantum fluctuations**, such as **virtual photons**, on the electrons. Nevertheless, the Schrödinger equation is accurate enough for most practical applications of quantum theory to chemistry, materials science, and biology, and it is the version of quantum theory that is usually adopted in treating those subjects.

Although one speaks of "the Schrödinger equation," what Schrödinger provided is not really a single equation, but rather a procedure for formulating equations that describe different situations where quantum mechanics can be applied.

One of the simplest Schrödinger equations is the equation for a single electron subject to the **electrical** attraction of a single **proton**. This gives a description of the hydrogen atom. Although it is formulated in a different universe of ideas—a universe where space-filling wave functions replace particles moving through orbits—the results that flow from the Schrödinger equation, in this case, largely vindicate Bohr's intuition about the meaning of hydrogen's spectrum. For an overview see **Spectra**.

We can also formulate Schrödinger equations that govern several electrons simultaneously. We must do that, of course, if we want to account for electrons' influence on one another. As explained in the entry **Wave function**, wave functions that fully describe the physical state of several electrons occupy spaces of very high **dimension**. The wave function for two electrons lives in a six-dimensional space, the wave function for three electrons lives in a nine-dimensional space, and so forth. The equations for these wave functions rapidly become quite challenging to solve, even approximately, and even using the most powerful computers. This is why chemistry remains a thriving experimental enterprise, even though in principle we know the equations that govern it, and that should enable us to calculate the results of experiments in chemistry without having to perform them.

### SPATIAL TRANSLATION SYMMETRY

Spatial translation is the transformation that moves the position of points in space through a common displacement. *Spatial translation symmetry* is the hypothesis that the laws of physics are unchanged or, as we say, **invariant** under such a transformation. Spatial translation symmetry is a **rigorous** way to formulate the idea that the laws of physics are the same everywhere. Spatial translation symmetry is closely connected, through Emmy Noether's theorem, to **conservation of momentum**.

### SPECIAL RELATIVITY

In his theory of *special relativity,* Einstein reconciled two ideas that appear to be contradictory.

- The observation of Galileo that overall motion at constant **velocity** leaves the laws of Nature unchanged
- The implication of **Maxwell's equations** that the speed of light is a consequence of the laws of Nature, and cannot change

There is tension between these two ideas, because experience with other objects suggests that the speed you observe them to have will change if you move at constant velocity. You can catch up to them, or outrun their pursuit. Why should light beams be different?

Einstein resolved that tension by critically analyzing the operations involved in synchronizing clocks at different places, and how that synchronization process is modified by overall motion at constant velocity. From this analysis, it emerges that the time assigned to an event by a moving observer is different from the time it is assigned by a fixed observer, in a way that depends on its position. In referring to common events, one observer's time is a mixture of the other's space and time, and vice versa. This "relativity" of space and time is the essential novelty Einstein's special theory of relativity brought into physics. Both of the theory's assumptions were already out there, and widely accepted, before his work—but no one had taken both seriously at the same time, and forced their reconciliation.

Special relativity is not only important in itself, but also because it introduced a new meta-idea for guessing and improving our laws of physics that has proved extremely fruitful and successful. This meta-idea is what we've called **symmetry**, poetically defined as Change Without Change. The two postulates of special relativity fit that description very well: the first tells us what kinds of changes to consider (namely, **Galilean transformations**) and the second tells us what they don't change (namely, the speed of light).

The theme of symmetry, or invariance—Change Without Change—plays out many times, with variations, in our meditation. At first it is tentative and muted, but it stands out ever more clearly, and heightens, until by the end we find it dominating our deepest understanding of Nature.

### SPECTRA (ATOMIC, MOLECULAR, AND OTHER)

Atoms of a given kind—for instance, hydrogen atoms—will absorb some colors of spectral light much more efficiently than others. (More generally, they will absorb **electromagnetic waves** that have some frequencies much more efficiently than those that have others. In this entry, I will use the less general, but more evocative, language of colors.) The same atoms, when heated up, will emit most of their radiation in those same spectral colors. The pattern of preferred colors is different for different kinds of atoms, and forms a sort of fingerprint through which we can identify them. An atom's pattern of preferred colors is called its *spectrum*.

A major achievement of **quantum theory** has been to provide a way to compute atomic spectra. The underlying idea is a lasting legacy of Bohr's atomic model. Bohr postulated that the electrons in an atom can take on only a discrete set of **stationary states**. The possible values of the electrons' energy, therefore, also form a discrete set. When an atom emits or absorbs a **photon**, it makes a transition between two stationary states. Because **energy** is **conserved** in this process, the energy of the photon is related to the difference in energy between those two stationary states. Finally, crowning Bohr's vision: the spectral color of a photon reveals its energy. Thus the spectrum of an atom encodes the energies of its possible states. (To be precise about this coding: the **frequency** of the color's electromagnetic wave, times Planck's constant, equals its **energy**. See **Photon / Planck-Einstein relation**.)

In modern quantum theory, we compute the possible stationary states and their energies by solving Schrödinger's equation, but the fundamental relationship between an atom's possible energies and its spectrum remains just as Bohr envisaged it. See **Schrödinger's equation**.

I have spoken of atoms, but the same logic applies to molecules, to solid materials, to **nuclei**, and even to **hadrons**. In nuclei we are concerned with **stationary states** of **nucleons**, and in hadrons we are concerned with stationary states of systems based on **quarks** and **gluons**, but in each case their spectra encode secrets of structure.

When light from the Sun, or from other stars, is analyzed into its spectral colors, one finds that some colors appear with heightened intensity (so-called "emission lines") and others with reduced intensity (so-called "absorption lines") compared with the average. The pattern of emission and absorption lines can be matched to the spectra, either measured or computed, of known atoms, molecules, and **nuclei**. They reveal what is in the stars' atmospheres, and the presence of hot or cold regions. They testify, in great and convincing detail, that matter throughout the Universe is made from the same stuff, and follows the same laws.

The use of the word "spectrum" in "electromagnetic spectrum" seems, on the face of it, rather different from its use in "spectrum of an atom." The former refers to the range of all possible forms of electromagnetic radiation, while the latter refers either to the particular colors (or **pure tones**, or *frequencies*) of light that the atom is capable of emitting. (This maps faithfully, as we have explained, onto the possible energies of its stationary states.) From a deeper perspective, however, it is quite correct to say that the **electromagnetic spectrum** is indeed the spectrum *of* something—namely, of the **electromagnetic fluid**! For the electromagnetic spectrum is the range of possible colors that the electromagnetic fluid can emit.

### SPECTRAL COLOR. SEE COLOR (OF LIGHT) AND ELECTROMAGNETIC SPECTRUM.

### SPIN

In common language, we say that an object has spin, or is spinning, if it is rotating around some axis. *Spin* means that in the quantum world too, but the concept assumes new importance, chiefly for two reasons.

- Many particles never cease spinning! For these particles, rotary motion around their center is an aspect of the spontaneous activity that is so characteristic of the quantum world. **Electrons**, **protons**, and **neutrons** all have that property. Whenever the rotary **angular momentum** is measured, its magnitude is found to be equal to one-half times the reduced **Planck's constant**. We say these particles have spin ½, or are spin -½ particles.
- Many particles, including, especially, electrons, act as little magnets. Like Earth, they generate **magnetic fields** whose structure is aligned with their direction of spin. The magnetic field associated with any single electron is quite small, but if many electrons have their spin axes aligned in the same direction, their fields add up. The magnetism of the classic "magnets"—bars of iron ores, basically—arises from the aligned fields of the spinning electrons they contain.

### SPINOR REPRESENTATION

*Spinors* are a kind of advanced version of **vectors**. They appear in the mathematical description of an electron's **spin**, in the **Dirac equation**, as the **property space** of **substance particles** in the *SO*(10) (Georgi-Glashow) unification scheme sketched in "Quantum Beauty IV," and in several other frontier subjects in physics. A description of the mathematics of spinors is well beyond the scope of this book, but I've indicated two readily available references on page 404 in the endnotes.

### SPONTANEOUS SYMMETRY BREAKING

Between accurate realization of **symmetry** and its total absence, there is an intermediate possibility, *spontaneous symmetry breaking,* that features prominently in our description of the world.

We say that we have spontaneous breaking of a system of equations, if

- The equations satisfy the symmetry, but
- The stable solutions of those equations do not.

In this way, an observed lack of symmetry excuses itself. The symmetry is there in the equations, but the equations themselves tell us that we will not observe it!

Example: In the basic equations that describe a lump of lodestone, any direction is equivalent to any other. But the lump forms a magnet, and in a magnet it is no longer true that all directions are equivalent. Each magnet has a polarity and can be used as a compass needle. The explanation of how rotation symmetry gets lost (or "broken") is simple, but profound. There are forces that tend to align the spins of electrons in the magnet with their neighbors.

In response to those forces, all the electrons must choose a common direction in which to point. The forces—and the equations that describe them—will be equally satisfied with any choice for that direction, but *a choice must be made.* So the stable solutions of the equations have less symmetry than the equations themselves.

In our **Core Theory** of the **weak force**, we have rotation symmetry among the directions in weak **color space** that is spontaneously broken by the existence of a space-filling **Higgs field**. The basic idea is very similar to what we just considered when discussing a common magnet. Just as the underlying equations for forces among electrons encourage the spin of neighboring electrons to align their directions, so the underlying equations encourage the Higgs field at neighboring points in space-time to align its directions in weak **property space**. A common direction must be chosen, and in that way the rotational symmetry (in weak property space) is spontaneously broken.

The success of these ideas in providing an excellent account of the weak force, and in predicting the existence of the Higgs particle, encourages us to further explore the possibility that the underlying symmetry of our world-equations is much larger than the symmetry we see in the world, by contemplating still larger underlying **symmetry groups**.

### STANDARD MODEL. SEE CORE THEORY.

### STANDING WAVE / TRAVELING WAVE

Wave oscillations in bounded regions are called *standing waves.* Thus the vibrations of strings in musical instruments, or of their sounding boards, are standing waves. Standing waves are often called vibrations, or oscillations.

Waves that are not confined to a finite region, but travel through space, are called *traveling waves.* In common language, and also in physics, when we speak of "sound waves," we usually mean traveling waves. The vibrations of a grand piano's sounding board, which are standing waves, push nearby air to and fro. The moving air exerts forces on its neighboring air, which exerts forces on its neighboring air, and so forth, resulting in a disturbance that takes on a life of its own. This is a traveling sound wave that we can detect—that is, hear—at great distances.

The **wave function** associated with an electron can be either a standing wave or a traveling wave.

The wave function of an electron bound to a proton to make a neutral hydrogen atom is considered a standing wave even though, strictly speaking, it extends through all space. Indeed, the electron's probability cloud, which reflects the magnitude of the wave function, diminishes very rapidly as we move far away from the proton, and never becomes significant outside of a small, fixed region near the proton. That is what we mean when we say the electron is bound to the proton. So, in effect, the wave function is confined to a bounded region of space, and should be considered a standing wave.

The wave function of an unbound electron, which moves freely through space, is a traveling wave.

See also **Schrödinger equation.**

### STATIONARY STATE

Historically, the term "stationary state" first arose in Bohr's atomic model. If one applies classical mechanics and electrodynamics to the problem of a negatively charged electron bound to a positively charged proton, one finds no stable solution. The electron will spiral into the proton while radiating **electromagnetic waves**. In order to avoid that catastrophe Bohr introduced a radical hypothesis, the hypothesis of *stationary states,* according to which the electron is allowed to take up only a few of the classically allowed orbits, which define its allowed "states." Within these particular orbitals the electron does not radiate, but is "stationary." Thus the allowed **orbitals** define stationary states.

Bohr's model has been superseded by modern quantum mechanics, but some elements of his picture, including the concept of stationary state, can be recognized in the modern theory. In modern quantum theory, an electron's state is described by a **wave function**. That wave function, and its associated **probability cloud**, evolves in time according to the fundamental **Schrödinger equation**. Among the solutions to the Schrödinger equation, there are some special ones whose probability clouds do not change in time at all. Those solutions have the properties in the quantum theory that Bohr postulated for stationary states in his model. So we say, in the quantum theory, that wave functions whose probability clouds do not change over time define *stationary states*. Here again, for more information, including pictures worth thousands of words, see "Quantum Beauty I," and also **Spectra**.

The special wave functions that define stationary states (i.e., wave functions whose probability clouds do not change in time) are extremely useful in thinking about problems in atomic physics and chemistry. In tribute to their origin in Bohr's allowed orbits, they are called orbitals.

The notion of stationary state is an approximate one because there are physical processes by which an electron can transition between these states. Specifically, an electron in a stationary state can transition into another stationary state by emitting or absorbing a **photon**. Bohr could not, within his model, provide a detailed picture or mechanism for this discontinuous change of orbit, and simply admitted it as an additional assumption: the possibility of **quantum jumps**, or **quantum leaps**.

In modern quantum theory transitions between stationary states occur as a logical consequence of the equations. Physically, they arise due to the interaction between electrons and the **electromagnetic fluid**. Because that interaction is fairly feeble, compared with the basic **electric forces** that bind electrons, we often do well to include it as a correction, while retaining the stationary states as a starting point. In this treatment, we find that the transitions are not true discontinuities, but do occur rapidly.

The emission process is particularly interesting, conceptually. In it, the electron gives birth to electromagnetic energy in the form of a photon, where initially there was none. This occurs when the electron encounters spontaneous activity in the electromagnetic fluid and, by imparting some of its own energy, amplifies that activity. In this way, the electron transitions to a state of lower energy, a **virtual photon** becomes a real photon, and there is Light.

### STRONG FORCE

The *strong force* is, together with **gravity, electromagnetism**, and the **weak force**, one of the four basic mechanisms through which Nature acts. The strong force is the most powerful force in Nature. It is responsible for holding atomic nuclei together, and governs most of what happens in the collisions studied at high-energy **accelerators** such as the **Large Hadron Collider**.

Soon after the discovery of atomic **nuclei** early in the twentieth century, physicists recognized that the forces known at that time, **gravity** and **electromagnetism**, could not account for their most basic properties, starting with their ability to hold together. This stimulated decades of intense research in subnuclear physics, both experimental and theoretical. The mature result of that work is the **Core Theory**, described at length in our main meditation. Within the Core Theory, the strong force is understood to be a manifestation of **quantum chromodynamics (QCD)**.

The use of "strong" and "force" together is potentially ambiguous because "strong force" could be taken to mean a powerful source of acceleration. Thus when speaking of the gravitational influence of a neutron star, or a black hole, one might say that gravity exerts a strong force on a nearby planet. To avoid ambiguity, in such cases I use terms like "powerful force" or "powerful interaction," avoiding "strong force" and "strong interaction."

The strong force is also called the strong interaction. See **Force**.

### SUBSTANCE PARTICLE

This is another way of referring, collectively, to fermions. In the **Core Theory**, these are the **quarks** and **leptons**.

If **supersymmetry** is a correct idea, then for each substance particle there is a related, "partner" **force particle**. The substance particle, by moving in a **quantum dimension**, becomes its partner force particle.

### SUPERCONDUCTIVITY / SUPERCONDUCTOR

A large variety of metals, and some other materials, exhibit qualitatively new behavior when they are cooled down to the absolute zero of temperature. Most dramatically, their resistance to the flow of electric charge abruptly drops to zero. For this reason, they are said to exhibit *superconductivity,* and to become *superconductors.*

Superconductivity was discovered experimentally by Kamerlingh Onnes in 1911. For many years it eluded theoretical explanation. A breakthrough came in 1957 when John Bardeen, Leon Cooper, and J. Robert Schrieffer proposed what we now call the BCS theory of superconductivity. Their work not only explained the emergence of superconductivity, but did it using ideas of great beauty and power that could be—and were—applied to other problems. In particular, it foreshadowed **spontaneous symmetry breaking** and the **Higgs mechanism.**

Inside superconductors, **photons** behave as if they have non-zero **mass.** The equations that describe this situation are essentially the same as the equations that we use in the **Core Theory**, to give non-zero mass to the **weakons** in the Higgs mechanism. I think it is both fair and poetic to say that the great lesson we can draw from the discovery of the **Higgs particle** is that we live inside a cosmic superconductor. (But it is superconductivity for the flow of weak charge, as opposed to **electric charge.**)

### SUPERSYMMETRY (SUSY)

*Supersymmetry* is a particular kind of **symmetry**. The transformations of supersymmetry involve displacement, or *translation,* in a **quantum dimension**. When a **force particle** (**boson**) moves into a quantum dimension it becomes a **substance particle** (**fermion**), and vice versa.

If we can convince ourselves that force and substance are the same thing, seen from different perspectives, we will have achieved a new level of unity and coherence in our fundamental understanding of Nature. At present, however, the evidence for supersymmetry, though impressive, is circumstantial.

### SYMMETRY / SYMMETRY TRANSFORMATION / SYMMETRY GROUP

In the mathematics and the mathematical sciences, we say that an object has *symmetry* if there are transformations that make changes in, or move, different parts of the object, while leaving the object as a whole unchanged, or **invariant**. Such transformations are called *symmetry transformations.*

The concepts of symmetry and symmetry transformations also apply to systems of equations. We say that a system of equations has symmetry with respect to a transformation if the transformation makes changes in the quantities that appear in the equations (typically by interchanging them, or by mixing them up in more complicated ways) without changing the meaning of the overall system.

Example: The equation $x = y$ has symmetry under the transformation that interchanges $x$ and $y$ because the transformed equation $y = x$ has exactly the same meaning as the original. The totality of transformations that leave an object invariant is called its *symmetry group.*

### SYNTHESIS

The process of assembling simple ingredients, or concepts, to produce more complex structures. See **Analysis and Synthesis.**

### TIME TRANSLATION SYMMETRY

Time translation is the transformation that moves the times of events through a common interval. *Time translation symmetry* is the hypothesis that the laws of physics are unchanged or, as we say, *invariant* under such a transformation. Time translation symmetry is a **rigorous** way to formulate the idea that the laws of physics are the same throughout history. Time translation symmetry is closely connected, through Emmy Noether's theorem, to *conservation of energy.*

### TONE / PURE TONE

The phrase "pure tone," as it is used in this book, means a simple wave disturbance that is periodic in both space and time. (Here "simple" has a definite technical meaning—the wave pattern is *sinusoidal*—but I will not spell that out here. Page 405 in the endnotes provides two easily available references.)

The most important examples of pure tones, for us, concern sound waves and **electromagnetic waves** (including, especially, light). In sound waves, it is pressure and density of air that varies; in electromagnetic waves, it is electric and magnetic fields.

A profound and satisfying insight that arises from the scientific study of Nature is that *pure tones,* defined in the preceding mathematical/physical fashion, correspond to simple perceptions. Pure audio tones are easily produced electronically, and may be familiar to you from hearing tests or primitive electronic music devices (as sometimes incorporated into greeting cards, for example), or tuning forks. Pure visual tones are the spectral colors of light that emerge in rainbows, or from sunlight dispersed by a prism, as in Newton's experiments. These complementary perspectives—sensory and conceptual—on pure tones beautifully exemplify our desired correspondence

Real ↔ Ideal

The *tones* produced by more traditional musical instruments, when you play a single "note," are far from pure. The details vary from instrument to instrument, but in all cases the note contains many pure tones sounding simultaneously, with different strengths. The most powerful of these is the pure tone that gives the name to the note, but the quality of the musical sound, which distinguishes different instruments, is largely a function of the additional so-called *overtones.*

These matters are discussed further in the main text. Also relevant is the entry **Spectra**.

#### TRANSLATION

Displacement of a system through a constant step, in either space or time. See **Space translation symmetry** and **Time translation symmetry**.

#### TRANSVERSE WAVE / POLARIZATION (OF LIGHT)

In our meditation, the most important *transverse waves* are **electromagnetic waves**, including light as a special case.

When an electromagnetic wave proceeds through space devoid of ordinary matter (see **Vacuum**) its electric and magnetic fields—both of which are directed, **vector** quantities—are perpendicular to the direction in which the wave is progressing. That—no more, and no less—is what we mean in saying that electromagnetic waves are *transverse.* Thus for a transverse wave, the activity the wave produces is perpendicular to the direction in which the wave progresses.

Sound waves, on the other hand, are not transverse waves. Their activity, the compression and rarefaction of air, takes place in the same direction as the direction the wave progresses. Waves of that kind are called *longitudinal* waves.

Even the simplest kinds of light waves, those associated with pure spectral colors, have an additional property besides their color and direction of progress. This property is called *polarization.* The simplest possibility is linear polarization. If a light wave is coming toward you, and its electric fields always point in the direction that joins your head to your toes, we say the light is linearly polarized in the direction joining your head to your toes. There are solutions to Maxwell's equations that correspond to linear polarization in any transverse direction—that is, in any direction perpendicular to the direction the wave progresses. There are other, more complicated possibilities too, where over time the electric fields describe circles, or ellipses, in the plane perpendicular to the direction of wave progress. These give circularly or elliptically polarized light.

Humans are not sensitive to light's polarization (with some minor exceptions), though many other animals, especially insects and birds, are.

#### TRAVELING WAVE. SEE STANDING WAVE / TRAVELING WAVE.

#### UBIQUITOUS

Pervasive; everywhere.

#### UNIFICATION

The *unification* of related ideas into a coherent whole is an aspect of economy of thought. Another, complementary aspect of unification is the reconciliation of apparent opposites. When opposites are reconciled, we see them as complementary aspects of an underlying unity.

Our Question poses the challenge to

unify beauty and physical embodiment, or Ideal and Real.

Unification, both in its aspect of combining related ideas and in its aspect of reconciling apparent opposites, has been a leading feature in many of the milestone achievements of natural philosophy:

- The systematic use of **coordinates**, pioneered by René Descartes (1596–1650) in his *La Géométrie* of 1637, unified algebra and geometry.
- Newton's universal law of gravitation, and his laws of motion, unified celestial astronomy and earthly physics. Galileo's telescopic observations, which revealed (among other things) the mountainous terrain of our Moon and the satellite system of Jupiter, contributed mighty images to this unification.
- **Maxwell's equations** of **electromagnetism** unified the description of electricity and magnetism. The same equations also provided an electromagnetic account of light, bringing all the phenomena of optics within this unification.
- Einstein's **special relativity** theory brought in **symmetry transformations** that mix space and time, allowing us to see those things as two aspects of a unified space-time.
- The **electromagnetic fluid** of Faraday and Maxwell, and the **metric fluid** of Einstein, by abolishing **Void**, unified space-time and matter.
- The concept of **quanta** of **quantum fluids**, typified by the photons of electromagnetic radiation (light), unified the description of particle and wave aspects of physical behavior.

At the frontier of physics today, there are tantalizing indications that new unifications may soon be fulfilled.

- Our **Core Theories** are all based on **local symmetry**, but the transformations envisaged in our theories of the **strong**, **weak**, and **electromagnetic forces** act independently on **property spaces**, while the transformations of our theory of the **gravitational** force act on space-time. We seek more encompassing local symmetry that makes these whole.

With **supersymmetry**, we might unify **substance** and **force**.

These are the ideas at play in "Quantum Beauty IV."

**UNIVERSE / VISIBLE UNIVERSE / MULTIVERSE**

Modern physics has opened up imaginative possibilities for cosmology that outrun the anticipations of ordinary language. To do them justice, we must both refine and expand everyday usage. In particular, the vague use of "universe" to mean "everything" will not do. Though the scientific literature is not yet entirely consistent on these issues, either, I think it is possible, and useful, to distinguish three concepts that reflect most recent scientific usage. They probably will emerge as standard usage.

The *visible Universe* consists of everything accessible to observation. Here, fundamental limitations arise from the finite speed of light, which (we assume) is the limiting speed for transmission of information, and the finite time that has passed since the Big Bang, which (we assume) is an event we can't see beyond. Limited by a finite speed and a finite time, we realize that we can access only out to a finite distance, the so-called horizon. Two things to note:

- The horizon grows as the time since the Big Bang gets longer. Thus the visible Universe was smaller in the past, and we can anticipate it will become larger in the future.
- If we discover that the speed of light is not a fundamental limitation to transmission of information, or if we learn to see past the Big Bang, we'll have to rethink what we mean by the visible Universe.

The visible Universe we see today seems to be broadly the same throughout. Astronomers have found the same sorts of stars, organized into the same sorts of galaxies, obeying the same sorts of physical laws, however far and in whatever direction they've looked. If we assume this pattern will continue as the horizon expands, we arrive at what is usually called the "Universe." The Universe, in this sense, is the conservative, logical extension of our past experience of the visible Universe into the indefinite future.

Modern physics has made it plausible, however, that the physical world can exist in **qualitatively** different forms, or phases, similar in spirit to how water can exist as ice, liquid water, or steam. In those different phases, space is permeated by different **fields** (or by the same fields with different magnitudes). See **Vacuum**. Because those fields largely determine the property of matter that moves through them, these different phases, in effect, implement different laws of physics. If such diverse regions of space exist, then the "Universe," as we've defined it, is not the whole of reality. We call the whole of reality the *multiverse*.

The idea that there is a multiverse, so that the laws of physics we observe are partly an accident of where we happen to be, figures prominently in **anthropic arguments**.

### VACUUM, VOID

The word "vacuum" is commonly understood to mean "empty space, devoid of matter." Thus one speaks of "creating a vacuum" by pumping air out of a container, or of "vacuum tubes," or of "the vacuum of interstellar space." This usage can become ambiguous, because

- What you find depends on how hard you're ready to look. The "vacuum" of interstellar space, for example, is permeated by the microwave background radiation, the sort of radiation our eyes would sense as starlight, cosmic rays, various neutrino streams, dark energy, and dark matter. On Earth, vacuum engineers can, with effort, exclude the first two of these things from a region of space, and most of the third, but not the last three. Fortunately, the reason it's so difficult is also the reason it makes no difference, for practical purposes: neutrino streams, dark energy, dark matter—and possibly other things we don't yet know about!—interact very feebly with **normal matter**.
- What you think is there depends on how hard you're ready to think. In our Core Theory, even ideally "empty" space is permeated by a variety of **quantum fluids**—the **electromagnetic fluid**, **metric fluid**, **electron fluid**, **Higgs fluid**, and so forth—as well as the metric and Higgs **fields**.

What people intend when they say "vacuum" is usually clear from the context, but in thinking about fundamentals one should be clear that the word "vacuum" does not refer unambiguously to a definite thing. In particular, the philosophical concept of *Void*—space as perfect nothingness—is quite different from any reasonable understanding of physical space anywhere in the present-day physical world.

In modern physical cosmology, it is important to take into account that space-filling fields, such as the **Higgs field**

- Have profound physical effects, both altering the behavior of matter and contributing to *dark energy*.
- Are present in any physically defined *vacuum* (because they are pervasive and inescapable).
- Can, in extreme conditions, change in magnitude.

Combining these observations, we reach the insight that there can be significantly different physical vacuums whose pervasive fields differ in magnitude. The behavior of matter within these different vacuums (or, in better Greek, *vacua*) can be drastically different, as can be the density of **dark matter** and **dark energy**.

It is suggestive, and pretty, to summarize this situation by saying that space itself is a sort of material that can exist in different phases, just as water can exist as liquid water, ice, or steam. See **Universe / Visible Universe / Multiverse**.

### VECTOR / VECTOR FIELD

Vectors can be defined either geometrically or algebraically.

Geometrically, a vector is a quantity that has both magnitude and direction. Examples:

- If we have two points, say *A* and *B*, then the straight-line displacement that takes *A* into *B* is a vector. Its magnitude is the distance between *A* and *B*, and its direction is the direction from *A* to *B*.
- The **velocity** of a particle is a vector.
- The **electric field** at any point is a vector.

Algebraically, a vector is simply a sequence of numbers.

The connection between these two definitions is made by introducing **coordinates**. In the examples above, the vectors are vectors in ordinary three-dimensional space, and correspond to triples of real numbers. Several interesting and important variations are mentioned in the entry for **Coordinate**.

When we have a vector quantity assigned

to each point in space, we say we have a *vector field*. Examples:

- If we have a body of water, its different parts will be in motion at different velocities. These velocities define a vector field.
- Electric and magnetic fields are vector fields.
- For each point on a computer screen, the intensities with which red, green, and blue are displayed at that point are a sequence of three numbers, and so define a vector. Thus, on the plane of the computer screen, we have a vector field of colors.

### VELOCITY

Intuitively, *velocity* is defined as the rate of change of position.

Thus to define the velocity of a particle, we consider its displacement $\Delta x$ during a small amount of time $\Delta t$, take the ratio $\Delta x/\Delta t$, and consider its limiting value as the interval $\Delta t$ is taken smaller and smaller. That limiting value, by definition, is the velocity.

See **Infinitesimal**, where some fundamental issues around this definition are discussed.

### VIRTUAL PARTICLE. SEE QUANTUM FLUCTUATION / VIRTUAL PARTICLE / VACUUM POLARIZATION / ZERO-POINT MOTION.

### W PARTICLE

A massive particle that plays a central role in the weak force. See **Weak force** and **Weakon**.

### WAVE FUNCTION

In classical mechanics particles occupy, at each time, some definite position in space. In quantum mechanics, the description of a particle is quite different. To describe an electron, say, in **quantum theory**, we must specify the electron's *wave function*. The electron's wave function governs its **probability cloud**, whose density in a region of space indicates the relative probability to find the electron there.

Here I will sketch a more precise description of electron wave functions. To benefit fully from this description, you will need at least passing familiarity with **complex numbers** and the mathematics of probability. The concluding remark in this subentry, indicated below with a star (*), is a highlight you should visit even if you decide to skim, or skip, the paragraphs that precede it.

The wave function of an electron assigns, to each point of space, at each time is a **field** of complex numbers. Thus to each point of space, at each time, the wave function assigns a complex number. That complex number is called the value, or sometimes the amplitude, of the wave function at that place and time. The wave function obeys a (relatively) simple equation, the **Schrödinger equation**, but in itself has no very direct physical meaning.

What does have direct physical meaning is a field of positive (or zero) real numbers that we obtain from the wave function by squaring its magnitude. This mathematical operation takes us from the electron's wave function to its associated probability cloud. The probability of finding an electron at a given position, at a given time, is proportional to the square of the magnitude of the value of the wave function at that place and time.

Though it is described by a function that fills space, one should not think that the electron is an extended object. When an electron is observed, it is always observed as a full object, with its full mass, electric charge, and so forth. The wave function carries information about the probability of finding a full particle, *not* about the distribution of parts of a particle.

The quantum mechanical description of two or more particles is also, naturally, based on wave functions. It brings in an important new feature: *entanglement*. The essential novelty arises already for two particles, and to keep things as concrete and simple as possible, I will focus on that case.

To put entanglement in context, let me begin by describing a guess about the description of two particles that might seem reasonable, but is wrong. One might guess that the wave function for two particles takes the form of a wave function for one particle, multiplied by a wave for the other particle. Starting from that guess, if we take the square to get the probability cloud, we find that the joint probability for finding the first particle at $x$ *and* the second particle at $y$ (say) is equal to the *product* of the probabilities for finding the first particle at $x$ and the second $y$. In other words, those probabilities are independent. That is not an acceptable result, physically, because we must expect that the position of the first particle affects the position of the second.

The correct description employs a wave function that is a field in a six-**dimensional** space whose **coordinates** are the three coordinates that describe the position of the first particle, followed by three coordinates that describe the position of the second particle. When we square this object to get the joint probability, we generally find that the particles are no longer independent. Measuring the position of one of them affects the probabilities for where we will find the other. We say, therefore, that they are entangled.

Entanglement is neither a rare phenomenon in quantum mechanics, nor an untested corner of that theory. It arises, for instance, when we compute the wave function for a helium atom's two electrons. The **spectrum** of helium has been both measured and computed with great accuracy, and we find that the highly entangled wave functions of quantum mechanics give results that match reality.

* It is near magical, in the context of our Question, to discover that six-dimensional space, a beautiful product of creative imagination, is embodied in something as specific and concrete as a helium atom. That atom's spectrum, when we learn to read it, sends us postcards from six dimensions! *

For additional perspective on wave functions, see especially the discussion in **Quantum theory**.

(Final comment, and warning: The phrase "wave function" is not the happiest choice for the concept it represents. "Wave," in general, suggests oscillation, and "wave function" suggests a function that oscillates, or a function that describes oscillations in some medium, but quantum mechanical wave functions need not oscillate, nor do they describe oscillations of something else. A better name might be "electron square root of probability field," but "wave function" is far too deeply engrained in our language and literature to consider changing it.)

### WAVELENGTH

Waves that repeat themselves or, as we say, vary **periodically** in space are particularly important, both because they occur naturally, and because they provide basic units from which we can build up more complex wave motions, in the spirit of **Analysis and Synthesis**. Pure musical tones, among sound waves, and pure spectral colors, among **electromagnetic waves**, are periodic in space as well as in time. See **Tone / Pure tone**.

The distance between repeats, in a simple wave, is called its *wavelength*. Thus *wavelength* expresses the same concept, for variation in space, as does **period** for variation in time. Examples:

- The lowest tones humans that can hear have wavelengths of around ten meters in air, while the highest tones humans can hear have wavelengths around a centimeter in air. Not coincidentally, the sizes of most musical instruments are near the middle of this range, because they are crafted to initiate sound waves humans can hear. The bass pipes of pipe organs, on the one hand, and of piccolos, on the other, explore the limits of this range. Dog whistles go a step beyond!
- Spectral colors that humans can see have wavelengths ranging from about 400 nanometers (equivalently $4 \times 10^{-7}$ meters, or .4 microns) on the blue side, to 700 nanometers on the red side. These small wavelengths are poorly matched to mechanical contrivances. The "musical instruments" of light are atoms and molecules.

It is, of course, possible to widen the doors of perception, with the help of appropriate devices.

### WEAK FORCE

The *weak force* is, together with **gravity**, **electromagnetism**, and the **strong force**, one of the four basic mechanisms through which Nature acts.

The weak force is responsible for a large variety of transformative processes, including some forms of nuclear radioactivity, the burning of nuclear fuel within stars, and the cosmological and astrophysical synthesis of all chemical elements (that is, their nuclei) starting from protons and neutrons.

The weak force is also called the weak interaction. See **Force**.

Within the Core Theory, the *weak force* is understood to result from the response of *W* and *Z* particles, the so-called **weakons**, to weak **color charge**. Like the other Core forces, the weak force is a manifestation of **local symmetry**.

The **Higgs mechanism** was proposed to explain aspects of the weak force: specifically, the non-zero mass of the weakons. This line of thought led to the discovery of the **Higgs particle**. The success of these ideas informs us of the existence of a **Higgs field** that permeates all space, and modifies the behavior of other particles in many ways.

The use of "weak" and "force" together is potentially ambiguous because "weak force" could be taken to mean an influence that is not very powerful. Thus when arguing against astrology, and speaking of the influence of gravity from a distant planet, or a star not the Sun, on human fortunes, we might say "It's such a weak force, it can't matter." To avoid ambiguity, in such cases I use terms like "feeble force" or "feeble interaction," avoiding "weak force" and "weak interaction."

### WEAKON

Part of the beauty of *weakons* is that they can be defined in several **complementary** ways:

- Weakons are the *W* and *Z* particles that are observed in detectors at **accelerators**.
- Weakons are the **quanta** of the weak force **fluid**, whose response to the motion of weak charges causes the **weak force**.
- Weakons are the embodiments of a particular **local symmetry**, the symmetry of rotations in the **property space** of weak charges. That is their most beautiful definition: It shows the kinship of weakons with color **gluons**, **photons**, and **gravitons**. All are embodiments of **local symmetry**. It reminds us of our Question, and its answer in the Core. We discover Real ↔ Ideal, as the objects and events of reality match the concepts we introduce to implement the anamorphic art of local symmetry.

### YANG-MILLS THEORY

In 1954 C. N. (Frank) Yang and Robert Mills discovered how to construct a large new class of theories in which **rigid symmetry** of a **property space** is expanded into **local symmetry**. In their honor, theories of this kind are often called *Yang-Mills theories*. Our **Core Theories** of the **strong** and **weak forces** incorporate their construction.

In passing from **special relativity** to **general relativity**, in 1915, Einstein had expanded **Galilean symmetry** from a rigid into a local form. Roughly speaking, Yang and Mills taught us how to make this kind of expansion, from rigid symmetry to local symmetry, for a wide class of possible **symmetry groups** acting among particles.

In the main text, we compare the passage from rigid to local symmetry to the passage from ordinary perspective, governed by **projective geometry**, to the freer possibilities of anamorphic art.

### Z PARTICLE

A massive particle that plays a central role in the **weak force**. See **Weakon**.

### ZERO-POINT MOTION. SEE QUANTUM FLUCTUATION / ZERO-POINT MOTION.

# NOTES

PYTHAGORAS II: NUMBER AND HARMONY

34 ***Why* do tones whose frequencies are in ratios of small whole numbers sound good together?:** Even the most basic facts about musical perception raise fascinating questions. Two simpler observations seem to me especially relevant to the puzzle Pythagoras bequeathed us: *Why* are tone-pairs whose frequencies are ratios of small numbers the ones we commonly perceive as harmonious?

**Abstraction**

When we speak of an octave based on middle C, for example, we mean that both middle C and the C just above, with twice the frequency, are sounding simultaneously. To simplify the phenomenon of *merging* to its essence, let us suppose that by electronic means we produce rigorously pure tones, and let us also suppose that the intensity (loudness) of both is equal. Those specifications still do not give us a unique prescription for the total wave form that the computer must produce and that arrives at our ear. For the two sine waves need not be in synchrony: the peaks of one might or might not be aligned with the peaks of the other. We say that there is a relative phase between the two tones. The total wave forms, plotted as a function of time, can look very different, depending on the value of that relative phase. But they don't sound different! I've tried this experiment, and many related ones, on myself. The response of the basilar membrane separates the tones spatially, but its response retains the relative phase information. (At least, that's my reading of a complex literature. Experiments on these inner ear structures are not easy, and they are essentially always made in vitro.) Yet somehow we lump together all these possibilities, at a low level of processing, and recognize the result as a C octave, full stop. We conflate signals representing a continuous range of physical properties into a single perception, to produce a useful abstraction.

The same principle holds good for octaves based on other tones, and for other two-note combinations, as long as the frequencies are not too close. (As a limiting case, we can add together two tones with the *same* frequency and intensity, but different relative phases—instead of an octave, a unison. Now as we vary that relative phase, we always obtain a combination tone at the unison frequency, but with variable phase *and intensity*. Changes in intensity are easily perceptible.)

The process of deliberate conflation or *abstraction* makes good sense as an information-processing strategy. In the natural world, and in the world of simple musical instruments (including voice), common sources often generate octaves with different, essentially random relative phases on different occasions. If those different wave forms resulted in different sensations, we would be burdened with mostly useless information, and might have more difficulty learning, recognizing, and appreciating the useful general concept "octave." Evolution, presumably, is pleased to lighten that burden.

Similarly, people who do not have perfect pitch—the vast majority—merge a wide range of physically distinct "octaves," based on different tones (but see the discussion of *retention,* immediately below). Thus they suppress both phase and absolute frequency information, but retain the relative frequency.

Given that it can be useful to suppress irrelevant information in order to construct a useful abstraction, the question arises of how to accomplish it. That is an interesting problem in reverse engineering. I can think of three simple, more or less biologically plausible, ways that it might be accomplished:

- Nerve cells (or small networks of nerve cells) that respond to oscillation at different parts of the basilar membrane could be mechanically, electrically, or chemically coupled to one another in such a way that their responses are synchronized in phase. That is the phenomenon known in physics and engineering as *phase locking.* A slight variant of this concept is that there could be a class of nerve cell that receives oscillatory signals from two such nerve cells (or directly from oscillating hair cells in the inner ear) and is driven to respond in a way that is independent of their relative phase.
- One could have banks of several nerve cells that respond to oscillations at any point in the basilar membrane with different phase offsets. Then when the two banks of outputs corresponding to two different locations are combined, there would always be some that are synchronized. A subsequent level of nerve cells, receiving input from those banks, could then respond most strongly to those synchronous pairs.
- One might have *standard representatives* for each frequency: nerve cells whose output is fixed to a global timing mechanism. Then the relative phase between standard representatives would always be the same, whatever the relative phase of the input signal.

I have not listed here the simple but drastic possibility of simply encoding the places where the basilar membrane is vibrating strongly, without resolving the time-structure of the peaks and troughs at all. (That is analogous to what happens with electromagnetic vibrations in vision.) That encoding certainly loses the phase information, but I think it goes too far. It leaves us at a loss to explain Pythagoras's discovery, since ratios of frequencies would no longer correspond to regularities in the encoded signal.

**Retention**

Benjamin Franklin was keenly interested in music. He perfected the glass harmonica, an ethereal instrument for which Mozart wrote a very beautiful piece (K. 356, freely available at several Internet sites). In a letter to Lord Kames (1765), Franklin made several insightful remarks on music, including this profound one:

> In common acceptation indeed, only an agreeable succession of sounds is called melody, and only the co-existence of agreeing sounds harmony. But since the memory is capable of retaining for some moments a perfect idea of the pitch of a past sound, so as to compare with it the pitch of a succeeding sound, and judge truly of their agreement or disagreement, there may, and does arise from thence a sense of harmony between present and past sounds, equally pleasing with that between two present sounds.

The fact that we can compare the frequencies of tones played at slightly different times strongly suggests that there are cell networks that reproduce and briefly retain received oscillatory patterns. This possibility fits well, I think, with our standard representative proposal, for such networks could incarnate standard representations. Here it is noteworthy that perception of relative pitch corresponds to simple *comparison* of standard representations, which is a different task from *recognition* of absolute pitch.

It is also notable, for this circle of ideas, that we are able to maintain a more-or-less fixed tempo over long periods of time. This again suggests the existence of tunable oscillatory networks in our nervous system, but now for significantly lower frequencies.

I do not have perfect pitch, which irritates me. I have tried to bypass my auditory abstraction of relative pitch by inducing a kind of artificial synesthesia. I wrote a program to play, at random, specific tones together with specific colors. Later I tested myself on one input or the other, trying to predict its mate. After many tedious sessions, I got some modest improvement over random guessing. There may be more efficient ways to do this, or it may be that it is easier for very young people.

To determine whether these particular ideas on harmony are on the right track would require hard experimental work. But it would be wonderful, after two and a half millennia, finally to get to the bottom of Pythagoras's great discovery, and thereby to honor the command of the Delphic oracle: *Know Thyself.*

## Plato I: Structure from Symmetry—Platonic Solids

41 **the five Platonic solids are the only finite regular solids:** It seems natural to ask whether we can further transcend the limitation we (or, rather, Euclid) discovered that limited the number of Platonic solids to five, by passing to Platonic surfaces in a more general way. Recall that we argued that no more than six triangles could be used at a vertex, because their angles would add up to more than 360 degrees, which is the most that space can accommodate at a vertex. With six, we got the plane as a Platonic surface.

With three, four, or five triangles, we get, by projecting from the center of our Platonic surface to a circumscribed sphere, regular dissections of spheres. That is possible because equilateral spherical triangles feature angles greater than 60 degrees, so we can surround a vertex with fewer than six of them. This is another way of thinking about both classes of Platonic bodies: as regular dissections of planes, or of spheres.

Thus we are led to ask, more concretely: Can we imagine a different kind of surface, where the angles get smaller? Then we might imagine Platonic surfaces with more than six triangles intersecting at a vertex.

We can indeed. What we need is a surface that results from warping a plane out, as opposed to curling it inward as we do to get a sphere. A saddle shape does the job. On a saddle, we can imagine uniform dissections based on vertices with seven triangles, or even more (in fact, an arbitrary number). More precisely, the mathematical figure known as a trochoid gives us the exact saddle shape we need to keep everything symmetric, so that each vertex and each triangle (or other figure) looks the same.

The ancient geometers knew more than enough about geometry to carry out all the necessary constructions. Further pursuit of this line of thought might have led clever people, around 0 BCE/CE, into the nineteenth-century concepts of non-Euclidean geometry, and to the sorts of graphic designs made popular by M. C. Escher in the twentieth century. Unfortunately, it didn't happen.

42 **you can see a display of five carved stones:** There is controversy over whether the Ashmolean stones, and their relatives, are convincing Platonic solids. See math.ucr.edu/home/baez/icosahedron.

## Newton III: Dynamic Beauty

116 **The great twentieth-century mathematician and physicist Hermann Weyl:** Hermann Weyl is one of my heroes. I grew up on his books and even now return to them frequently. I never got to meet him in person, since I was a very small child when he died. But the beautiful passage quoted in the text has opened up an opportunity for us to collaborate, which I take up here. It always struck me as poetic, and so it occurred to me: Why not take the next step, and make it a poem?

Here is that poem. The first line is also the title.

The world simply is.

In my consciousness
Tethered to my brain and body
Fleeting images come to life—
Of the world, samples only.

The world simply *is*
It does not *happen.*

MAXWELL I: GOD'S ESTHETICS

130 **some excellent, free Web sites where you can explore the Maxwell equations interactively:** The Web site maxwells-equations.com is a comprehensive entry-level introduction to Maxwell's equations, including a video tutorial. The Wikipedia entry en.wikipedia.org/wiki/Maxwell%27s_equations is very good. In consulting that article, I recommend that you start with the "Conceptual Descriptions" section, which follows much the same lines as our main text, and work outward. There is also a beautifully clear little movie of the field pattern of an electromagnetic wave moving through space, which I highly recommend: en.wikipedia.org/wiki/Maxwell%27s_equations#mediaviewer /File:Electromagneticwave3D.gif.

MAXWELL II: THE DOORS OF PERCEPTION

155 **This ability seems to be rare, and has not been much studied:** It is not implausible, however, that tetrachromacy is common among the mothers and daughters of color-blind men. If color-blind men carry a defective receptor, so that their green and red receptors are closely similar—but not identical—which is carried on their X chromosome, then their daughters will also receive it. Together with the normal receptors from their mother, then, those daughters will have four distinct receptors (though two will be similar). If this is correct, then tetrachromacy is not terribly rare, but its consequences may be subtle. For similar reasons, one would also expect the mothers of color-blind men to be tetrachromats.

QUANTUM BEAUTY I: MUSIC OF THE SPHERES

194 **anthropic arguments beg many questions:** The general nature of anthropic arguments is discussed explicitly in "Terms of Art." They have their own entry, and also appear importantly in the entry on dark matter and dark energy. I decided not to interrupt the main text with them.

QUANTUM BEAUTY III: BEAUTY AT THE CORE OF NATURE

232 **the *properties* that the gluons respond to were also christened *colors*:** The literature contains several different choices for the names of the three strong color charges. Like any other choice, the one made here (RGB) is basically arbitrary, but it dovetails nicely with our earlier discussion of spectral colors and their mixing, as you'll see.

I've left the description of color property spaces in the text a little loose, because a precise description is a bit more complicated, and involves complex numbers. Thus the strong color space is a property space with three complex dimensions, and likewise for the weak and electromagnetic color property spaces. In each case the symmetry transformations do not change the overall distance from the origin, so the property spaces of what we've called entities (particles related to one another by symmetry transformations) are spheres of various dimensions. In the case of the strong interactions we start with three complex dimensions, which are six real dimensions, and so the property space of a quark entity is a sphere with five real dimensions. For electromagnetic charge we have one complex dimension, two real dimensions, and finally a one-dimensional sphere, also known as a circle. The radius of that circle is the magnitude of the electric charge.

240 **The historical origin of the term "gauge symmetry" is quite interesting:** In 1919 Hermann Weyl, in his paper "Eine neue Erweiterung der Relativitätstheorie" ("A New Extension of Relativity Theory"), proposed a brilliant theory to explain the origin of electromagnetism. Though that theory, in its original form, is quite incorrect, it introduced ideas that have proved extremely fruitful. Indeed, it was the first attempt to go beyond

Einstein, and invoke *local symmetry* as a fundamental creative principle for nongravitational interactions. As we've discussed, that strategy—implemented with different tactics—leads to our Core Theory.

The term "gauge symmetry" is a relic of Weyl's original theory.

As we have discussed, the essential idea of local symmetry is to require that many different images of the world represent the same physical content. If we want a wide variety of "distorted"' arrangements of space, time, and substance to be valid—i.e., if we want the behavior that each of them depicts to be physically possible—then we must bring in a medium that enables, or you might say "creates," the distortions. (See plate EE and figure 33 on page 242 for a visual equivalent of this idea.) The kind of medium we will need is closely connected to the kind of distortions we choose to implement.

Weyl, in his original theory, postulated *local scale symmetry*. That is, he postulated that one could change the size of objects independently at every point in space-time—and still get the same behavior! To make that outrageous idea viable, he had to introduce a "gauge" connection field. The gauge connection field tells us how much we must adjust our scale of length, or re-*gauge* our rulers, as we move from one point to another. Weyl made the remarkable discovery that this gauge connection field, in order to do its job of implementing local scale symmetry, must satisfy the Maxwell equations! Dazzled by this apparent miracle, Weyl proposed to identify his *ideal* mathematical connection field with the *real* physical electromagnetic field.

Unfortunately, although Weyl's connection field is a necessary ingredient of local scale symmetry, it is not sufficient to ensure that symmetry. Other properties of matter, such as the dimensions of a proton, give us objective scales of lengths that don't change as we move from point to point.

Einstein and others did not fail to notice the shortcomings of Weyl's theory. Despite its visionary brilliance, that theory seemed destined for oblivion.

The situation changed, however, with the emergence of quantum theory. In that context electric charge is associated with a one-dimensional property space, living on top of space-time, as we have discussed in the main text.

In 1929 Weyl exploited this opening, reviving his "gauge" theory in a modified form. In the new theory, the local symmetry transformations are not space-time dependent changes in the scale of length, but rather rotations of the electric property space. After that modification, we get a satisfactory theory of electromagnetism!

Decades later it emerged that implementing local (space-time dependent) symmetry under rotations in other, larger property spaces gives us our satisfactory theories of the strong and weak interactions, too. In homage to Weyl's early vision, physicists call all theories of this kind *gauge* theories.

244 **nuclei are collections of protons and neutrons, bound together:** Isolated neutrons are unstable, but by binding with other neutrons and with protons, neutrons become stable within atomic nuclei.

261 **What Lee and Yang proposed:** As a historical matter, their original proposal was not quite this specific, but later work refined it.

262 **it might be possible to describe that force . . . as an embodiment of local symmetry:** For experts: the current × current form of the total interaction, and the universal strength of the coupling, are both characteristic of gauge theory couplings.

SYMMETRY III: EMMY NOETHER—TIME, ENERGY, AND SANITY

286 **no less a scientist than Niels Bohr, in the 1920s:** These proposals by Bohr and by Landau were made after Noether's theorem. Bohr and Landau both anticipated radical changes in the foundations of physics that would render Noether's theorem inapplicable. But both quantum theory in general (not available to Bohr) and the Core Theories in particular (not available to Landau) are built on the same principles as Noether used to prove her theorem—namely, the principles of Hamiltonian mechanics. As mentioned in the main text, it would be very desirable to have a more conceptual, less technical foundation.

QUANTUM BEAUTY IV: IN BEAUTY WE TRUST

320 **Blessed are those who believe what they see:** The unification of forces and the unification of force with substance are very well-advanced theoretical programs. As we've discussed, they have achieved significant explanatory power, and they suggest essentially new effects accessible through concrete, doable experiments—suggestions that are now being tested. There are two other unifications in fundamental physics that I think would be most desirable, but where existing ideas are less mature.

One is the unification of our description of matter and of information. The former is based, speaking very broadly and roughly, on equations that describe flows of energy and charge. These equations are derived, formally, by manipulation of a quantity called action. Action has some interesting connections to entropy, and entropy has close connections to information, so the possibility of a unified theory is tantalizing. Such a theory might well provide a more conceptual understanding of Noether's theorem, and strengthen its foundations. The other is the unification of dynamics with initial conditions, mentioned several times in our main meditation.

At the boundary of physics, but important to any discussion of ultimate unification, is what Francis Crick called the "Astonishing Hypothesis": that consciousness, also known as Mind, is an emergent property of Matter. As molecular neuroscience progresses, encountering no boundaries, and computers reproduce more and more of the behaviors we call intelligence in humans, that hypothesis looks inescapable. But the particular go of it remains, to say the least, obscure.

A BEAUTIFUL ANSWER?

324 **Walt Whitman**, in the famous lines from *Leaves of Grass* we've recalled, anticipated complementarity. In the spirit of this concluding section, I would like to take them further in that direction:

The world is large—
It contains multitudes.
I look with all-embracing eyes
And tell you what I see.
Do I contradict myself?
Very well, I contradict myself.
If you are not bedazzled yet:
Look differently, and marvel.

TERMS OF ART

341 **the analysis of functions by studying their variation over small ranges, as in (differential) calculus:** Mathematically, the simplest periodic motions are those in which a particle moves at constant speed around a circle. If we look at the height of a particle moving that way, we get the simplest periodic motion you can realize on a line. It is called a sinusoidal oscillation. At www.youtube.com/watch?v=mitioODQYgI you can see an artistic presentation of sinusoidal wave motion, to the music of Bach.

At http://www.mathopenref.com/trigsinewaves.html you can find a more straightforward presentation that also contains an animation of an important physical realization of this sort of motion, shown in the vibrations of a weighted spring around an equilibrium. If you unfold this motion in time—that is, plot the height as a function of time—you get the sine function. Sinusoidal waves appear in the description of sound waves associated with a pure tone and light waves associated with a pure spectral color. In a pure tone, the variation of density and of pressure in space (relative to their average) takes the form of a sinusoidal wave, as does the variation, at a fixed point in space, of those quantities with time. Similarly, in a light of a pure spectral color, the electric and magnetic fields vary sinusoidally.

Thus when our ear resolves an incoming chord into its constituent tones, or when a prism disperses an incoming light beam into spectral colors, they are performing a kind of analysis that is quite different, mathematically, from that based on carefully studying behavior over small time intervals and building up from there. In general, the mathematical analysis of functions into sine functions of different wavelengths or frequencies is called Fourier analysis, after the French mathematical physicist Joseph Fourier (1768–1830). Fourier analysis, and the corresponding synthesis, is a powerful tool complementary to the infinitesimal analysis of calculus.

360 **there is no compelling theory explaining why Nature indulges in her threefold family repetition at all:** The distinction among families can be considered as another property, analogous to strong or weak color charge. One can define a property space associated with the family property. In this way, the different families would be distinguished by an additional set of colors, with the first family being (say) chartreuse, the second lavender, and the third peony. Anthony Zee and I, among others, have speculated that this property space might also support local symmetry. But since there is no hint, in any existing experiment, of the sort of transformations the gauge bosons of that hypothetical symmetry would induce, any "family symmetry" of this sort must be badly broken, and its gauge bosons very heavy.

364 **But the other three forces respond to charges that can have either sign:** There is an interesting question: why (and whether) the Universe is, on large scales, electrically neutral. If it were not, then electrical forces could not be canceled accurately, and they—rather than gravity—might dominate astronomy. We may also pose questions about overall angular momentum. If it were not zero, the Universe would divide up into aligned vortex-like structures. Whatever the reason, the Universe does seem to be balanced in charge and angular momentum. On the other hand, it is crucial to the emergence of humans as physical beings that the Universe is not balanced between baryons and antibaryons. There are plausible ideas about how that asymmetry might have arisen in the early stages of the Big Bang, starting from maximally symmetric conditions, and then frozen in. For an account of this, see frankwilczek.com/Wilczek_Easy_Pieces/052_Cosmic_Asymmetry_between_Matter_and_Antimatter.pdf.

364 **gravity leads to attraction between bodies:** The possibility of what is now called "dark energy" was anticipated by Einstein. He noted that the metric fluid might have a characteristic energy density, which is basically his "cosmological term." In order that the density is invariant under Galilean transformations, it must be accompanied by a pressure of the same magnitude but opposite sign. Thus a positive density of metric fluid is associated with negative pressure. In this case, we say there is a positive cosmological term. And, closing this circle of logic, negative pressure encourages expansion. Thus a positive "dark energy" density is associated with a tendency to expansion. In that sense, it generates a repulsive gravitational force.

It is also possible to contemplate a negative cosmological term: if the energy density of the metric fluid is negative, we get positive pressure and a tendency toward contraction.

In later years physicists realized that not only the metric fluid, but also the other fluids that pervade our description of Nature might have finite energy density, either positive or negative. Galilean symmetry then insures that they too exert pressure of the opposite sign. "Dark energy" refers to the totality of such effects, whereas "cosmological term" refers more specifically to the metric fluid. Physicists do not know how to calculate the magnitude of these densities, or even if it makes sense to speak of them as separate quantities. (See **Renormalization**.)

The literature on these topics is confused and (therefore) confusing. You can find more information at en.wikipedia.org/wiki/Cosmological_constant, en.wikipedia.org/wiki/Dark_energy, and scholarpedia.org/article/Cosmological_constant. The basic definitions and the description of observations are uncontroversial, but beyond that the theoretical terrain becomes treacherous and unstable.

366 **There is a complicated relationship among the weak force, hypercharge, and electromagnetism:** The position of electromagnetism with our Core Theory is a little complicated, as it gets tied in with the weak interaction. The problem is that the gauge bosons that act most simply on property spaces differ from the ones that have simple physical properties. The fundamentally simple ones are usually called *B* and *C*. *B* responds to the

difference between yellow and purple weak charges, while *C* responds to hypercharge. Hypercharge is closely related to electric charge, but not equal to it. The photon and *Z* boson are, mathematically, combinations of *B* and *C*. The photon, which has zero mass, gives us electromagnetism, while the *Z* boson, which weighs nearly as much as one hundred protons, was first observed experimentally in 1983, and plays a very limited role in the natural world.

The hypercharge of an entity is the *average* electric charge of the particles it represents. (Sometimes an extra factor of two is inserted, for historical reasons.) Because the weak interaction connects particles within an entity, and can change electric charge, we can't assign a definite electric charge to the entity, and hypercharge is the appropriate stand-in.

*The Theory of Almost Everything* by Robert Oerter (Plume) is a fine presentation of the ideas of our Core strong and electroweak force theories for general readers, complementary to ours.

The article arxiv.org/pdf/hep-ph/0001283v1.pdf by S. F. Novaes is very far from light reading, but its second section contains the basic equations in about as simple a way as it is possible to present them, while the first section has a useful timeline and background material.

369 **A technical discussion of precisely how the magnetic field is defined:** The relation between magnetic fields and the forces they generate is complicated. The magnetic force on a moving charged particle is proportional to the magnitude of the field, the magnitude of the charge, and the velocity of the particle. The force's direction is perpendicular to the plane formed by the velocity and the direction of the magnetic field vector. Finally, the direction of force is given by a right-hand rule, as if spinning from the direction of the velocity toward the direction of the magnetic field vector. This is described in en.wikipedia.org/wiki/Lorentz_force. You can find much more material on magnetic fields in the excellent article en.wikipedia.org/wiki/Magnetic_field. *Principles of Electrodynamics* (Dover), by Nobel Prize winner Melvin Schwartz, is a modern, clear textbook.

376 **The usual right-hand rule for breaking that ambiguity:** Neutrino physics is a world of its own, dominated by heroic experiments in exotic locations. The Web site of the Ice-Cube experiment—an experiment that involves inserting long strings of photodetectors deep into Antarctic ice—contains a wide-ranging discussion of the field, with loving description of experimental techniques, an extensive timeline, and a nice collection of links to other sources, at www.icecube.wisc.edu/info/neutrinos.

The Wikipedia article en.wikipedia.org/wiki/Neutrino is also good, though less self-contained.

387 **A description of the mathematics of spinors:** Spinors appear in several different places in physics and allied areas.

One can define spinors in any number of dimensions, and their finer properties depend on the dimension in interesting ways.

In some ways the most impressive use of spinors—because it is so basic and geometrical—is their application in computer graphics. Spinors provide the most compact, efficient way of dealing with rotations in three-dimensional space. If you need to calculate a lot of rotations in a short amount of time, say in constructing an interactive game, it pays to use spinors.

The most basic use of that kind of spinor in physics is to describe the spin degree of freedom of electrons and other spin-½ particles. Another form of spinor—the type appropriate to four-dimensional space-time—occurs in the Dirac equation for relativistic electrons. Yet another form of spinor, associated with ten-dimensional space, occurs in the entity that represents substance in the *SO*(10) unification scheme. Other forms of spinor occur in the theory of error correction for quantum computers. What, if anything, connects those three last appearances of spinors is at present unclear. Herein perhaps lies another opportunity for unification seekers.

Although I'd love to be proved mistaken about this, I'm afraid it's beyond human intuition, unaided by specialized experience and algebra, to understand the meaning of spinors in any depth. The Wikipedia article en.wikipedia.org/wiki/Spinor is very well done, but it does not perform that miracle. The great modern mathematician Michael Atiyah gave a lecture, "What Is a Spinor?," which you can find on YouTube at youtube.com /watch?v=SBdW978Ii_E.

The lecture alternates between entertaining anecdotes and general wisdom on the one hand, and very advanced math on the other.

One thing that spinors reveal is that the act of rotation through 360 degrees is not the same as no rotation at all, whereas rotation through twice that—that is, 720 degrees—is. That distinction also emerges in an experiment you can do at home, after watching here: youtube.com/watch?v=fTlbVLGBm3Q.

390 **Here "simple" has a definite technical meaning:** Two references mentioned earlier, www.youtube.com/watch?v=mitioODQYgI and http://www.mathopenref.com/trigsinewaves .html, are relevant here again. I will add here two classics of acoustics by master physicists: *On the Sensations of Tone*, by H. Helmholtz, and *Theory of Sound*, by Lord Rayleigh. Both are available free online, and also in attractive Dover editions.

# RECOMMENDED READING

This is a brief list of recommendations for further exploration of major themes touched by our meditation. I have made recommendations in three categories: classics, quantum theory, and modern developments. Each of the quoted items has meant a lot to me.

## CLASSICS (PRE-QUANTUM)

Nothing can replace the experience of direct communication with great thinkers at their best. So although the technical, scientific content of the works here listed has been superseded, I have no hesitation in recommending them to your attention. Some of this material is in the public domain and can be found on the Internet, if you know what you're looking for. But well-produced books are an attractive and mature technology with advantages in portability, and tactile and esthetic qualities, that you might want to consider as an alternative.

Plato, *The Collected Dialogues of Plato, Including the Letters,* edited by Edith Hamilton and Huntington Cairns, translated by Lane Cooper (Princeton University Press). See especially *Timaeus.*

Bertrand Russell, *The History of Western Philosophy* (Simon & Schuster). See especially Book 1 ("Ancient Philosophy") and Book 3, part 1 ("From the Renaissance to Hume").

Galileo Galilei, *The Starry Messenger* (Levenger).

Isaac Newton, *The Principia: Mathematical Principles of Natural Philosophy*

(University of California Press). This masterpiece should not be attempted without a guide. Fortunately this magnificent recent edition features a new translation (from the original Latin into English) by I. Bernard Cohen and Anne Whitman, with an excellent introduction and guide by Cohen.

Isaac Newton, *Opticks* (Dover Publications). This is much more approachable Newton. This is a very special, inexpensive edition featuring a foreword by Albert Einstein, an introduction by Sir Edmund Whittaker, a preface by I. Bernard Cohen, and a helpful analytical table of contents by Duane Roller.

John Maynard Keynes, "Newton, the Man." This remarkable short essay, the tribute of one genius to another, very different one, is available at www-history.mcs.st-and.ac.uk/Extras/Keynes_Newton.html.

James Clerk Maxwell, *The Scientific Papers of James Clerk Maxwell*, edited by W. D. Niven (Dover Publications).

Albert Einstein, H. A. Lorentz, H. Weyl, and H. Minkowski, *The Principle of Relativity*, with notes by A. Sommerfeld (Dover Publications). This is an extraordinary collection! It includes Einstein's founding papers on special and general relativity, his brief note on the conversion of mass into energy, Minkowski's address introducing the modern concept of space-time, and Weyl's early attempt at a unified field theory, wherein the concept of "gauge invariance" first appeared. These papers are research work, and general readers should not expect to follow all the mathematical details, but many of them feature conceptual discussions and passages that are memorable as literature.

## SOME QUANTUM THEORY

Here it gets more difficult to approach the original works without substantial prior background in mathematics and physics. But general readers may enjoy the early parts of texts by the masters, and of the discovery papers.

P. A. M. Dirac, *The Principles of Quantum Mechanics* (Oxford University Press). The early parts are conceptual, and give a (correct) impression of profundity.

R. P. Feynman, R. Leighton, and M. Sands, *The Feynman Lectures on Physics* (Addison-Wesley). The third volume is devoted to quantum theory, and its early parts are conceptual. The early parts of volume 1 survey physics as a whole (pre-Core) and then give a conceptual introduction to mechanics; the early parts of volume 2 present a conceptual introduction to electromagnetism. All feature Feynman's unique combination of insight and enthusiasm.

Henry A. Boorse, ed., *The World of the Atom* (Basic Books). This is a very well-thought-out collection of excerpts from original works going back to Lucretius and up to the pioneering era of particle physics, with helpful commentaries. It shows how the analysis of matter guided people to create something as weird and wonderful as quantum theory.

## MODERN DEVELOPMENTS

The Nobel Foundation Web site, nobelprize.org, is a very rich resource. It contains detailed descriptions of the prizewinning work going back to 1901, as well as the acceptance lectures by the winners.

The Particle Data Group Web site, pdg.lbl.gov, is mostly intended for professionals, but the "Reviews, Tables, Plots" section contains many broad reviews of frontier physics, whose introductory sections will repay study. Most important: By poking around this site, you will gain an impression of the imposing, detailed empirical evidence for our Core Theories.

New research contributions to physics usually first appear on the Web site arXiv.org. You may want to take a peek to see what physics in the making looks like. Of course, only a small fraction of this work will survive the test of time.

The *Stanford Encyclopedia of Philosophy* at plato.stanford.edu has many fascinating, mind-bending articles.

Although it mainly deals with pure mathematics, the *Princeton Companion to Mathematics*, edited by Timothy Gowers (Princeton University Press), will be of interest to anyone who has enjoyed *A Beautiful Question*. I am presently editing the *Princeton Companion to Physics*, scheduled to appear in 2018.

# ILLUSTRATION CREDITS

In-Text Figures

*Frontispiece:* Printed by permission of He Shuifa.
*Page 21:* Courtesy of the author.
*Page 23:* Courtesy of the author.
*Page 28:* Woodcut from Franchino Gaffurio, *Theorica Musice, Liber Primus* (Milan: Ioannes Petrus de Lomatio, 1492).
*Page 38:* Albrecht Dürer, *Melancholia I,* copper plate engraving, 1514.
*Page 40:* Courtesy of the author.
*Page 41:* Courtesy of the author.
*Page 42:* © Ashmolean Museum, University of Oxford.
*Page 44:* From Ernst Haeckel, *Kunstformen der Natur,* 1904. Plate1, Phaeodaria.
*Page 51:* Model of Johannes Kepler's Solar System theory, on display at the Technisches Museum Wien (Vienna), photograph © Sam Wise, 2007.
*Page 68:* Courtesy of the author.
*Page 69:* www.vertice.ca.
*Page 70:* Filippo Brunelleschi, perspective demonstration, 1425.
*Page 84:* Abell 2218, Space Telescope Science Institute, NASA Contract NAS5-26555.
*Page 88:* Diary of Isaac Newton, University of Cambridge Library.
*Page 89:* Sir Godfrey Kneller, portrait of Isaac Newton, oil on canvas, 1689.
*Page 102:* Galileo Galilei, *Sidereus Nuncius,* 1610.
*Page 104:* Sir Isaac Newton, *A Treatise of the System of the World,* 1731, p. 5.
*Page 107:* Courtesy of the author.
*Page 109:* Sir Isaac Newton, *The Mathematical Principles of Natural Philosophy,* vol. 1, 1729.
*Page 122:* Newton Henry Black and Harvey N. Davis, *Practical Physics* (New York: Macmillan, 1913), figure 200, p. 242.
*Page 125:* James Clerk Maxwell, "On Physical Lines of Force," *Philosophical Magazine,* vol. XXI, Jan.–Feb. 1862. Reprinted in *The Scientific Papers of James Maxwell* (New York: Dover, 1890), vol. 1, pp. 451–513.
*Page 128:* © Bjørn Christian Tørrissen, "Spiral Orb Webs Showing Some Colours in the Sunlight in a Gorge in Karijini National Park, Western Australia, Australia," 2008.
*Page 144:* James Clerk Maxwell with his color top, 1855.
*Page 172:* Courtesy of the author.
*Page 174:* Hans Jenny, *Kymatic,* vol. 1, 1967.
*Page 214:* Courtesy of the author.
*Page 216, above:* © D&A Consulting, LLC. *Below:* Care of Wikimedia contributor Alexander AIUS, 2010.
*Page 219, above:* Wikimedia user Benjahbmm27, 2007. *Below:* Harold Kroto, © Anne-Katrin Purkiss, reprinted by permission of Harold Kroto.
*Page 242:* Printed by permission of István Orosz.
*Page 256: Mechanic's Magazine,* cover of vol. II (London: Knight & Lacey, 1824).
*Page 258:* Andreas S. Kronfeld, "Twenty-first Century Lattice Gauge Theory: Results from the QCD Lagrangian," *Annual Reviews of Nuclear and Particle Science,* March 2012. Reprinted by permission of Andreas Kronfeld.
*Page 267:* Courtesy of the author.
*Page 288:* Emmy Noether, 1902.

*Page 296, above and below:* Created by Betsy Devine.
*Page 307:* Courtesy of the author.
*Page 316:* Courtesy of the author.
*Page 324:* Wikimedia.
*Page 326:* NASA Mars Rover image, NASA/JPL-Caltech/MSSS/TAMU.

## Color Plates

*A:* Printed by permission of He Shuifa.
*B:* Detail of Pythagoras from Raphael, *Scuola di Atene*, fresco at Apostolic Palace, Vatican City, 1509–11.
*C:* Courtesy of the author.
*D:* RASMOL image of 1AYN PBD by Dr. J.-Y. Sgro, UW-Madison, USA. RASMOL: Roger Sayle and E. James Milner-White. "RasMol: Biomolecular Graphics for All," *Trends in Biochemical Sciences* (*TIBS*), September 1995, vol. 20, no. 9, p. 374.
*E:* Salvador Dalí, *The Sacrament of the Last Supper*. Image courtesy of the National Gallery of Art, Washington, D.C.
*F:* Camille Flammarion, *L'atmosphère: météorologie populaire*, 1888.
*G:* Pietro Perugino, *Giving of the Keys to St. Peter*, fresco in Sistine Chapel, 1481–82.
*H:* Courtesy of the author.
*I:* Fra Angelico, *The Transfiguration*, fresco, c. 1437–46.
*J:* © Molecular Expressions.
*K:* William Blake, *Newton*, pen, ink, and watercolor on paper, 1795.
*L:* William Blake, *Europe a Prophecy*, hand-colored etching, 1794.
*M:* "Phoenix Galactic Ammonite," © Weed 2012.
*N:* Courtesy of the author.
*O:* Courtesy of the author.
*P:* Spectrum image by Dr. Alana Edwards, Climate Science Investigations project, NASA. Reproduced by permission.
*Q:* Courtesy of the author.
*R:* R. Gopakumar, "The Birth of the Son of God," digital painting print on canvas, 2011. Via Wikimedia Commons.
*S:* William Blake, *The Marriage of Heaven and Hell*, title page, 1790.
*T:* Courtesy of the author.
*U:* Courtesy of the author.
*V:* Claude Monet, *Grainstack* (*Sunset*), oil on canvas, 1891. Juliana Cheney Edwards Collection, Museum of Fine Arts, Boston.
*W:* Courtesy of the author.
*X:* Photographs by Jill Morton, reproduced by permission.
*Y:* Image created by Michael Bok.
*Z:* Mantis shrimp by Jacopo Werther, 2010.
*AA:* Image created by Michael Bok.
*BB:* Courtesy of the author.
*CC:* Courtesy of the author.
*DD:* Via Wikimedia Commons.
*EE:* Printed by permission of István Orosz.
*FF:* Via Wikimedia Commons. Created by Michael Ströck, 2006.
*GG:* Photograph by Betsy Devine; effects by the author.
*HH:* Winter Prayer Hall, Nasir Al-Mulk Mosque, Shiraz, Iran.
*II:* Courtesy of the author.
*JJ:* Courtesy of the author.
*KK:* Amity Wilczek photographed by Betsy Devine; effects by the author.
*LL:* Photograph by Mohammad Reza Domiri Ganji.
*MM:* Typoform, The Royal Swedish Academy of Sciences.
*NN:* © CERN image library.
*OO:* © Derek Leinweber, used by permission.
*PP:* © Derek Leinweber, used by permission.
*QQ:* Courtesy of the author.
*RR:* Courtesy of the author.
*SS:* Courtesy of the author.
*TT:* Courtesy of the author.
*UU:* Courtesy of the author.
*VV:* Courtesy of the author.
*WW:* Courtesy of the author.
*XX:* © Derek Leinweber, used by permission.
*YY:* Caravaggio, *The Incredulity of St. Thomas*, oil on canvas, 1601–2.
*ZZ:* Leonardo da Vinci, *Vitruvian Man*, ink and wash on paper, c. 1492.
*AAA:* Via NASA.

# INDEX

Page numbers in **boldface** refer to entries in "Terms of Art."


absorption, **339**
abstraction, 397*nn*
abundance/reduction complementarity, 325
acceleration, 6, **339–40**, 344–45, 366
accelerators, **340**
Achilles and the tortoise (Zeno's paradoxes), 58–59, 107–8, 107*fig.*, 340
acoustics. *See* musical harmony; musical instruments; sound
action, 402*n*
action at a distance, 118–20, 124, 134, **340**, 358–59, 368
Adams, John Couch, 318, 353
alchemy, 85–86, 96
algebra, 229, 287, 383–84, 392
All Things Are Number, 4, 20, 21, 26, 30, 57, 324, 327, 340
alpha particles, **340**
Ampere-Maxwell's law, 132–33, **340**, 361, 370
Ampere's law, 129, 132, **340**, 370, plate N
anachromy, 222, plates GG, HH
analog quantities, **340–41**, 379
analysis, 5–6, **341**, 345, 402*nn*
Analysis and Synthesis, 43, 45, **341**, 345–46
  as Newton's strategy, 5–6, 43, 45, 78, 80–81, **341**
  seen as reductive, 112–14
anamorphic symmetry. *See* local symmetry
anamorphy, 206, 221–22, 239–40, 242*fig.*, plates EE, HH, LL
Anderson, Carl, 354
Anderson, Philip, 335
angular momentum, 194, **341–42**, 387, 403*n*
  conservation of, 103, 286, 341, 349
anthropic arguments/anthropic principle, 194–95, **342**, 353, 393, 400*n*
antibaryons, 365–66
antielectrons. *See* positrons
antileptons, 368
antimatter, **343**
antiparticles, 261, 319, **342–43**
antiprotons, 342
antiquark-quark pairs. *See* mesons
antiquarks, 252, 268, 365–66, plate NN
antisymmetry, **343**
appearances, vs. hidden/unseen/ultimate reality, 55, 61–62, 141–42
  *See also* Cave allegory
Aristotle, 6, 46–47, 58, 66–67, 78
  "All humans are mortal" syllogism, 290–92
Arnold, Vladimir, 110
*Art Forms in Nature* (Haeckel), 43, 44*fig.*
Artisan
  constraints on, 199–200
  Plato's demiurge as, 5, 48, 64
  visual representations of, 241, 242*fig.*, plates K, L
  *See also* Creator
artistic perspective, 67–72, **378**, plates G, H
astrology, 64
astronomy and cosmology, 13, 31, 50, 64–67
  origins of the Solar System's features, 193–94
  significance of spectral lines, 98–99, 203, 386–87

astronomy and cosmology (*Cont.*)
*See also* celestial mechanics; Copernicus, Nicolaus; Galilei, Galileo; Kepler, Johannes; planetary motion
asymptotic freedom, 255, 305–7, 336, 337, 381, 384
basics of, 253–54, **343**
Atiyah, Michael, 404*n*
atomic clocks, 190
atomic engineering, 189–90
atomic nuclei, 226, **373**
components/structure of, 244–45, 365, 401*n*
discoveries and related questions, 176–77, 243–46, 333
isotopes, 95, 245, 343, **367**
spectra of, 386
*See also* decay; neutrons; nuclear forces; protons; quantum chromodynamics
atomic number, **343**
atomic spectra, 180–81, 359, 373, 376, **386–87**
atoms, 176–77, 189–90
as objects of beauty, 169, 191, plate CC
in Platonic theory, 46–49, 57, 191, 319
as stable structures, 176–77, 178–81, 191–92, 195–96, 388, 389
*See also* stationary states
auditory perception, 31–32, 33–36, 157, 158, 362, 395, 397–99*nn*
*Autobiographical Notes* (Einstein), 23
axial currents, **344**
axial vectors, 341, 344, 375
axions, 271, 272–73, 337, **344**, 352, 381

*b* (bottom) quarks, 270, 271–72, 361
Babylonian astronomy, 64–65
Bardeen, John, 335, 390
baryons, 246, 247, 256, 365, plate PP
*See also* hadrons
BCS theory of superconductivity, 335, 390
beauty, 1–9, 111–12, 200, 318–23
austerity/gorgeousness complementarity, 189
beauty/ugliness complementarity, 328
desire for, as constraint on creation, 200
enlarging our sense of, 166
human perception of (Mind/Matter unity), 11–15
varieties of, 10–11
vs. truth, 63–66, 318
*See also* complementarity; Real/Ideal duality/complementarity
Békésy, Georg von, 33
beliefs, justifications for, 289–94
Berkeley, George, 289
biblical scholarship, 85–86
birds, 13–14, 155
*Birth of the Son of God, The* (Gopakumar), plate R
black, 92, 143
Blake, William, 113, 141–42, plates K, L, S
bleaching rules, 250, 256, 301, plate QQ
blueshift, 96, 202, 203
Bohr, Niels, 52, 190
embrace of complementarity, 16, 236, 324–25, 324*fig.*, 347
energy conservation violation proposal, 286, 401*n*
*See also* stationary states
boosts, **344**, 367
*See also* Galilean transformations
Bose, Satyendra, 310, 333, 345
Bose's inclusion principle, 310, 345
bosons (force particles), 310–11, **344–45**, **362**, 389, 390
*See also specific types*
Boswell, James, 289
bottom (*b*) quarks, 270, 271–72, 361
Brahe, Tycho, 80, 357
branching ratios, 269, **345**
Brout, Robert, 264, 335
Brunelleschi, Filippo, 5, 70–71, 70*fig.*, 77, 331
buckminsterfullerenes (buckyballs), 218, 219*figs.*, 220, 295, **345**
Bunsen, Robert, 98

*c* (charmed) quarks, 270, 275, 319, 336, **346**, 361
Cabibbo angle, 360
calculus, 87, 108, 341, **345–46**, 366–67, 383–84
calendars, 64
Campbell, Lewis, 162, 163–64
Caravaggio, Michelangelo Merisi da, *The Incredulity of St. Thomas*, 320, plate YY
carbon and carbon compounds, 212–20
buckminsterfullerenes (buckyballs), 218, 219*fig.*, 220, 295, **345**
carbon atoms, 212–13
diamonds, 213, 214–15, 214*fig.*, 216*fig.*

graphene and graphite, 213–14, 214*fig.*, 215, 216*fig.*, 217, 218, **364**
nanotubes, 217–18, **372**, plate FF
Cave allegory (Plato's *Republic*), 5, 7, 56–57, 61–64, 141, plate F
Cay, Charles, 162
celestial mechanics, 6–7, 167, 193–95, **346**, 364
*See also* astronomy and cosmology; planetary motion
celestial spheres, 50, 51*fig.*, 80
Music of the Spheres, 30, 51
CERN. *See* Large Hadron Collider
Chadwick, James, 244
change, 58–59, 327
*See also* dynamical laws and equations; motion
Change Without Change, symmetry as, 73–74, 137–38, 167, 327, 384, 386
chaos theory, 115
charmed (*c*) quarks, 270, 275, 319, 336, **346**, 361
chemistry and chemical elements, 46, 95–97, 220, 245, **376**
spectral lines and, 98–99, 386–87
*See also specific elements*
chords, 34, 157, 174–75
*See also* musical harmony
*Chronology of Ancient Kingdoms Amended, The* (Newton), 85
circles, 39, 50, 137, 357
circular-motion theories of planetary motion, 6–7, 50, 65–66
circulation, 131–32, 340, **346**, 361
Clarke, Arthur C., 175
Cleve, Per, 99
color (of light), **346–47**, 359
anachromy, 222, plates GG, HH
as infinite-dimensional, 151–52, 161
polarization and, 94–95, 152–53, 378, 391
*See also* color perception; perceptual color; spectral colors
color (of pigments), 143, 147
color blindness, 145, 154, 155, 159–60, 400*n*, plates U, X
color boxes, 146
color charges, 234, 237, 253, 304, 352, 379, 396, 400*n*
basics, 249–50, **347**
color currents, 352
color gluons, 303, 319, 337, **364**
as avatars of local symmetry, 240, 241
behavior of, 242–43, 250, 253–54, 310, 344, plate OO
as elementary particles, 356, 373
as force particles, 310
as gauge particles, 363
jets as avatars of, 252, plate NN
as massless particles, 263, 270
color perception, 34, 93, 142–64, 347
color as representation of material, 156–57
color as temporal information, 151–52, 204
color receptors and varieties of color vision, 145, 153–56, 159–60, 400*n*, plates U, X, Y
disregarded or ambiguous color information, 151–53
expanding our color perception, 147–48, 159–61, 190, 204, plates AA, BB
imagery vs. color, 150–51
Maxwell's related work, 7, 142, 143–47, 144*fig.*, plate U
motion and spectral color shifts, 96, 202–3
by nonhumans, 154, 155–56, plates X, Y
observers' motion and, 202–3
pigment colors, 143, 147
relativity and, 96, 202–3, plate DD
spectral and nonspectral colors perceived as identical, 142–43
wavelengths visible to humans, 150, 395
*See also* perceptual color; spectral colors
color photography, 144, 147
color property spaces, 230–32, 242–43, 274, **378–79**, 400*n*, plates II, JJ, KK, RR, SS, TT, UU
colors (in quantum chromodynamics), 232
color tops, 143–46, 144*fig.*, plate U
complementarity, 16, 97, 113–14, 185, 391
basics, 74–75, **347**
as essential to quantum theory, 74, 185, 382–83
as wisdom, 324–28
yin-yang complementarity, 235–36, 237, 324–25, 324*fig.*, plate A
*See also* Real/Ideal duality/complementarity
complex dimensions, **347**, 350
complex numbers, **347–48**, 350, 382, 394
computer displays, 147, 160, 230, 394

confinement, 248, 253, 304, 305, **348**, plate MM
consciousness, 402*n*
*See also* Mind/Matter
conservation laws, 341, **348–49**, 354
as products of symmetries, 279, 281, 285–86, 293, 341, 348, 372
conserved quantities, 279, 341, 347, **348–49**
angular momentum, 103, 286, 341, 349
energy, 180, 205, 281–85, 286, 293, 349, 359, 369, 386
mass, 284, 285, 357–58, 369, 372
momentum, 285, 293, 349, 372, 385
consistency, **349**
continuous groups (of transformations), 349, **365**, **390**
continuous symmetry, **349**
contradiction, **349**
Cooper, Leon, 335, 390
coordinates, 347, **349–50**, 353, 392, 393
Copernicus, Nicolaus, 6–7, 50, 77, 103, 331
Core Theories, 227–29, 234–35, **350–51**
beauties of, 8–9, 276, 297
census of forces and entities, 273–74, plates RR, SS
as embodiments of symmetry, 8–9, 273, 298, 382, 392
flaws and imperfections, 9, 277, 297–98, 321, 344, 350–51, 352–53, 354
mantras for, 233–37
visual summary of, plates RR, SS, TT, UU
yin-yang complementarity of, 235–36
*See also* unification(s); *specific forces*
cosmic inflation, 323
cosmic rays, **351**, 354
cosmological term, 352, 403*n*
cosmology. *See* astronomy and cosmology; Universe
Coulomb's law, 354–55
Cowan, Clyde, 335
Creator and creation, 2–3
Blake's Urizen, plate I
Einstein on the Creator's agency, 199–200
Maxwell on, 133–34, 192
Newton on, 84–85, 116
*See also* Artisan
Creutz, Michael, 337
Crick, Francis, 402*n*
cubes, 40, 40*fig.*, 46
*See also* Platonic solids
current, **351–52**
*See also* axial currents; electric current

*d* (down) quarks, 273–74, 360, 361, plates RR, SS, TT, UU
Dalí, Salvador, *The Sacrament of the Last Supper*, 53, plate E
dark energy, 9, 351, **352–53**, 364–65, 369, 393
Einstein's "cosmological term," 352, 403*nn*
dark matter, 9, 337, 344, 351, **352–53**, 364–65, 369, 373, 393
Davy, Humphry, 121
decay, 243, 246, 261, 343, 395
branching ratios, 269, 345
weak decays, 334
deep truths, 52
demiurge. *See* Artisan
*De Revolutionibus Orbium Coelestium* (*On the Revolutions of the Celestial Spheres*) (Copernicus), 77, 331
Descartes, René, 77, 123–24, 392
determinism, 185, 327, 382
diamonds, 213, 214–15, 214*fig.*, 216*fig.*
digital quantities, 323–24, 340–41, **353**, 379
dimensions (in space), 230, 347, 350, **353**
*See also* property spaces; quantum dimensions
dimensions (measurement), 353
Dimopoulos, Savas, 312, 313–15, 337
Dirac, Paul, 319, 333–34, 380
Dirac equation, 333, 342, **354**, 385, 404*n*
*See also* wave functions
discreteness, 170, 171–72, 195, 383
*See also* digital quantities; quantization; stationary states
dodecahedra, 40, 40*fig.*, 46, 47, 218, 295–98, plates D, E
*See also* Platonic solids
dogs, 13, 154, plate X
Doubting Thomas, 319–20, plate YY
down (*d*) quarks, 273–74, 360, 361, plates RR, SS, TT, UU
Dürer, Albrecht, *Melancholia I*, 37, 38*fig.*
dynamical laws and equations, 6–7, 101, 170, 191–96, **354**, 367
Newton's laws as, 111–12, 114–15, 354
*See also* initial conditions

"Dynamical Theory of the Electromagnetic Field, A" (Maxwell), 117, 124, 129, 133, 162, 332
*See also* Maxwell's equations
Dyson, Freeman, 264, 334

Ecclesiastes, 280
economy, 11, 15, 49, **354**
Einstein, Albert, 118, 383
anticipation of dark energy, 352, 403*n*
on Bohr's stationary states hypothesis, 180–81
calculation of Sun's bending of light, 83
photon hypothesis, 177–78, 333
Planck-Einstein relation, 359, **376**, 377, 386
proof of Pythagoras's theorem, 23–24, 23*fig.*
*See also* general relativity; special relativity
electrical resistance, 283
electric charge, 234, 237, 249, 304, **354–55**, 403*n*
antimatter/antiparticles, **342–43**
as determined by atomic number, 343
Gauss's electric law, 130–31, plate N
hypercharge, 274*n*, **366**, 403–4*nn*
subatomic, 244, 245
*See also* antiparticles; color charges; hypercharge; *specific particle types*
electric current, 340, **351–52**, **355**
electric fields/fluids, 123, 125–27, 125*fig.*, **355**, 361
Gauss's electric law, 130–31, plate N
*See also* Faraday's law; Maxwell's equations
electric force, 354–55
electricity, **355**
Faraday's insights, 123
photoelectric effect, 176, 177–78
electric property space, 240, 274, 401*nn*
electrochromics, 160
electrodynamics. *See* electromagnetism; Maxwell's equations; quantum electrodynamics
electromagnetic fields/fluids, 237, 306, **355–56**, 380, 387, 392
Maxwell's model and its implications, 125–27, 125*fig.*
photon generation, 188–89
Planck-Einstein relation and Planck's constant, 359, **376–77**, 386, 387
electromagnetic property space, 234–35, plate KK
electromagnetic spectrum, 387
*See also* color (of light); spectral colors
electromagnetic waves, 149, 355, **356**, plate W
light as, 97, 127, 148–50, 176, 346–47, 356, 392, plates P, W
as pure tones, 390
speed of, 127
as transverse waves, 391
electromagnetism, 226, 237, 297, 304, 340, **355**, 382
conservation of energy and, 283
in Core Theories, 234
Faraday and his work, 120–24, 126, 340, **360**
hypercharge, the weak force and, 366, 403–4*nn*
*See also* Maxwell, James Clerk; Maxwell's equations; quantum electrodynamics
electron clouds, 175, 187*fig.*, 394
electron fluid, 306, **356**, 380
electron-positron collisions, 252, plate NN
electrons, **356**, 380
arrangements in carbon compounds, 212–20
discovery of, 176
electric charges of, 355, 356
as elementary particles, 356, 373
as fermions, 333
fundamental properties of, 210–11, 310, 359, 387
interchangeability of, 293–94
in quantum dimensions, 313
two-electron atoms, 196–97, 394–95
*See also* stationary states
electron wave functions, 210, 385, 388, 394–95
Dirac equation, 333, 342, **354**, 385, 404*n*
*See also* Schrödinger equation; wave functions
electroweak theory, 336
elementary particles, **356**, 369, 373, 380
*See also specific types*
elements. *See* chemistry and chemical elements; *specific elements*
*Elements* (Euclid), 26, 43, 45, 331, 377

ellipses, **357**
planetary orbits as, 52, 80, 103, 357
empty space, 305, 351, 366, plate XX
*See also* fluids; media; Void
energy, 205, 341, **357–59**
atoms as energy starved, 196
conservation of, 180, 205, 281–85, 286, 293, 349, 359, 369, 386
of electrons, as discrete, 180
forms of, 282, 283, 284, 357, 358–59
as origin of mass, 259–60, 369–70
quantization of, 379–80
energy currents, 352
energy-momentum, 233, 234, 237
energy of motion. *See* kinetic energy
Englert, François, 264, 335
entanglement, 394–95
entropy, 402*n*
equations
atoms as embodiments of, 191
power in, 135–36
*See also* symmetry of equations
eternity/transience complementarity, 327
Euclid, 4, 227–28
*Elements*, 26, 43, 45, 377
exclusion principle, 211, 212, 214, 310, 333, 345, **359**

falsifiability, 308, **359–60**
families, 273–76, 298, **360**, 403*n*, plates RR, SS, TT, UU
familons, 272
family property space, 403*n*
family transitions, 360
Faraday, Michael, 7, 120–24, 340, 368
Faraday's work as basis for Maxwell's, 124–27, 340
Faraday's law, plate N
basics, 131–32, 340, 355, **360**, 361
in Maxwell's model and equations, 126, 129, 131–32, 340, 356, 370
Fermi, Enrico, 310, 333, 334
fermions (substance particles), 309–10, 333, **344–45**, **389**, plates RR, SS
behavior and properties of, 310–11, 343, 345, 354, 359
families of, 273–76, 298, **360**, 403*n*, plates RR, SS, TT, UU
supersymmetry and, 389, 390
*See also specific types*
Feynman, Richard, 200, 334
field energy, 283, **358–59**
fields, 7, 126, 230, **360–61**, 393
vs. fluids, 123, **355**
wave functions of, 382–83
*See also specific field types*
fields and field theory, spiders and, 128–29
fish-eye lenses, 241, plate LL
flavors (of quarks), 247, **361**
*See also specific flavors*
fluids, **361**, 363
accounting for fluid fluctuations, 305–8, 307*fig.*
as essential to local symmetry, 239–40, 400–401*n*
vs. fields, 123, **355**
*See also* media; *specific fluid types*
flux, 131, 132, 340, **361–62**
flux tubes/distribution, 253, plates OO, PP, QQ
force, **362**
quantifying, 339
force particles. *See* bosons; *specific types*
forces, **362**
strength inequities among, 304–8, 314–15, 336
*See also* electromagnetism; gravity; strong force; weak force
Fourier analysis, 402*n*
fractals, 113–14, **362**, plate M
fractional (fractal) dimensions, 353
Franklin, Benjamin, 398*n*
Fraunhofer, Joseph von, 98
Fredkin, Ed, 25
freedom/determinism complementarity, 327
free energy, 349
free parameters, 354
frequencies, **362**
natural (resonant) frequencies, 171–75, 172*fig.*, 173–75, 174*fig.*, **372–73**
of spectral colors, 346
*See also* musical harmony; pure tones
friction, 14, 190, 194, 282, 283
Friedman, Jerome, 251, 336
Fuller, Buckminster, 218, 220
functions, **362–63**

Galilean symmetry (Galilean invariance), 168, 200–204, 309, **363**, 381
local Galilean symmetry, 205–8, 297, 363
Galilean transformations, 167–68, 203, 309, 344, **363**, 386

Galilei, Galileo, 3, 10, 77, 281
*Sidereus Nuncius* (*Starry Messenger*), 101, 102*fig.*, 103, 332
gamma rays, 135, 351, 356, plate P
gauge connection field, 401*nn*
gauge particles, 298–99, **363**
*See also specific types*
gauge symmetry(ies), 240–41, 400*n*
*See also* local symmetry
gauge theories, 304, 400*n*, 401*n*
Gauss, Carl, 371
Gauss's laws, 129, 130–31, 362, **363**, plate N
Geiger, Hans, 243, 333
Geim, Andre, 215, 217
Gell-Mann, Murray, 247, 335, 383
general covariance, 208, 238, 240, **363**
general relativity, 204–8, 333, 340, **363**, 365
dark matter and, 352, 403*n*
Einstein's calculation of the Sun's bending of light, 83
as embodiment of local symmetry, 205–8, 238, 240, 273, 363, 396
general covariance, 208, 238, 240, 241, **363**
mantras for, 232–35, 237
reconciling with Core Theories, 350–51
*See also* gravity; space-time
geodesics, 233, **363–64**
geometry, 26, 229, 392
Euclid's *Elements*, 26, 43, 45, 331, 377
as a fluid, 236–37
geometric mantras, 233–36
Pythagorean theorem, 4, 20–21, 21*fig.*, 23*fig.*, **379**, plate C
*See also* projective geometry
Georgi, Howard, 299, 336
Georgi-Glashow unification scheme, *SO*(10), 299–308, 336, 387, 404*n*, plates VV, WW
Ginzburg, Vitaly, 335
*Giving of the Keys to Saint Peter* (Perugino), plate G
Glashow, Sheldon, 262, 263, 299, 335, 336
*See also* Georgi-Glashow unification scheme
Glaucon, 56
*See also* Cave allegory
global symmetry. *See* rigid symmetry
glue balls, 366
gluon fields/fluids, 305–6, **364**, 369–70, 380, plate XX
gluon fusion, 267*fig.*, 268–69, 270–72
gluons, 49, 232, 348, **364**, 366, 373, 380
*See also* color gluons
Goldstone, Jeffrey, 335
Gopakumar, R., *The Birth of the Son of God*, plate R
*Grainstack* (*Sunset*) (Monet), plate V
graphene, 213–14, 214*fig.*, 215, 216*fig.*, 217, 218, **364**
graphite, 215, 217
Grassmann numbers, 311, 353, **364**, 380
gravitational fluid. *See* metric fields/fluids
gravitons, 207, 208, 263, 310, 356, 363, **364**
as avatars of local symmetry, 240, 241
gravity, 226, 297, 339–40, **364–65**
and bending of light, 82–83, 84*fig.*
dark matter/dark energy and, 352–53, 364–65
force unification and, 315–17, 316*fig.*, 350–51
Newton's Mountain, 103–6, 104*fig.*, 108, 109
questions and challenges raised by Newton's theory of, 118–19, 204–5, 318, 345–46
strength relative to that of other forces, 304–5, 316–17, 364
*See also* general relativity; Newtonian motion and mechanics
gravity waves, 236, 364, 381
Gross, David, 256–57, 336
ground state, 196
groups (of transformations). *See* symmetry groups
guitars, resonant frequencies of, 173–75, 174*fig.*
Guralnik, Gerald, 264, 335

hadrons, 246, 348, **365–66**, 367
calculating masses of, 257–59, 258*fig.*, 337
Dirac equation's application to, 354
electric charges of, 355
jets of, 252, 337, 348, **367**, plate NN
Large Hadron Collider, 254, 318, 337, **367–68**
masses of, 257–59, 258*fig.*, 337
quark model and its shortcomings, 247–48, 251, 335, 365–66, **383**
spectra of, 386
*See also* baryons; mesons; quantum chromodynamics

Haeckel, Ernst, 43, 44*fig.*
Hagen, Carl, 264, 335
handedness, 262, 274, **374–75**
*See also* parity and parity transformations
harmony, 30, **366**
of mind and matter, 11–15, 47
*See also* musical harmony
hearing. *See* auditory perception; musical harmony; musical instruments; sound
heat, as product of energy loss, 284
Heisenberg, Werner, 188, 333
Heisenberg's uncertainty principle, 185, 383
helium, 99, 196–97, 395
Hertz, Heinrich, 134–35, 318, 332, 362
He Shuifa, vi, plate A
hexagons, 39, 41, 41*fig.*, 125, 125*fig.*, 137, 218
hidden/unseen/ultimate reality, vs. appearances, 55, 61–62, 141–42
*See also* Cave allegory
higglets, 272
Higgs, Peter, 264, 335
Higgs fields/fluids (Higgs mechanism), 335, 351, 352, 380, 388, plates TT, UU
basics, 264–66, 275, **366**
as weak force solution, 265–66, 298, 303, 366, 387–88, 396
Higgs particles (Higgs bosons), 266–73, 310, 319, 356, **366**, 380, 390
discovery/production of, 267–69, 267*fig.*, 270–71, 303, 337, 368
high-energy collisions and particles, 245–46, 254, 304, 308, 317, 340, 343
jets as result of, 252, 337, 348, **367**, plate NN
Large Hadron Collider, 254, 318, 337, **367–68**
*See also* hadrons; quantum chromodynamics; strong force
high-energy particles, 246
cosmic rays, **351**
*See also* hadrons; *specific types*
Hilbert, David, 287
Hippasus, 24
Hooft, Gerard 't, 264, 336
human body, as reflection of the cosmos, 321, 322
Hume, David, 289
Hunt, Leigh, 91
Huxley, Aldous, Guido's proof, 22, plate C
hydrogen atoms, 196, 210–11, 244, 385, plate CC
in Bohr's stationary states model, 178–79, 385, 386
probability clouds for, 187, 187*fig.*
wave functions of, 185–89
hypercharge, 274*n*, **366**, 403–4*nn*

Ice-Cube experiment, 404*n*
icosahedra, 40, 40*fig.*, 43, 46, plate D
*See also* Platonic solids
Ideal/Real duality. *See* Real/Ideal duality/complementarity
Iliopoulos, John, 336
imagery, 150–51, 159
imaginary unit, 347–48
Impressionist painting, 147, plate V
inclusion principle, 310, 345
induction. *See* Faraday's law
inertia, 369
infinite decimals, 383, 384
infinitesimal quantities, 6, **366–67**, 384
motion analyzed into, 110
information
digital quantities and, 323–24, 340–41, **353**
unification with matter, 402*n*
infrared radiation, 135, 155, 347, 356, plates P, Y
initial conditions, 114–16, **367**, 402*n*
and Solar System's attributes, 193–95
insect vision, 155
instantons, 271
intensity (of light), **367**
interactions, 362
*See also* forces; *specific types*
invariance, 74, 344, **367**
*See also* symmetry(ies); *specific types*
Islamic art and architecture, 222–23, plates HH, LL
isotopes, 95, 245, 343, **367**

Janssen, Pierre, 98–99
jets (of particles), 252, 337, 348, **367**, plate NN
Johnson, Samuel, 289
Jona-Lasinio, Giovanni, 335
Joule, James Prescott, 284

Kavli Institute for Theoretical Physics, 312
Keats, John, 113, 114, 127
Kelvin, William Thomson, Baron, 319
Kendall, Henry, 336

Kepler, Johannes, 50–52, 77, 79–80, 193, 331
  Kepler's Solar System model, 50–52, 51*fig.*, 79–80, 112, 319, 331
Kepler's laws, 79, 103, 108–9, 322, 357
  first law, 52, 103
  second law, 103, 286, 341
  third law, 103, 105
Keynes, John Maynard, "Newton, the Man," 85–86
Kibble, Tom, 264, 335
kinetic energy, 281–82, **357**, **358**
  *See also* motion
Kirchhoff, Gustav, 98
knots, mathematical, 319
Koyré, Alexandre, 80–81
Kroto, Harold, 219*fig.*, 220
Kusch, Polykarp, 334

Lamb, Willis, 334
Landau, Lev, 286, 335, 401*n*
Langlet, Nils, 99
Large Electron-Positron Collider, 337, 343
Large Hadron Collider, 254, 318, 337, **367–68**
lasers, 310, 345
Lavoisier, Antoine, 284
learning, 14–15, 36, 61–62
*Leaves of Grass* (Whitman), 324, 402*n*
Lee, Tsung-Dao, 261–62, 335, 375
left-hand rules, 374
  *See also* handedness
Leibniz, Gottfried Wilhelm von, 366–67
Leonardo da Vinci, 77, 321, 331, plate ZZ
leptons, 309–10, 356, **368**, 373, 389
  in census of forces and entities, 273–76, plates RR, SS, TT, UU
  *See also* fermions
leverage, 255, 256*fig.*
Le Verrier, Urbain, 318, 353
Lie, Sophus, 299, 332, 365
Lie groups, 299, **365**
*Life of Samuel Johnson* (Boswell), 289
light, 13, 322
  bending of, 82–83, 84*fig.*
  chemistry of, 95–97
  as electromagnetic waves, 97, 127, 148–50, 176, 346–47, 356, 392, plates P, W
  emission/generation of, 188–89, 190, 196, 389
  as encoder of spatial information, 158
  intensity of, **367**
  Maxwell equations and, 97, 127, 148–50, 176, 201, plate W
  Newton's work, 92–95, 97, 119–20, 203, 332
  particle theories of, 95, 97, 119, 177, 376
  photoelectric effect, 176, 177–78
  polarization of, 94–95, 152–53, 378, **391**
  speed of, 83, 127, 201, 367, 385
  *See also* color; color perception; photons; vision and visual perception; *specific types of light*
light-quanta. *See* photons
lines of force, 121–23, 122*fig.*, 130–31, **368**
  of color gluons, 253, plate OO
  *See also* Faraday's law; Gauss's laws
local scale symmetry, 401*n*
local symmetry, 228, 238–41, **368**
  anamorphic art as model of, 239–40, 241, 242*fig.*, 368, plate EE
  Core Theories as embodiments of, 8–9, 273, 297, 382, 392
  fluids as essential to, 239–40, 400–401*n*
  four subtypes and avatars of, 240–41
  Galilean, 205–8, 297, 363
  general covariance, 208, 238, 240, **363**
  massive particles and, 335
  quarks and, 335
  weak force as embodiment of, 262–63, 396
  Weyl's gauge symmetry theory, 400*nn*, 401*nn*
  Yang-Mills equations as embodiments of, 240, 263, 334, **396**
Lockyer, Norman, 98–99
logic, geometry as, 26
London, Fritz, 335
London, Heinz, 335
Lucretius, 119

machines, 283–84
magnetic fields/fluids, 125, 125*fig.*, 340, **369**, 404*n*
  Ampere's law, **340**
  Faraday's law, 126, 129, 131–32, 340, 355, **360**
  Gauss's magnetic law, 131, 363, plate N
  Maxwell's model and its implications, 125–27, 125*fig.*
  *See also* Maxwell's equations
magnetism, **368**, 387, 404*n*
  Faraday's lines of force, 121–23, 122*fig.*, 130–31, **368**

Maiani, Luciano, 336
Malley, Jim, 176, 348
mandalas, 139, 197, plate R
mantis shrimps, 155–56, plates Y, Z
many worlds issue, 185, 382
*Marriage of Heaven and Hell, The* (Blake), 141–42, plate S
Marsden, Ernest, 243, 333
mass, 205, 339, **369–70**, 372
  conservation of, 284, 285, 357–58, 369, 372
  of hadrons, 257–59, 258*fig.*, 337
  origins of, 259–60, 352–53, 369–70, 373, 379
mass currents, 352
mass energy, **357–58**, 369
massive particles, 265, 269, 335
  heavy quarks and quark mesons, 270, 272, 336
  *See also* Higgs particles; weakons
massless particles, 263, 265–66, 270, 271, 272
Mass Without Mass, 232, 259, 370
mathematical knots, 319
mathematics, 10, 113–14
  All Things Are Number, 4, 20, 21, 26, 30, 57, 324, 327, 340
  *See also specific mathematical branches, terms, and tools*
matter, 9, 209, 233–35, 343, **373**
  normal vs. dark, 352, 369
  six fundamental entities, 273, 298, 300, plates RR, SS, TT, UU
  unification of matter and information, 402*n*
  *See also* dark matter; fermions
matter/space-time complementarity, 233–34, 235
"Matthew Effect," 41–42
Maxwell, James Clerk, 3, 7, 144*fig.*, 162–64, 332
  color vision work, 7, 142, 143–46, 332, plate U
  on the Creator and stability of matter, 192–93
  death of, 134, 163–64
  and Faraday's work, 121, 124–27, 340
Maxwell's contradiction, 132–33, plate O
Maxwell's equations, 7, 129–39, 204, 340, 361–62, **370**
  beauties of, 117–18, 130, 134–39
  Dirac's quantized version, 333
  exploring further, 400*n*
  foundations in Faraday's work, 124–27, 340
  Hertz's confirmation of, 134–35, 176, 332
  influence/significance of, 117–18, 134–36
  pictorial representations of, 118, 130, plate N
  predictions of, 318
  symmetry in, 118, 136–39
  as unification, 392
  as wave theory of light, 97, 127, 148–50, 176, 346–47, 356, plate W
  Weyl's gauge theory and, 401*n*
  *See also* Faraday's law; Gauss's laws
Maxwell's law, 126–27, 133, 355, 356, **370**, plate N
Maxwell's model and its predictions, 125–28, 125*fig.*
Maxwell term, 340, **370**
  *See also* Maxwell's law
media, **370**
  Faraday's lines of force as evidence of, 121–22
  *See also* fluids; Void; *specific fluid types*
*Melancholia I* (Dürer), 37, 38*fig.*
Merton, Robert, 41
mesons (quark-antiquark pairs), 246, 247, 251, 365, plate OO
  masses of, 257–59, 258*fig.*
  *See also* hadrons
metaphysics, 58, 62
method of fluxions, 108
metric fields/fluids, 207, 236–37, 239, **371**, 392, 403*nn*
metrics, 236, **370–71**
microcosmos/macrocosmos, 322–23, 325–26, plate AAA
microwave background radiation, **371–72**
microwaves, microwave radiation, 135, 322–23, 356, **371–72**, plates P, AAA
Mills, Robert, 334
  Yang-Mills equations, 240, 249, 263, 298–99, 334, **396**
Mind/Matter, 11–15, 47, 402*n*
Molière, 229–30
momentum, 285, 341, **372**, 382
  conservation of, 285, 293, 349, 372, 385
  *See also* angular momentum
Monet, Claude, *Grainstack* (*Sunset*), plate V
Moon, 194
  Galileo's observations and drawings, 101, 102*fig.*, 103

motion
  Newtonian analysis and description of, 108–9, 204, 345–46
  spectral color shifts and, 96, 202–3
  time as dimension of, 106–8
  Zeno's paradoxes of, 58–59, 107–8, 107*fig.*, 340
  *See also* kinetic energy; Newtonian motion and mechanics; planetary motion
multiverse, 326, 353, **392–93**
muons, 275, 373
musical harmony, 27–36, **366**
  perception of, 33–36, 397–99*nn*
  physics of, 31–32
  Pythagoras's work, 4, 27–30, 28*fig.*, 79
musical instruments
  atoms as, 169
  harmony and, 28–30
  physics of sound production by, 171–75, 172*fig.*, 174*fig.*
  quantum dots as, 190
Music of the Spheres, 30, 51
mutatrons, 303, 317, **372**
*Mysterium Cosmographicum* (Kepler), 50–51, 80, 331

Nambu, Yoichiro, 335
nanotubes, 217–18, **372**, plate FF
natural (resonant) frequencies, 171–75, 172*fig.*, 173–75, 174*fig.*, **372–73**
natural numbers, 341, **373**
Neptune, 318, 353
neutrinos, 274, 275, 335, 343, 352, 368, **373**, 404*nn*
neutrons, 244–45, 246, **373**, 387, 401*n*
  as composite structures, 260–61, 356, 365, 373
  proton/neutron conversions, 260–61
  *See also* atomic nuclei
Newton, Isaac, 3, 5–7, 77–90, 89*fig.*
  Analysis and Synthesis as Newton's strategy, 5–6, 43, 45, 78, 80–81, **341**
  background, education, and early work, 86–90, 88*fig.*
  Blake's depiction of, plate K
  concern with theology and ethics, 84–86
  emphasis on scientific precision, 78, 80–82
  ill health and later life, 89–90
  on light and color, 91, 119–20
  mathematical work, 5, 87, 108, 383–84
  range of hypotheses and Queries, 81–83, 84–86
  on Solar System's stability, 50–52, 79, 80, 112, 192
  work seen as reductionist, 112–14
  *See also* calculus; Newtonian motion and mechanics; Newtonian optics and color theory
"Newton, the Man" (Keynes), 85–86
Newtonian motion and mechanics, 87, 103–12, 124, 318, 322, 332, 365, 392
  basics, 339, 369
  challenges and questions raised by, 118–19, 204–5, 318, 345–46
  conservation of energy theorem, 282
  Kepler's laws and, 357
  laws of, as dynamic, 111–12, 114–15, 354, 357
  motion analyzed and described, 108–9, 204, 345–46
  Newton's anagram, 110
  Newton's Mountain, 103–6, 104*fig.*, 108, 109
  predictions based on, 110–11, 120
  *See also* conservation; gravity; motion
Newtonian optics and color theory, 82–83, 87, 91–99, 203, 318, 332
Noether, Emmy, 279, 287–88, 288*fig.*, 348
Noetherian rings, 348
Noether's theorem, 401*n*, 402*n*
  basics, 279–81, 285–87, 333
  as tool for discovery, 285–87, 348
  *See also* conserved quantities
normal matter, 369, **373**
  *See also* matter
Novoselov, Konstantin, 215, 217
nuclear forces, 243–44, 245–46, 247, 248, 251
  *See also* confinement; quantum chromodynamics; strong force
nuclei. *See* atomic nuclei
nucleons, **373**
  origins of nucleon masses, 259–60, 369–70, 373
  *See also* atomic nuclei; neutrons; protons
numbers. *See* mathematics; *specific number types*

object/person complementarity, 326
observation, as active/interactive process, 74–75

octahedra, 40, 40*fig.*, 46
*See also* Platonic solids
one world/many worlds complementarity, 325–26
"On Faraday's Lines of Force" (Maxwell), 124, 332
Onnes, Kamerlingh, 390
"On Physical Lines of Force" (Maxwell), 124–25, 129, 332
*Opticks* (Newton), 82, 83, 95, 332
*See also* Newtonian optics and color theory
orbitals, 187*fig.*, 211, **374**, 388
in carbon compounds, 212–20
*See also* stationary states
orbits, 103–6, **373**
*See also* planetary motion
organ of Corti, 33, 34
Orphism, 60–61
oscillations, **374**

parity and parity transformations, **374–75**
parity violation, 261–62, 335, **375**
Parmenides, 58, 59, 116, 327
particle accelerators, 340
particle position, 184, 185
*See also* property spaces
Pati, Jogesh, 336
Pauli exclusion principle, 211, 212, 214, 310, 333, 345, **359**
Peccei, Roberto, 336
Peccei-Quinn symmetry, 272, 336, 337, 344
pentagons, 39, 40, 53, plate E
Penzias, Arno, 372
perception
appearances vs. hidden/unseen/ultimate reality, 55, 61–62, 141–42
*See also* auditory perception; Cave allegory; color perception; sensory perception; vision and visual perception
perceptual color, 347, 371
relativity and, 96, 202–3, plate DD
as synthesis of three base colors, 145–47, 152, 161, 347, 378–79, plate T
*See also* color perception
periodic table
of chemical elements, 95–97, **376**
of light, 95–97
periods, periodic processes, **375**, 395
*See also* frequencies
perpetual motion, 284
person/object complementarity, 326
perspective, 378
artistic perspective, 67–72, **378**, plates G, H
symmetry and, 73–74, 205–6, 239
Perugino, *Giving of the Keys to Saint Peter*, plate G
photoelectric effect, 176, 177–78
photons, 351, **376**, 380
as avatars of local symmetry, 240, 241
as bosons, 310, 333
color perception and, 93, 95, 153, 158
as electrically neutral, 253, 342
as elementary particles, 209, 356, 373
emission/generation of, 188–89, 196, 389
as gauge particles, 363
and Higgs particle, 267*fig.*, 268
as massless particles, 263, 265
particle theories of light, 95, 97, 177–78, 333, 376
property space of, 232
superconductivity and, 335, 390
*See also* light; Planck-Einstein relation
Planck, Max, 178, 333, 376
Planck-Einstein relation, 359, **376**, 377, 386
Planck's constant, **376–77**, 387
planetary motion
ancient theories of, 6–7, 65–66
Copernicus's theory of, 103
Kepler's model, 50–52, 51*fig.*, 79–80, 112, 319, 331
orbits as ellipses, 52, 80, 103, 357
problems related to dark matter and energy, 352–53
*See also* celestial mechanics; Kepler's laws; Newtonian motion and mechanics
Plato, 4–5, 6, 55–56
allegory of the Cave, 5, 7, 56–57, 61–64, 141, plate F
*Timaeus*, 46, 47–48, 331
Platonic Ideals, 60–61, 63–64, 65–66, 79
Platonic solids, 4, 7, 37–53, 331, **377**, 399*nn*
anticipations and biological embodiments of, 42–43, 42*fig.*, 44*fig.*, 49, 399*n*, plate D
in Euclid's *Elements*, 45
five solids, 40, 40*fig.*

as foundation of Kepler's Solar System model, 50–52, 79–80, 112, 331
viewed as building blocks of Universe, 46–49, 57, 191, 295
*See also specific solids*
Platonic surfaces (Platonic prodigals), 40–41, 41*fig.*, 216*fig.*, **377**, 399*nn*
points at infinity, 68–70, 68*fig.*, 69*fig.*, 367, **377**
polarization
of light, 94–95, 152–53, 378, **391**
vacuum polarization, **381**
Politzer, David, 256, 336
Polyakov, Alexander, 337
polygons, 39, **377**
*See also specific polygons*
polyhedra, **377**
*See also* Platonic solids; *specific types*
Popper, Karl, 308, 360
positrons, 342, 351, 354, 356, **377**
electron-positron collisions, 252, plate NN
potential energy, 282, **358–59**
powerful equations, 135–36
powerful theories, **360**
pressure, **377**
*Principia* (Newton), 45, 111, 115–16, 119–20, 332, 349
motion analysis, 108–9, 109*fig.*
Newton's Mountain, 103–6, 104*fig.*, 108, 109
probabilities, 184, 185
probability clouds, 170, 183, **378**, 388, 394
electron clouds, 175, 187*fig.*, 394
wave function vibrations and, 186–87, 187*fig.*
projections, 350, **378**
projective geometry, 5, 67–75, 239, 331, 367, **378**
artistic perspective, 67–72, **378**, plates G, H
human visual perception and, 12–13
relativity and, 73
property spaces, 228, 273, 297, 309
basics, 229–32, 231*fig.*, 234, **378–79**
geometry of, 234, 238, 273–74, plates RR, SS, TT, UU
particle behavior as function of location within, 242–43, 262–63
quark positions encoded as color, 250
symmetries and symmetry groups of, 273, 299, 315, 365
*See also specific types*
protons, 244, 245–46, 257, 336, 343, **379**
atomic number and, 343
components/structure of, 251–52, 260–61, 356, 365, 379
and Higgs particle, 267*fig.*, 268
properties of, 251–52, 365, 369–70, 379, 387, plates PP, QQ
proton/neutron conversions, 260–61
*See also* atomic nuclei
Ptolemy, 6, 50
pure tones, 171, 347, 372, **390–91**, 395, 402*n*, 405*n*
*See also* spectral colors
pure waves, 149
Pythagoras, 4, 6, 17–20, 331, plate B
All Is Number credo, 20, 21, 26, 30, 57, 324
work on musical harmony, 4, 27–30, 28*fig.*, 79
Pythagorean theorem, 4, 20–26, 21*fig.*, 23*fig.*, **379**, plate C

QCD. *See* quantum chromodynamics
QED. *See* quantum electrodynamics
qualitativeness, **379**
quanta, 240, **380**, 381, 392
*See also specific particle types*
quantitativeness, **379**
quantization, 333, **379–80**
*See also* discreteness
quantum chromodynamics (QCD), 248–54, 336, 337, 384–85
basics of, 249–50, 260, **380**
beauties of, 254–55, 257, 380
bleaching rule, 250, 256, 301, plate QQ
*See also* asymptotic freedom; color charges; color gluons; hadrons; strong force
quantum dimensions, 309, 310–11, 353, 364, **380**
quantum dots, 190, **380**
quantum electrodynamics (QED), 249–50, 254–55, 260, 264, 333–34
quantum engineering, 189–90
quantum fields/fluids, 268, 356, 380, **381**, 384
*See also specific types*

quantum fluctuations, 334, 351, 352, **380–81**, 383, plate XX
in early history of Universe, 323
force strength inequities and, 306–8, 307*fig.*, 315–18, 316*fig.*
renormalization, 384
*See also specific field/fluid types*
quantum jumps or leaps, 180, 188–89, **381**, 389
quantum theory (quantum mechanics), 181–91, 255, 368, 379–80
basics and key principles, 74–75, 172, 181–82, 184–85, **381–83**
conflicts with general relativity, 350–51
early history and developments, 170, 176–82, 188, 211, 332–35
music/sound as analogous to, 170, 185–89
new frontiers and possibilities, 189–90
timeline, 332–37
wave functions and probability clouds, 181–85, 382–83
as web of ideas, 182, 382
*See also* quantum chromodynamics; quantum electrodynamics; stationary states
quark-antiquark pairs. *See* mesons
quark fluid, 306
quark mesons. *See* mesons
quark model, 247–48, 251, 335, 365–66, **383**
quarks, 49, 251, 335, 336, **383**
in census of forces and entities, 273–76, plates RR, SS, TT, UU
as elementary particles, 356, 373
as fermions, 309–10, 389
as hadron components, 247–48, 365–66
heavy quarks in gluon fusion process, 267*fig.*, 270–72
jets as avatars of, 252, plate NN
properties of, 247, 248, 250, 348, 355
as proton components, 251–52
proton/neutron conversions and, 260–61
quark flavors, 247, **361**
quark triads, 247, 256, plate PP
*See also* confinement; fermions; hadrons; *specific quark types*
*Queries* (Newton), 82–83, 84–85
Quinn, Helen, 336
Peccei-Quinn symmetry, 272, 336, 337, 344
Raby, Stuart, 314–15, 337
radioactivity. *See* decay
radiolaria, 43, 44*fig.*
radio waves and technology, 134, 135–36, 318, 347, 356, plate P
Ramsay, William, 99
Raphael, *The School of Athens*, plate B
Rayleigh, John William Strutt, Baron, *The Theory of Sound*, 170
real dimensions, 347
Real/Ideal duality/complementarity, 63–64, 78–81, 209–10, 376, 396
Aristotelian focus on the Real, 66–67, 78
Kepler's work and, 79–80, 357
mathematics and, 113–14
Newton's work and, 111
Noether's theorem as expression of, 279, 285
Platonic Ideals, 60–61, 63–66, 79
pure tones as expressions of, 391
symmetry of equations as, 138–39
*See also* unification(s)
reality, hidden/unseen/ultimate, vs. appearances, 55, 61–62, 141–42
real numbers, 341, 347, 349–50, **383–84**
redshift, 96, 202–3
reduced Planck's constant, 377
reduction/abundance complementarity, 325
reductionism, 112–14, 341, **384**
regular polygons, 39, **377**
*See also specific polygons*
Reines, Frederick, 335
relativity and relativistic mechanics, 73, 74–75, 340, 342, 354, **384**
*See also* general relativity; special relativity
renormalization and renormalization groups, **384**
*Republic* (Plato), allegory of the Cave, 5, 7, 56–57, 61–64, 141, plate F
resonant (natural) frequencies, 171–75, 172*fig.*, 173–75, 174*fig.*, **372–73**
Riemann, Bernhard, 371
right-hand rules, 361–62, 374
"Rigid Body" (Maxwell), 162–63
rigid symmetry, 238–39, 333, 334, 363, 368, **384**
Yang-Mills theories of expansion, 240, 249, 263, 298–99, 334, **396**
rigor, **384–85**

rotational symmetry, 73–74, 137, 285–86, 387–88
Rubbia, Carlo, 337
Ruskin, John, 92
Russell, Bertrand, 19, 58, 189, 289, 292
Rutherford, Ernest, 243–44, 332, 333, 340

*s* (strange) quarks, 275–76, 361, plates TT, UU
*Sacrament of the Last Supper, The* (Dalí), 53, plate E
Salam, Abdus, 262, 263, 335, 336
*School of Athens, The* (Raphael), plate B
Schrieffer, J. R., 335, 390
Schrödinger, Erwin, 371
Schrödinger equation, 188, 189, 354, 376, 386
  as analog to equations describing music, 185, 186
  basics, 185, 333, **385**
  *See also* wave functions
Schwinger, Julian, 334
scientific method and discovery
  beauty as motivation for discovery, 3, 9
  Noether's theorem as tool for, 285–87, 348
  *See also* Analysis and Synthesis
selectrons, 313
semiquantitativeness, 379
sensory perception, 12–15
  of animals and insects, 13–14, 128–29, 154, 155–56, plates X, Y, Z
  vision vs. hearing, 157–58
  *See also* auditory perception; color perception; vision and visual perception
shadows, 136, plate Q
  Plato's Cave allegory, 5, 7, 56–57, 61–64, 141, plate F
Shakespeare, William, 280
*Sidereus Nuncius* (*Starry Messenger*) (Galileo), 101, 102*fig.*, 103
sinusoidal wave patterns, 390, 402*nn*
*SO*(10) (Georgi-Glashow) unification scheme, 299–308, 336, 387, 404*n*, plates VV, WW
Socrates, 56
  *See also* Cave allegory
solar systems
  diversity of, 193–94, 195
  Kepler's model of our Solar System, 50–52, 51*fig.*, 79–80, 112, 319, 331
  *See also* planetary motion
sound, 31–32, 158, 170–71, 388, 390, 391
  auditory perception, 31–32, 33–36, 157, 158, 395, 397–99*nn*
  natural (resonant) frequencies, 171–75, 172*fig.*, 174*fig.*, **372–73**
  pure tones, 171, 347, 372, **390–91**, 395, 402*n*, 405*n*
  *See also* musical harmony; musical instruments
space
  empty, 305, 351, 366, plate XX
  as Void, 117, 119–20, 133–34, 392, 393
  *See also* fluids; media
space-time, 107–8, 116, 233–35
  geometry of, 233, 236, 238, 371
  matter/space-time complementarity, 233–34, 235
  *See also* general relativity
spatial translations, 280, 367, 372, **391**
spatial translation symmetry, 285, 293, 294, **385**
special relativity, 206, 240, 255, 334, 341, 355, 357–58
  basics of, 200–202, 333, 367, 381–82, **385–86**
  color perception and, 202–4
  symmetries and, 182, 201–2, 240, 333, 363
  as unification, 392
spectra, 180–81, 359, 373, 376, **386–87**
spectral colors, 92–95, **346–47**, 391, plate J
  combinations of, 142–46, plate T
  motion and spectral color shifts, 96, 202–4
  Newton on, 91
  and photoelectric effect, 177
  Planck-Einstein relation, 359, **376**, 377, 386
  polarization and, 94–95, 152–53, 378, 391
  wavelengths of, 395, plate P
  *See also* color perception; spectra
spectral lines, 98–99, 181, 203, 293, 386–87
spherical cow joke, 172*n*
spiders and spiderwebs, 128–29, 128*fig.*
spin, 342, 377, **387**
  handedness, 262, 274, **374–75**
spinors, 299, **387**, 404–5*nn*
Spinoza, Benedictus de, 288
spiritual cosmology, 2–3

spontaneous symmetry breaking, 264, 271, 312–13, 335, 336, **387–88**
  applied to supersymmetry, 313–18
squares, 39, 41, 41*fig.*
  drawn in perspective, 71–72, plate H
  *See also* Pythagorean theorem
stability
  atoms as stable structures, 176–77, 178–81, 191–92, 195–96, 388, 389
  as result of dynamical laws and equations, 170, 191–92, 193–96
  *See also* change
Standard Model, 8, 227
  *See also* Core Theories
standing waves, 170–75, 172*fig.*, **388**
*Stanford Encyclopedia of Philosophy*, 18, 19
Stanford Lineal Accelerator, 251
stationary states, 359, 373, 377, 386, plate CC
  basics, 178–81, 186–89, **388–89**
  transitions between, 180, 188–89, 196, 376, **381**, 386, 389
  *See also* orbitals
strange (*s*) quarks, 275–76, 361, plates TT, UU
string theory, 317
strong bleaching rule, 301
strong color charges, 234, 237, **347**
  *See also* color gluons
strong fluid, 237
strong force, 226, 228, 243–59, 297, 304
  basics, 232, 234, **389**
  discovery of atomic nuclei and related questions, 243–46, 389
  quantum chromodynamics as theory of, 249, 256–57
  *See also* asymptotic freedom; nuclear forces; quantum chromodynamics
strong property space, 234–35, 241, 274, plate KK
*SU*(2) gauge symmetry, 241
*SU*(3) gauge symmetry, 241
subatomic particles
  in Platonic theory, 49
  *See also* atomic nuclei; *specific particle types*
substance
  unification of force with substance, 315–17, 316*fig.*, 392, 401–2*n*
  uniformity of, 293–94
substance particles. *See* fermions; *specific types*
superconductivity and superconductors, 265, 335, **389–90**
supersymmetry (SUSY), 319, 380, 392
  basics, 308–9, 311, 336, 337, **390**
  particles predicted by, 312, 313, 317, 318, 352, 389
  potential and challenges of, 312–18, 316*fig.*, 390, 392
symmetry(ies), 11, 15, 49, 165–66
  antisymmetry, **343**
  basics, 73–74, 343, **390**
  as Change Without Change, 73–74, 137–38, 167, 327, 384, 386
  conservation laws as products of, 279, 281, 285–86, 293, 341, 348, 372
  as fundamental principle, 165, 202, 386
  gauge particles as embodiments of, 298–99
  perspective and, 73–74, 205–6, 239
  relativity as, 384
  structure as dictated by, 48
  *See also specific forms of symmetry*
symmetry breaking, 264, 270, 271, 312
  *See also* spontaneous symmetry breaking
symmetry groups, 349, **365**, **390**
  Lie groups, 299, **365**
symmetry of equations, 238, 309, 390
  in Maxwell's equations, 118, 136–39
  *See also* local symmetry; spontaneous symmetry breaking; supersymmetry
symmetry transformations, **365**, **390**
synthesis, 6, 110–11, 179, **341**, 345, **390**
  *See also* Analysis and Synthesis
*System of the World, The* (Newton), 111

*T* (time reversal) symmetry violation, 344
*t* (top) quarks, 268, 270, 272, 275, 361
Taiji, vi, 235, plate A
Tait, Peter, 319
tauons, 262, 275, 373
Taussky, Olga, 288
Taylor, Richard, 251, 336
telescopes, 97
television, 147
tetrachromacy, 154–55, 400*n*
tetrahedra, 40, 40*fig.*, 46
  diamond structure as tetrahedral, 213, 216*fig.*
  *See also* Platonic solids

*Theaetetus* (Plato), 42, 331
  *See also* Platonic solids
*Theory of Sound, The* (Rayleigh), 170
θ parameter/problem, 272, 336, 354
Thomson, J. J., 176, 356
*Timaeus* (Plato), 46, 47–48, 331
time
  as dimension, 106–8
  sound as temporal information, 157, 158
  vision and temporal information, 151–52, 157–58, 204
  *See also* change; space-time
time reversal (*T*) symmetry violation, 344
time translation symmetry, 280, 309, 359, **390**
  justifying our belief in, 289–94
Tomonaga, Sin-Itiro, 334
tones, 174, 347, **390–91**
  pure tones, 171, 347, 372, **390–91**, 395, 402*n*, 405*n*
  *See also* musical harmony
top (*t*) quarks, 268, 270, 272, 275, 361
trajectories, 106–8
transformation groups. *See* symmetry groups
transformations, **365**, **390**
  *See also* symmetry(ies); *specific symmetry types*
transience/eternity complementarity, 327
translations, **391**
  *See also* spatial translations; time translation symmetry
transverse waves, **391**
traveling waves, 170–71, **388**
triangles, 39, 40, 41, 41*fig.*, 49, 137
  *See also* Pythagorean theorem
trochoids, 399*nn*
*T* symmetry violation, 344

*U*(1) gauge symmetry, 240
*u* (up) quarks, 273–74, 360, 361, plates RR, SS, TT, UU
ubiquity, **391**
ugliness/beauty complementarity, 328
ultraviolet radiation, 135, 155, 177, 347, 356, plates P, Y
unification(s), 226, 255, **391–92**, 401–2*nn*
  of forces, 247–48, 254–55, 299–308, 314–18, 316*fig.*, 381, 392, 401–2*n*
  of force with substance, 315–17, 316*fig.*, 392, 401–2*n*
  Maxwell's work as, 7, 127
*SO*(10) Georgi-Glashow unification scheme, 299–308, 336, 387, 404*n*, plates VV, WW
uniformity of substance, 289, 292–94
uniformity through space. *See* spatial translation symmetry
uniformity through time. *See* time translation symmetry
universal gravitation. *See* gravity; Newtonian motion and mechanics
Universe, **392–93**
  expansion of, 196, 202–3, 323, 364–65, 369, 392
  one world/many worlds complementarity, 325–26
  origins and early development of, 254, 351, 371–72, 403*n*
  in Platonic and Aristotelian theory, 4–5, 46–47, 295
  vastness of, 325–26
up (*u*) quarks, 273–74, 360, 361, plates RR, SS, TT, UU
Uranus, 318, 353

vacuum, **393**
  *See also* Void
vacuum polarization, **381**
vanishing point. *See* points at infinity
vector fields, 346, 361, **393–94**
vector quantization, 156
vectors and vector quantities, 341, 375, **393–94**
velocity, 6, 339, 345–46, 366, **394**
  *See also* acceleration; motion
velocity fields, 125, 131, 360
Veltman, Martinus, 264, 336
virtual particles, 305, **380–81**, 389
  *See also* quantum fluctuations
viruses, 43, 49, plate D
visible Universe, **392**
vision and visual perception, 12–14, 145, 150–51, 153, 362, 378
  Newton's experiment on his own eye, 87, 88*fig.*, 89
  unseen/disregarded visual information, 151–53, 158
  vision as space sense, 151–52, 157–58
  *See also* color perception; light; perspective
*Vitruvian Man* (Leonardo), 321, plate ZZ
Void, 117, 119–20, 133–34, 392, **393**

Ward, John, 262, 263, 335
wave functions, 170, 182–89, 382–83, **394–95**
  amplitudes of, 183
  of hydrogen atoms, 182–85
  natural vibrations of, 185–89
  of two-electron atoms, 197, 394–95
  *See also* Dirac equation; electron wave functions; Schrödinger equation; stationary states
wavelengths, 158, **395**, plate P
waves, 388
  *See also specific wave types*
weak bleaching rule, 301
weak color charges, 234, 237, 304, **347**, 390
  *See also* weakons
weak color transitions, 360
weak fluid, 237
weak force, 226, 228–29, 260–73, 304
  basics of, 234, 260–62, 332, **395–96**
  as embodiment of local symmetry, 262–63, 297, 396
  as embodiment of spontaneous symmetry breaking, 264
  handedness and parity violation, 261–62, 274, **374–75**
  hypercharge and, 366, 403–4*nn*
  proton/neutron conversions, 260–61
  *See also* Higgs fields/fluids; Higgs particles
weakons, 337, 347, 390, **396**
  as avatars of symmetry, 240, 241, 263, 396
  as elementary particles, 356
  as force particles, 310, 344–45
  as gauge particles, 363
  properties of, 263, 303, 366
  *See also W* particles; *Z* particles
weak property space, 234–35, 241, 262–63, 273–74, 387–88, plate KK
Weinberg, Steven, 165, 264, 272, 336, 337
Wess, Julius, 308, 336
Weyl, Hermann, 116, 165, 240, 287, 334, 399*n*, 400*n*
  gauge theory, 400*nn*, 401*nn*
Wheeler, John, 232–33, 363
white, white light, 91, 92–93, 94, 142, 347, plates I, J, T
Whitehead, Alfred North, 55–56
Whitman, Walt, 324, 402*n*
whole numbers, 25–26
Wigner, Eugene, 334
Wilczek, Frank, 269–73, 336, 337
Wilson, Kenneth, 337
Wilson, Robert, 372
winds, 122–23
Wittgenstein, Ludwig, 52
Wolfram, Stephen, 25
*W* particles, 263, 265, 266, 319, 337, **394**
  *See also* weakons

X-rays, 135, 347, 356, plate P

Yang, C. N. (Frank), 165, 261–62, 334, 335, 375
  Yang-Mills equations, 240, 249, 263, 298–99, 334, **396**
yin-yang complementarity, 235–36, 237, 324–25, 324*fig.*, plate A
  *See also* complementarity
"Young Archimedes" (Huxley), 22

Zee, Anthony, 403*n*
Zeno's paradoxes, 58–59, 107–8, 107*fig.*, 340
zero-mass particles, 263, 265–66, 270, 271, 272
zero-point motion, 305, 342, **381**
*Z* particles, 263, 265, 266, 319, **396**
  *See also* weakons
Zumino, Bruno, 308, 336
Zweig, George, 247, 335, 383